D0713086

Emulsion Polymerization

Irja Piirma, EDITOR
University of Akron

John L. Gardon, EDITOR
M&T Chemicals, Inc.

An international
symposium sponsored
by the Division of
Polymer Chemistry, Inc.
at the 169th Meeting
of the American Chemical
Society, Philadelphia,
Penn., April 8–10, 1975

ACS SYMPOSIUM SERIES **24**

AMERICAN CHEMICAL SOCIETY
WASHINGTON, D. C. 1976

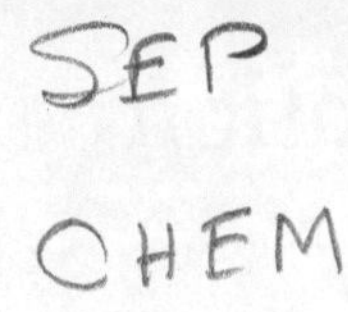

AMERICAN CHEMICAL SOCIETY
1876-1976

Library of Congress CIP Data

Emulsion polymerization.
(ACS symposium series; 24 ISSN 0097-6156)

Includes bibliographical references and index.

1. Additional polymerization—Congresses.
I. Piirma, Irja, 1920- . II. Gardon, J. L. III. American Chemical Society. Division of Polymer Chemistry. IV. Series: American Chemical Society. ACS symposium series; 24.

QD281.P6E46 547'.28 75-44458
ISBN 0-8412-0309-1 ACSMC 8 24 1-407

PRINTED IN THE UNITED STATES OF AMERICA

ACS Symposium Series

Robert F. Gould, *Series Editor*

FOREWORD

The ACS SYMPOSIUM SERIES was founded in 1974 to provide a medium for publishing symposia quickly in book form. The format of the SERIES parallels that of the continuing ADVANCES IN CHEMISTRY SERIES except that in order to save time the papers are not typeset but are reproduced as they are submitted by the authors in camera-ready form. As a further means of saving time, the papers are not edited or reviewed except by the symposium chairman, who becomes editor of the book. Papers published in the ACS SYMPOSIUM SERIES are original contributions not published elsewhere in whole or major part and include reports of research as well as reviews since symposia may embrace both types of presentation.

CONTENTS

PREFACE

The progress in synthetic latex technology has been phenomenal. Because of the pioneering work in the thirties, the rubber industry converted to synthetic latexes during World War II. Subsequently the emulsion polymerization technology expanded rapidly so that in addition to styrene, butadiene, and isoprene, many monomers became commonly used; acrylics, vinyl acetate, vinyl chloride chloroprene, and acrylonitrile represent a large volume. The industrial end-uses of emulsion polymers now include finishes for paper, textiles, leather and wood, adhesives, sealants, air-dry trade sale paints, metal decorating coatings, high impact plastic composites, and structural plastics. The science of emulsion polymerization and emulsion polymers grew parallel with this technology, and this book is a good indication of this growth.

In the past the really well understood particle formation and kinetics involved the single-charge homopolymerization of "styrene-like" monomers in the presence of micellar surfactants and of water soluble initiators with long half lives. Even for these systems the theory was predictive only at relatively low conversions. Now the kinetics are well understood even at high conversions.

The mechanism of particle formation at submicellar surfactant concentrations was established several years ago. New insight was gained into how the structure of surfactants influences the outcome of the reaction. The gap between suspension and emulsion polymerization was bridged. The mode of popularly used redox catalysts was clarified, and completely novel catalyst systems were developed. For "non-styrene-like" monomers, such as vinyl chloride and vinyl acetate, the kinetic picture was elucidated. Advances were made in determining the mechanism of copolymerization, in particular the effects of water-soluble monomers and of difunctional monomers. The reaction mechanism in flow-through reactors became as well understood as in batch reactors. Computer techniques clarified complex mechanisms. The study of emulsion polymerization in nonaqueous media opened new vistas.

The research on the structural behavior of emulsion polymers has been greatly influenced recently by colloid chemistry. Surface charges, particle morphology, film formation, flow properties, and interactions with organic solvents and monomers have been studied. Of particular interest is the interaction of tagged latex particles with proteins leading to new applications in medical diagnosis. The editors hope that the reader will share their excitement for tremendous diversity of this field and will find challenges for future work.

Institute of Polymer Science
The University of Akron
Akron, Ohio 44325

I. PIIRMA

M&T Chemicals Inc.
26701 Telegraph Rd.
Southfield, Mich. 48076

J. L. GARDON

November 25, 1975

1

Emulsification and Emulsion Polymerization of Styrene Using Mixtures of Cationic Surfactant and Long Chain Fatty Alcohols or Alkanes as Emulsifiers

A. R. M. AZAD, J. UGELSTAD, R. M. FITCH, and F. K. HANSEN

Laboratory of Industrial Chemistry, The University of Trondheim, The Norwegian Institute of Technology, N-7034 Trondheim-NTH, Norway

In some recent papers Ugelstad et al.(1,2,3) have reported results on the emulsion polymerization of styrene with a mixed emulsifier system consisting of Na hexadecyl sulphate and hexadecanol. Under given conditions for emulsification and polymerization it was found that the monomer droplets became the main loci for initiation of polymerization. The first part of the present work describes emulsification experiments using a mixture of a cationic emulsifier, octadecyl pyridinium bromide (OPB) with n-fatty alcohols of different chain lengths, applying ordinary stirring equipment. In order to get a comparison of the monomer droplet size with that of the particles in the final latex, it was necessary to develop a method for obtaining electron microscope pictures of the monomer emulsions (4).

Polymerizations of the monomer emulsions were carried out with oil-soluble initiators. Oil-soluble initiators have often been employed in emulsion polymerization recipes and are generally used in suspension polymerization. Whereas in the latter case the initiation naturally takes place in the monomer droplets, the locus of initiation and growth of particles in emulsion polymerization with oil-soluble initiators has been open to some doubt. However, the fact that the particle size and size distribution is not very different from the results with water-soluble initiators and that the particles are generally much smaller than the droplets in the monomer emulsions indicates that with

---

This work is part of the thesis by A.R.M.A. at the Norwegian Institute of Technology, Trondheim, Norway

ordinary emulsifier recipes initiation of particles takes place mainly in the aqueous phase. An additional part describes some results in which the emulsification of the monomer is carried out with a Manton Gaulin homogenizer. Some experiments were also carried out where the hexadecanol was replaced by hexadecane.

Experimental part

Materials: The styrene monomer was distilled twice, the second time immediately prior to emulsification. A small amount of inhibitor (p-benzoquinone, 500 mg/kg monomer) was added to avoid polymerization in the emulsification experiments. OPB was synthesized from octadecyl bromide (pure, Koch Light Lab. Ltd., England) and pyridine (pure, Merck, Germany); the product was washed four times with dry ether and recrystallized twice from acetone, m.p. 75 ^{o}C. Hexadecanol (HD) (Hyfatol 16, Aarhus Oliefabrik) and octadecanol (OD) (pure, Fluka, Switzerland) were distilled twice in vacuum. 2,2'-azobisisobutyronitrile (AIBN) (pure, Fluka), was crystallized twice from 96% ethanol. Eicosanol, ES (Arachidic alcohol, pure Koch Light Lab. Ltd., England), tetradecanol, TD (pure, Schuchardt, Munich), hexadecane (puris, Koch Light Lab.) benzoyl peroxide, BP (97% Merck), cumene hydroperoxide, CHP (70% in cumene, Merck-Schuchardt), osmium tetraoxide (Merck) and dioctyl sodium sulfosuccinate (pure, Merck) were used without further purification. Redistilled water was used.

Osmium tetroxide staining of monomer emulsion. 0.01 cm^3 of the emulsion was diluted with 0.5 cm^3 water saturated with styrene. To this was added gradually with mixing a saturated solution of OsO_4 in water prepared immediately before use. The amount of OsO_4 solution added was 0.13 cm^3, corresponding to a styrene :OsO_4 molar ratio of 1:1.5. An immediate blackening of the emulsion took place. Samples (2-3 μl) were withdrawn after 5 and 10 minutes placed on a formvar and carbon-coated grid and allowed to dry. The grids were examined in a Siemens electron microscope. Some experiments were carried out in which the OsO_4:styrene molar ratio and time of reaction were varied. Inferior electron microscope pictures were obtained when the molar ratio of OsO_4:styrene was less than 1:1 or higher than 2:1. Prolonged reaction times greater than 30 minutes also invariably gave inferior results. After drying on the grids the particles were stable and preparations could be left for a week without any noticeable change in the electron microscope pictures. In some experiments the stained particles were shadowed with a 80/20 Pt/Pd alloy at an angle of 30° to ascertain that the particles were spherical.

Apparatus and procedure

The emulsification experiments were carried out in a 500 cm^3 glass vessel with a paddle stirrer fitted with thermometer, manometer and equipment for charging and sampling. Hot water was passed through the outer jacket of the reactor. OPB, fatty alcohol and H_2O were first mixed with stirring at 70-80 °C, the temperature depending upon the chain length of the alcohol. After cooling to 60 °C, the styrene was added and the stirring continued at 600 rpm.

Samples were withdrawn at intervals through a bottom stopcock and analysed for emulsifier in the aqueous phase after centrifugation as described in a previous paper (2) with the exception that 0.002 M dioctyl sodium sulfosuccinate was used for the titration. From some of the samples of the monomer emulsions electron micrographs were obtained in the manner described above. Emulsions from the Manton Gaulin homogenizer were subsequently stirred at 60 °C and analysed in the same manner.

The polymerization experiments were carried out as described previously (2). The initiator was dissolved in the monomer prior to emulsification. Samples were withdrawn through the bottom valve into 25 cm^3 of a short stop solution of p-benzoquinone in methanol. The conversion was determined by evaporation of the volatiles at 60 °C.

Electron micrographs of the latex particles were obtained in the usual way.

Results and discussion

Effect of n-fatty alcohol as additive: As found previously in the anionic mixed emulsifier systems with fatty alcohols, the method of preparing the emulsions is crucial when moderate stirring is employed. Thus, if the fatty alcohol was added to the monomer prior to mixing with an aqueous solution of the emulsifier, only very coarse emulsions resulted, which separated within a few minutes. With the present method there took place a rapid emulsification, with the result that at moderate initial emulsifier concentrations more than 97.5 % of the emulsifier was adsorbed within 15-20 minutes. Table I gives some results of the amount of emulsifier adsorbed on the droplets after 15 minutes stirring with 2 g/dm^3 H_2O of OPB and with various molar ratios of hexadecanol (HD) to OPB. It appears that a HD:OPB ratio of 2:1 is sufficient to bring about practically total adsorption of emulsifier on the droplets.

Fig. 1 gives the results of some experiments with OPB and hexadecanol in which the amount of OPB adsorbed on the droplets was followed as a function of time at 60 °C and with stirring at 600 rpm. It appears that in all cases the amount of OPB adsorbed on the droplets rapidly reaches an optimal value. As the stirring is continued the OPB is gradually transported back to the aqueous phase, indicating a gradual degradation of the emulsion. This degradation could be followed by electron micrographs of the emulsion after OsO_4-staining as described above.

Fig. 2 gives an electron micrograph of one of the emulsions after 15 minutes stirring. The droplet sizes are in the range of 0.4 to 1.5 μm.

Fig. 3 shows an electron micrograph of the same emulsion after 21 h stirring at 60 °C. The drastic increase in droplet size is clearly apparent.

Fig. 4 gives some results with fatty alcohols of different chain length. It appears that with the C_{14} fatty alcohol the emulsification is poor and the emulsion is relatively unstable. As the chain length of the fatty alcohol is increased the stability of the emulsion gradually increases. With the C_{20} alcohol the amount of emulsifier adsorbed on the droplets after 20 h stirring at 60 °C is only reduced from 95 to 90 %.

For comparison some experiments were carried out in which the fatty alcohol was dissolved in the monomer phase. The results of these experiments are given in Table II. It appears that in this case the amount of OPB adsorbed on the droplets after 15 minutes is very low. After 20 h stirring the amount adsorbed is still only ca. 50 % which, when compared to the results in Fig. 1, seems to correspond to an "equilibrium" value at the given temperature and stirring rate.

The pronounced dependence of the degree of emulsification upon the order of mixing with fatty alcohols is not yet satisfactorily explained. Several possibilities may be advanced. a) During the stirring, in regions of high shearrate in the neighbourhood of the stirrer, there will be formed continuously small droplets which, without stabilization, will continuously coalesce with the larger droplets inthe bulk of the mixture. In the presence of the mixed emulsifier these droplets may be

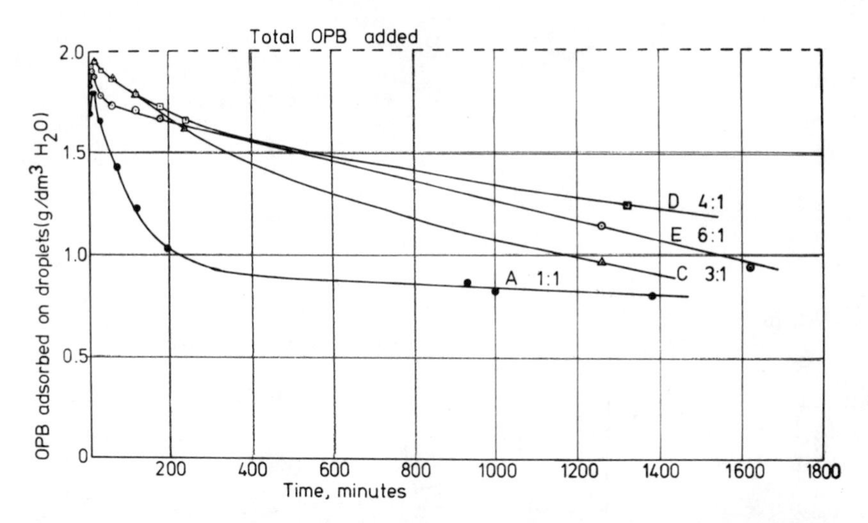

Figure 1. OPB adsorbed on the monomer droplets as a function of time at different molar ratios HD:OPB. Styrene = 83.3 g, H_2O = 250 g, OPB = 2.0 g/dm³ H_2O. Temp. = 60° C. Stirring = 600 rpm.

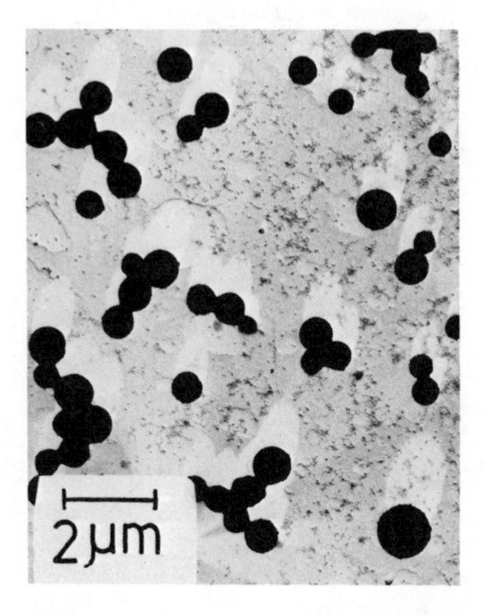

Figure 2. Electron micrograph of monomer emulsion after 15 min stirring for HD:OPB. Molar ratio 6:1 (Figure 1, curve E).

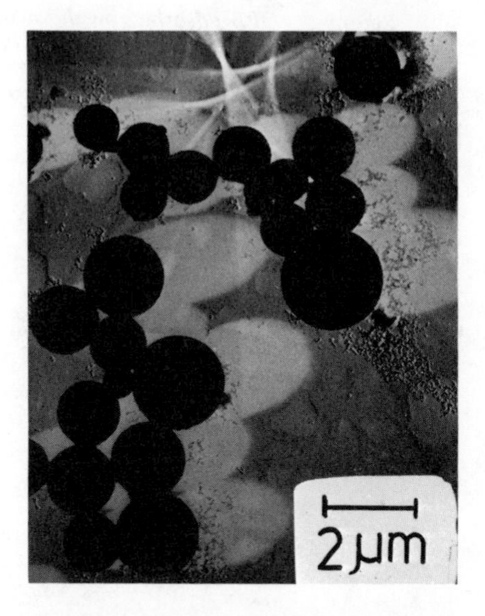

Figure 3. Electron micrograph of monomer emulsion of Figure 2, after 21 hr stirring at 600 rpm and 60°C

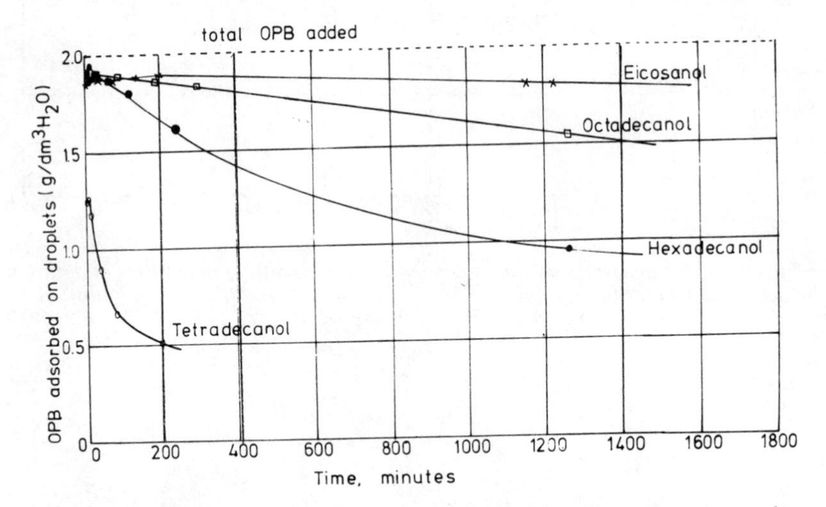

Figure 4. OPB adsorbed on the monomer droplets as a function of time for different long chain fatty alcohols. Styrene = 83.3 g, H_2O = 250 g, OPB = 2.0 g/dm^3 H_2O. Temp. = 60°C. Stirring = 600 rpm. Molar ratio fatty alcohol:OPB = 3:1.

Table I. OPB adsorbed on the monomer droplets after 15 minutes of stirring. HD dissolved in the water phase prior to emulsification. Temp. = 60 °C. Stirring = 600 rpm. OPB = 2.0 g/dm^3 H_2O, H_2O = 250 g. Styrene = 83.3 g.

Molar ratio HD:OPB	Amount of OPB in the aqueous phase. g/dm^3 water	Amount of OPB adsorbed on monomer droplets. g/dm^3 water
0:1	1.88	0.12
1:1	0.21	1.79
2:1	0.06	1.94
3:1	0.05	1.95
4:1	0.06	1.94

Table II. OPB adsorbed on styrene droplets after different stirring times where HD is dissolved in the monomer. Temp. = 60 °C, Stirring = 600 rpm. OPB = 2.0 g/dm^3 H_2O. H_2O = 250 g. Styrene = 83.3 g.

Molar ratio HD:OPD	Time of stirring (min)	Amount of OPB in the aqueous phase g/dm^3 water	Amount of OPB adsorbed on monomer droplets g/dm^3 water
4:1	15	1.82	0.18
	1230	1.06	0.94
10:1	15	1.76	0.24
	1330	0.95	1.05

rapidly covered with a complex layer of emulsifier and fatty alcohol, which will prevent the coalescence for the time necessary for dispersion of all the monomer to take place. b)During stirring, fresh oil-water interfaces are created by the paddle stirrer. The alcohol and emulsifier present in the water phase diffuse rapidly to this freshly formed interface, resulting in a momentary high concentration of alcohol at the interphase. This may cause a local lowering of the interfacial tension which may drastically facilitate the emulsification. As shown by Davies and Haydon (6), a high concentration of fatty alcohol in addition to emulsifier at the interface will lower γ to approximately zero value which may lead to spontaneous emulsification. Davies and Haydon have added the fatty alcohol to the oil phase prior to mixing with the water solution of the emulsifier. In this case, therefore, they had to apply a relatively very high concentration of fatty alcohol.c)It is also possible that the development of transient interfacial tension gradients on the monomer droplets formed may facilitate emulsification with fatty alcohol present in the aqueous phase. Immediately after a droplet is separated into two droplets, the interfacial tension, γ, will tend to be higher at the points of closest approach than at the more distant parts of the interfaces. The ensuing gradient in γ tends to suck aqueous solution between the newly formed droplets forcing them apart and hence providing them with time to stabilize themselves against coalescence after the interfacial tension gradient has vanished (5).

Fig. 5 gives electron micrographs of latexes prepared with AIBN as initiator from emulsions with a constant styrene:H_2O weight ratio = 1:3, a constant amount of OPB and with HD:OPB molar ratios of 0, 1:1, 3:1 and 4:1 respectively. In the first case the latex contains only small particles with a relatively broad particle size distribution. With a HD:OPB ratio of 1:1 the particles are somewhat larger and more monodisperse. In both cases the particle nucleation apparently takes place completely in the aqueous phase. The larger particles in the second case probably stem from the fact that the amount of emulsifier in the water phase is lower, due to the fact that a relatively large part is adsorbed on the monomer droplets. However there is still enough emulsifier present in the aqueous phase to bring about practically complete particle nucleation in that phase. In the cases of HD:OPB = 3:1 and 4:1 the situation is completely different. We have in both cases a bimodal distribution. The major part

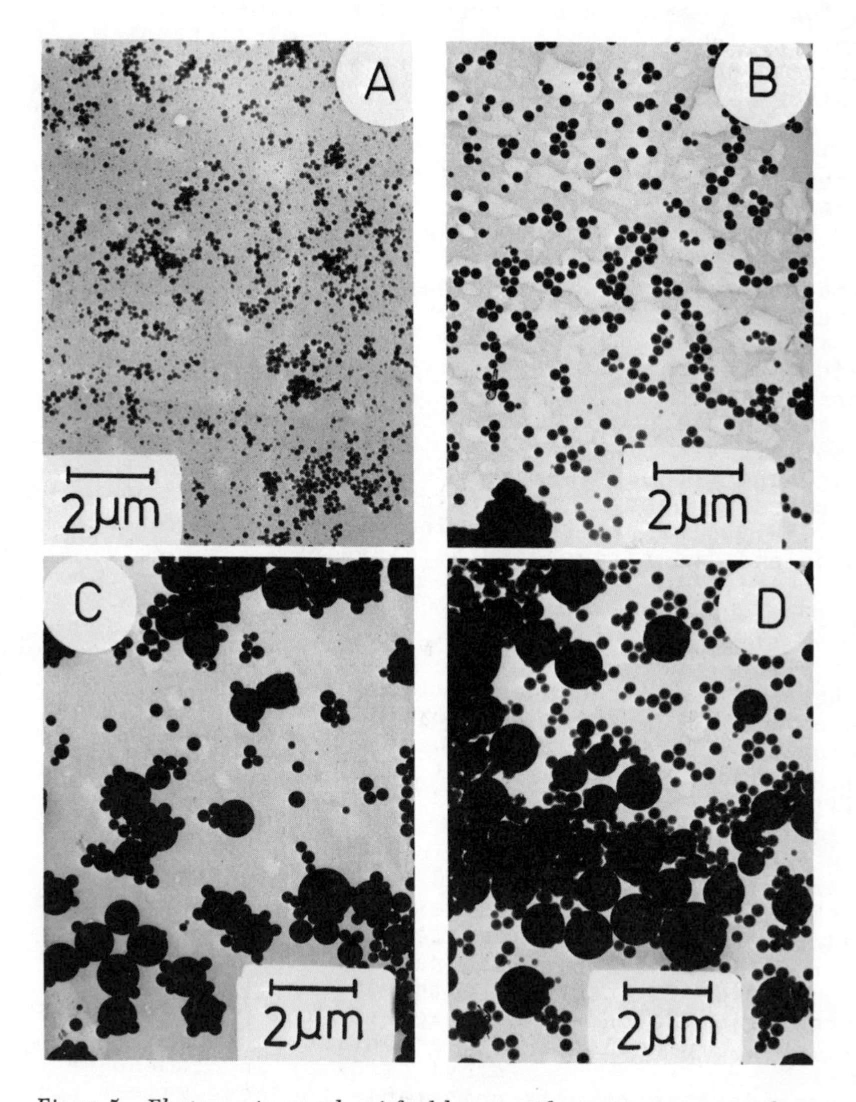

Figure 5. Electron micrographs of final latexes with varying amounts of hexadecanol (HD). Styrene = 166.7 g, H_2O = 500 g, OPB = 2.0 g/dm³ H_2O. Temp. = 60°C. AIBN = 1.0 g in 166.7 g styrene. Molar ratios HD:OPB = (A): 0, (B): 1:1, (C): 3:1, and (D): 4:1. OPB remaining in the water phase after emulsification of monomer (A) = 1.78, (B) = 0.35, (C) = 0.08 g/dm³ H_2O.

by weight consists of particles of diameter 0.4 to 1.5 μm. By comparing with the electron micrographs of the corresponding monomer emulsions in Fig. 6, it is apparent that these particles stem from initiation in monomer droplets. In addition, there are a considerable number of particles of about 0.2 μm not present in the monomer emulsion. These particles therefore most probably stem from nucleation in the aqueous phase.

In Fig. 7 are given the results of kinetic measurements of the latexes given in Fig. 5. It appears that in the case of pure emulsifier the kinetic results are in agreement with common experience in emulsion polymerization of styrene with a water soluble initiator. The rate is approximately constant up to high conversion.

In case C and D, however, the situation is completely different. The initial rate is relatively high and decreases significantly up to about 30 % conversion when it starts increasing slowly.

With the reaction taking place in the monomer droplets the rate should be given by

$$r = k_p [M]_M \left(\frac{k_i [I]_M}{k_t}\right)^{\frac{1}{2}} V_M \qquad (1)$$

where $[M]_M$ is the concentration of monomer in the droplets, $[I]_M$ the concentration of initiator in the droplets, V_M the total volume of monomer droplets per 1 H_2O and r is given in mol sec^{-1} per 1 H_2O. With literature values of k_p = 300 l/mole-sec, $k_t = 10^8$ l/mole-sec, $k_i = 1.0 \times 10^{-5}$ sec^{-1} and $[M]_M$ = 8.4 mole/l, V_M = 0.383 l/l H_2O, the calculated initial rate is approximately 20 g PS/l H_2O/h. The experimental initial rate is found to be 20 g PS/l H_2O/h, in good agreement with the calculated value.

In the beginning the rate decreases with increasing conversion in accordance with equation (1) ($[M]_M$ decreases). Beyond 30 % conversion the rate increases slightly, probably due to a continuous decrease in k_t.

Benzoyl peroxide and cumene hydroperoxide gave results comparable to those with AIBN: initiation both in monomer droplets and in the aqueous phase, although to varying degrees, was observed, as shown in Figs. 8 and 9. Under the same conditions of emulsifier and fatty alcohol, benzoyl peroxide gave the highest degree of monomer droplet initiation, while cumene hydroperoxide led to more nucleation in the aqueous phase.

A more finely dispersed monomer droplet emulsion could be achieved by increasing the concentration of

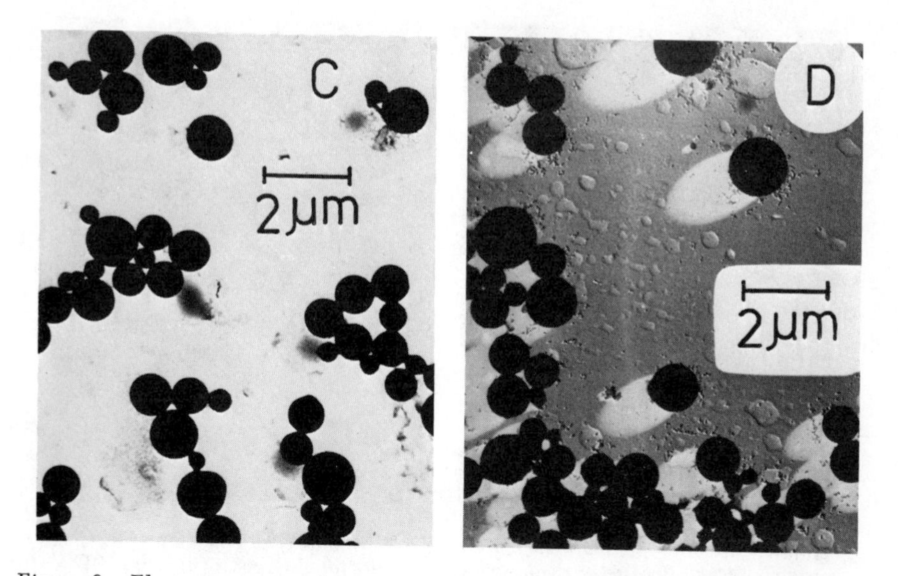

Figure 6. Electron micrographs of monomer emulsion obtained by OsO_4 staining after 15 min of stirring from the experiments given in Figure 5. (Letters (C) and (D) refer to Figure 5.)

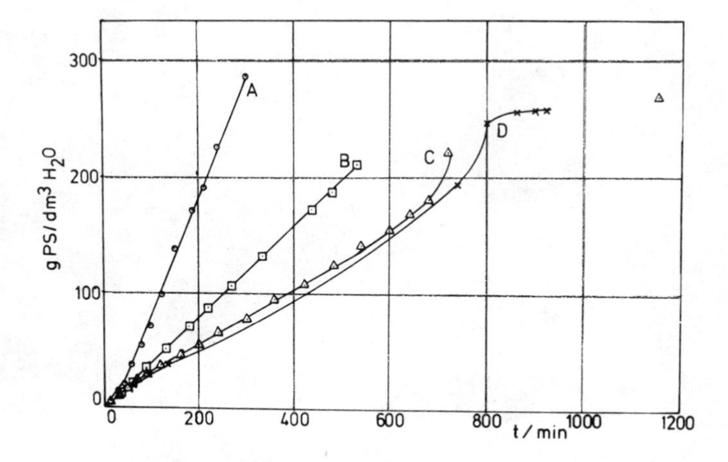

Figure 7. Polymer formed as a function of time with varying amounts of HD for the experiments given in Figure 5. (Letters (A), (B), (C), and (D) refer to Figure 5.)

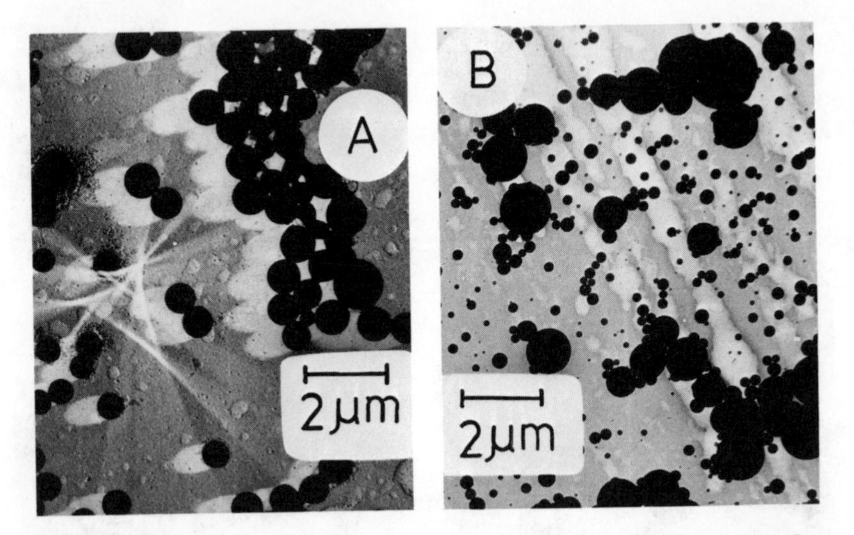

Figure 8. Electron micrographs of monomer emulsion (A) and final latex (B) when benzoyl peroxide was used as the oil soluble initiator. Styrene = 166.7 g, H_2O = 500 g, HD:OPB = 4:1, OPB = 2.0 g/dm³ H_2O. Temp. = 60°C. BP = 2.0 g in 166.7 g styrene.

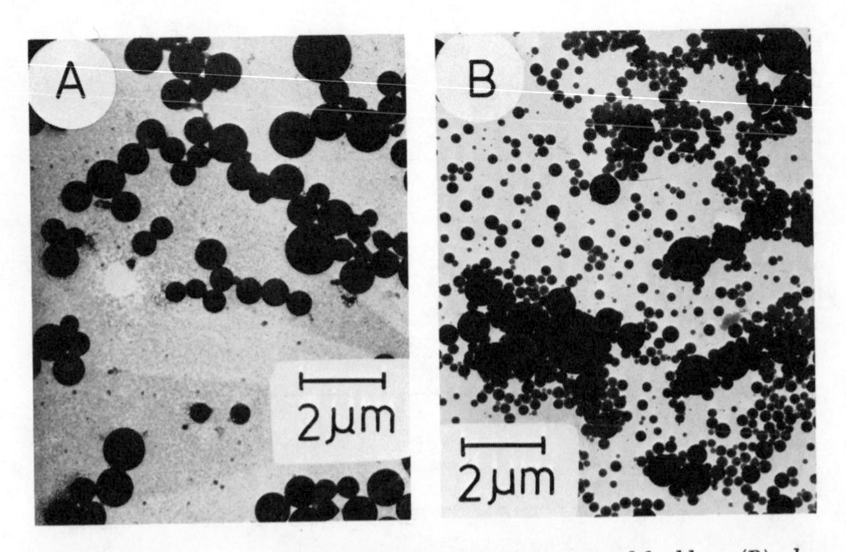

Figure 9. Electron micrographs of monomer emulsion (A) and final latex (B) when cumene hydroperoxide was used as the oil soluble initiator. Styrene = 166.7 g, H_2O = 500 g, HD:OPB = 4:1, OPB = 2.0 g/dm³ H_2O. Temp. = 60°C. CHP = 2.5 g in 166.7 g styrene.

the mixed emulsifier system (Fig. 10). However, the monomer droplet surface does not increase proportionately to the amount of emulsifier added so that the concentration of emulsifier left in the aqueous phase after emulsification is inthis case increased. It turns out that this fact leads to that one gets a predominant nucleation in the aqueous phase (Fig. 11). The results are in accordance with previous results with mixed emulsifier systems of Na hexadecyl sulphate and hexadecanol with $K_2S_2O_8$ as initiator (2). Also in the present case with oil soluble initiator and with OPB it is therefore a necessary condition for obtaining a high degree of monomer droplet initiation that not only a fine dispersion of monomer is achieved but that at the same time the concentration of emulsifier left in the aqueous phase is very low.

Effect of hexadecane as additive: In a series of papers Hallworth and Carless (7,8,9,10) have investigated the effect of the nature of the internal phase on the stability of oil in water emulsions as well as the effect of addition of long chain fatty alcohols with sodium dodecyl sulphate or sodium hexadecyl sulphate as the ionic emulsifier. They found that light petroleum and chlorobenzene emulsions prepared only with sodium hexadecyl sulphate were much less stable than those produced using the longer chain paraffins, white spirit and light liquid paraffins. Most interestingly, however, they found that addition of small amounts of light liquid paraffins to the light petroleum or chlorobenzene led to an increase in stability which even surpassed that which was obtained with long chain fatty alcohols. It should be noted, however, that Hallworth et al. in preparing their emulsions applied the usual method of addition of the additives to the main component of the internal phase before mixing with the water solution of the anionic emulsifier. As shown above and discussed in previous papers, this procedure does not lead to rapid emulsification with fatty alcohols when ordinary, moderate stirring is applied for preparing the emulsions. In order to investigate the possible applications of the results of Hallworth and Carless on emulsion polymerization we have carried out some experiments with hexadecane as additive. It turned out that with ordinary stirring equipment, addition of hexadecane did not give the rapid emulsification which could be obtained with the long chain fatty alcohols. Therefore a series of experiments were carried out in which the emulsions after premixing were homogenized with a Manton Gaulin

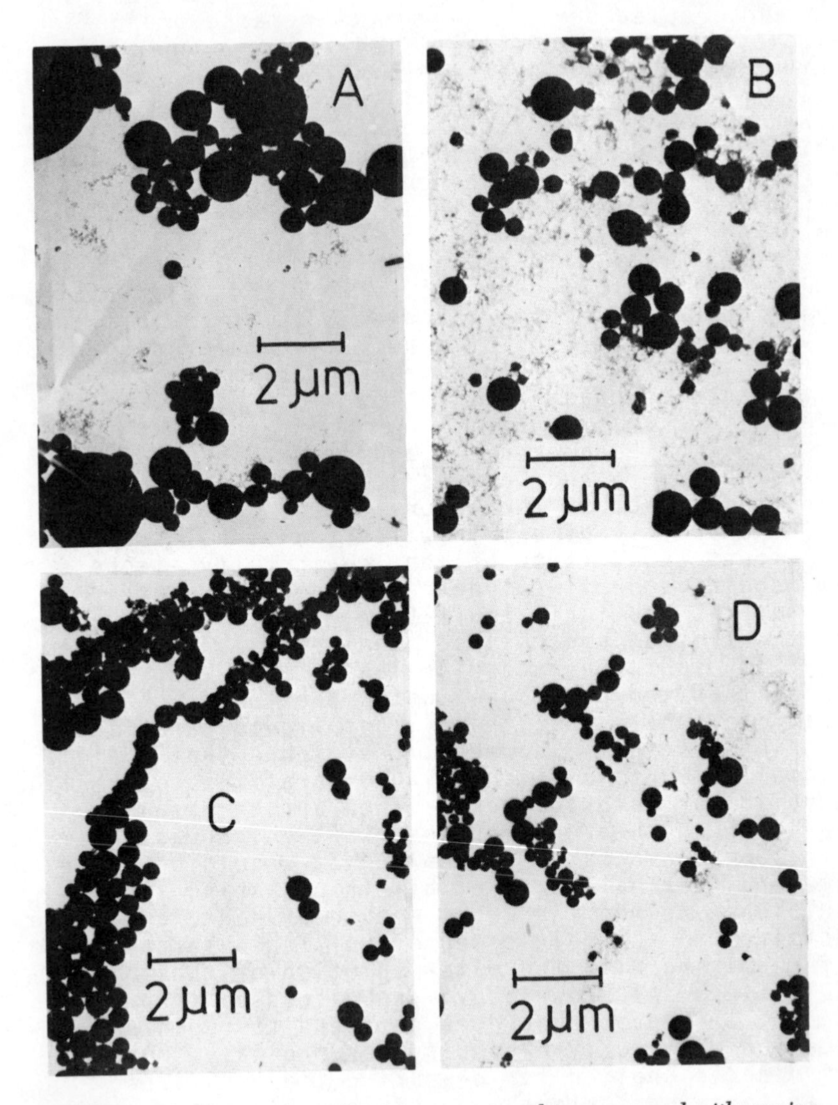

Figure 10. Electron micrographs of monomer emulsions prepared with varying concentrations of OPB. Styrene = 166.7 g, H_2O = 500 g. Temp. = 60°C. Molar ratio HD:OPB = 3:1. OPB_3 = (A) 2.0, (B) 4.0, (C) 6.0, and (D) 8.0 g/dm^3 H_2O. OPB remaining in the water phase after emulsification of monomer, (A) = 0.10, (B) = 0.53, (C) = 1.12, and (D) = 3.14 g/dm^3 H_2O.

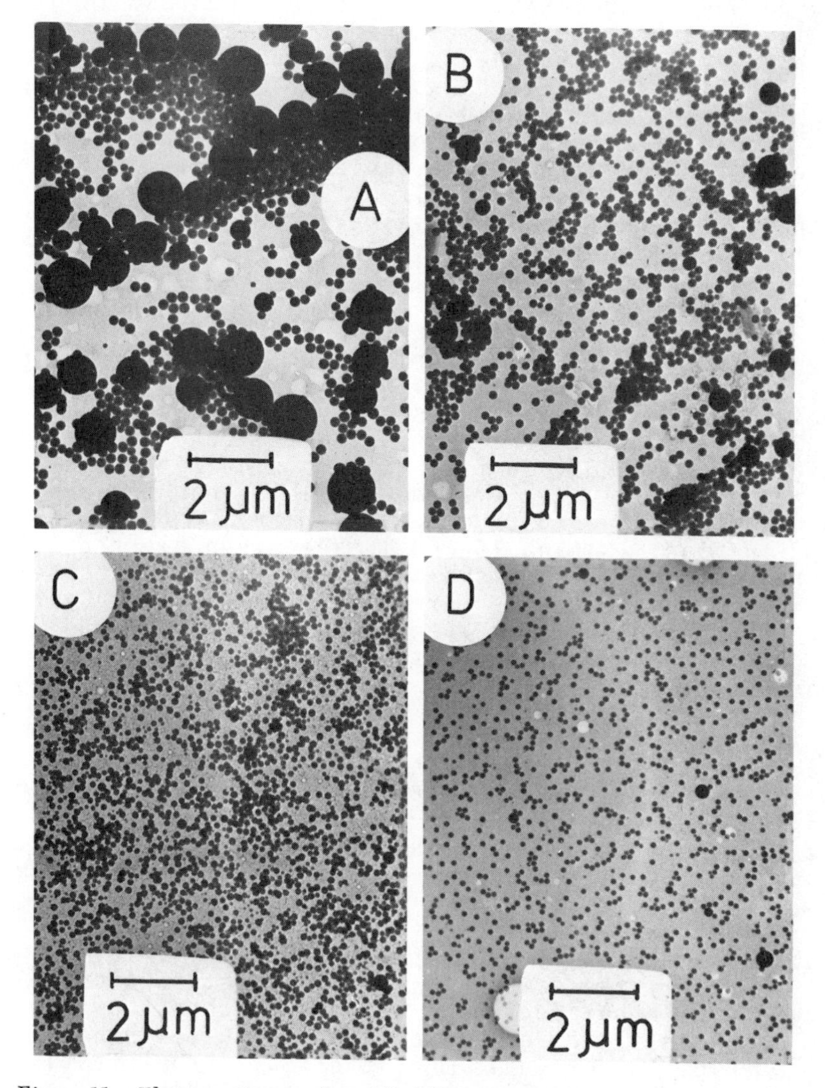

Figure 11. Electron micrographs of final latexes from the experiments given in Figure 10 when polymerized using AIBN as initiator. AIBN = 4.0 g in 166.7 g styrene. (Letters (A), (B), (C), and (D) refer to Figure 10.)

laboratory homogenizer. As expected, the order of mixing had no detectable influence on the resulting emulsions even with long chain fatty alcohols, when such an effective homogenizing device was applied.

In Fig. 12 are given some results where the effect of hexadecane is compared with that of hexadecanol on the stability of the emulsions using the Manton Gaulin homogenizer and with the cationic OPB emulsifier. The figure also includes a result of an experiment with OPB without any additive, which as expected led to a very unstable emulsion.

As shown, the application of the homogenizer for preparing the emulsions did not lead to any increase in the stability of the emulsion with hexadecanol as additive as determined by measuring the amount of adsorbed OPB as a function of time with stirring at 60 °C. Addition of hexadecane leads to an extremely stable emulsion even at the relatively severe conditions of stirring at 60 °C. In fact, the emulsion with hexadecane as additive is even more stable than the one obtained with n-eicosanol.

In Figs. 13, 14, 15 and 16 are given electron micrographs of the emulsions with hexadecanol and hexadecane immediately after preparation and after about 20 hrs stirring at 60 °C. When comparing Fig. 13 and Fig. 2, both with hexadecanol, it appears that the application of the homogenizer has resulted in the formation of a larger number of small droplets in the range of 0.2-0.3 μm. This does not, however, lead to a more stable emulsion. After 20 h stirring at 60 °C all the small droplets have disappeared and the electron micrograph of the emulsion, Fig. 14, is rather similar to the one shown in Fig. 3. This is in agreement with the results of the measurements of adsorption of OPB (Fig. 12).

With hexadecane the electron micrographs of the emulsion immediately after preparation (Fig. 15) show approximately the same size and size distribution as obtained with hexadecanol. With hexadecane, however, the electron micrograph taken after 23 h stirring at 60 °C (Fig. 16), reveals that the small droplets to a large extent are still present and that only a relatively small number of larger droplets in the range of 1 μm have been formed.

Fig. 17 gives results of a polymerization experiment where AIBN was added to the monomer before homogenizing the mixture. The electron micrograph of the latex should be compared with that of the monomer (Fig. 15). It appears that we again have initiation both in the monomer droplets and in the aqueous phase.

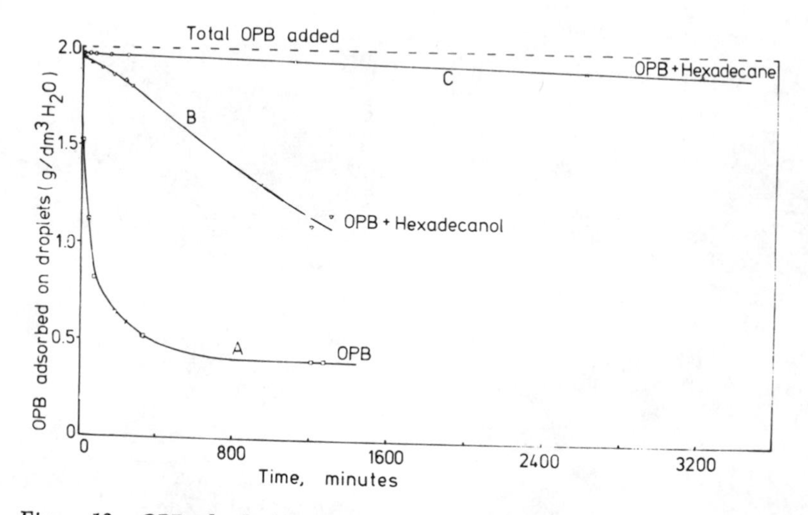

Figure 12. OPB adsorbed on the monomer droplets as a function of time for pure OPB, OPB + hexadecanol (HD), and OPB + hexadecane as emulsifier, homogenized with the Manton Gaulin homogenizer and afterwards stirred at 60°C using the paddle stirrer (600 rpm). Styrene = 333.3 g/dm^3 H_2O, OPB = 2.0 g/dm^3 H_2O. Temp. = 60°C. Molar ratio hexadecanol:OPB = 4:1, hexadecane:OPB = 4:1.

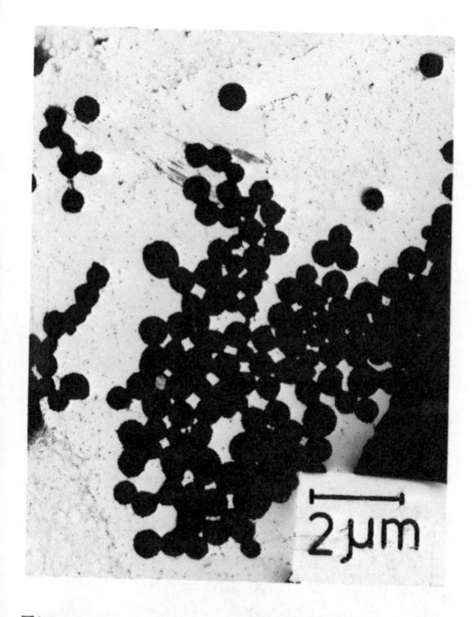

Figure 13. Electron micrograph of monomer emulsion of Figure 12 with OPB + hexadecanol (HD) immediately after homogenization

Figure 14. Electron micrograph of monomer emulsion of Figure 13 after 21 hr stirring at 60°C and 600 rpm

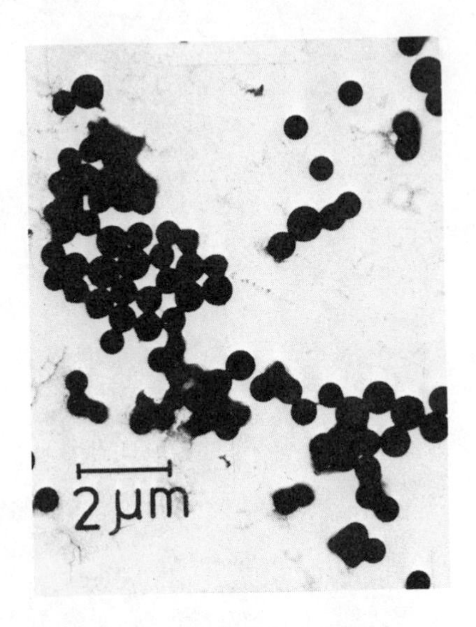

Figure 15. Electron micrograph of monomer emulsion of Figure 12 with OPB + hexadecane immediately after homogenization

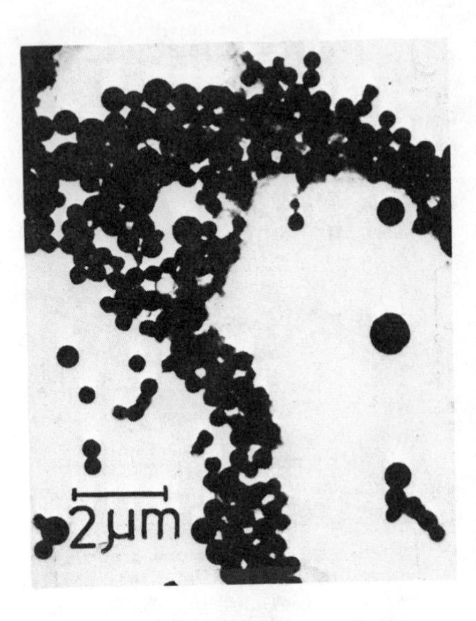

Figure 16. Electron micrograph of monomer emulsion of Figure 15 after 23 hr of stirring at 60°C and 600 rpm

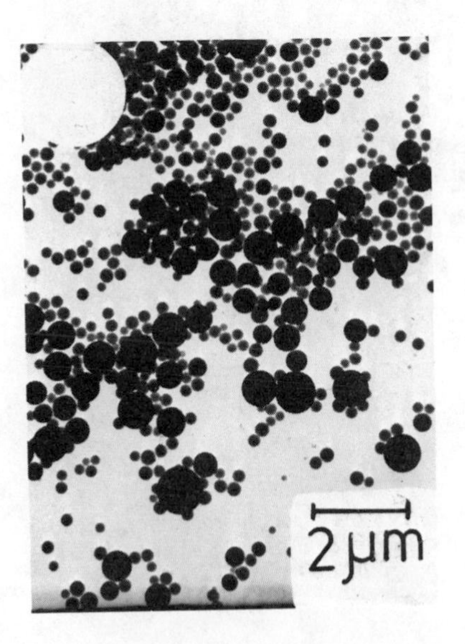

Figure 17. Electron micrograph of final latex from an emulsion prepared as the one in Figure 15 with AIBN added to the styrene before homogenization, homogenized, and polymerized. Styrene = 250 g, H_2O = 750 g, OPB = 2.0 g/dm³ H_2O. AIBN = 6.0 g in 250 g styrene. Molar ratio hexadecane:OPB = 4:1. Temp. = 60°C.

Hallworth and Carless (10) discuss several possibilities for the effect of light liquid paraffin on the stability of emulsions with light petroleum or chlorobenzene as the main components. They seem to prefer an explanation previously advanced by them and several other authors for the effect of fatty alcohol, namely that the increased stability is due to the formation of an interfacial complex between the additive and sodium hexadecyl sulphate. The condenced mixed film will resist coalescence primarily by virtue of its rheological properties. With mixed films of the present type, the importance of the film viscoelasticity lies in its ability to maintain electrical repulsion between approaching droplets by preventing lateral displacement of the adsorbed ions. The effective paraffinic oil has chains at least as long as those of the alkyl sulphate and will be associated by van der Waals' forces with the hydrocarbon chain of the alkyl sulphate at the interface.

Davies and Smith (11) measured the stability of a series oil in water emulsions including benzene and hexane as the main component of the internal phase and with addition of small amounts of hexadecane and long chain fatty alcohols. They measured the change in droplet size with time and found a remarkable increase of the stability by addition of hexadecane which was far more effective than hexadecanol. Davies and Smith reject the explanation of Hallworth and Carless of the effect of hexadecane on the stability of hexane emulsions on thermodynamic grounds. The activity coefficient of an alkanol in alkane is in the region of 15 at 25 ^{o}C, and is approximately independent of chain length. The non-ideality of alcohol as solutes in alkanes can therefore be attributed entirely to the hydroxyl group. Similarly, the adsorption of alkanols from the alkane phase to the o/w interface is dominated by the change in the environment of the hydroxyl group and not to the methylene groups. On the other hand, the long chain alkane (C_{16}) will behave almost ideally in a shorter chain oil (C_6) and there should be no tendency for the longer chain fraction to concentrate at the interface.

Davies and Smith suggest that the effect of addition of small amounts of hexadecane on stability may be due to a prevention of emulsion degradation by molecular diffusion. This approach to emulsion instability was first presented by Higuchi and Misra (12), and was based on the fact that small droplets will demonstrate deviations in physical properties as compared to larger droplets or plane surfaces.

For the case of an oil droplet in water one has for low oil solubilities:

$$C_r = C_\infty \exp (2\gamma M)/(r\rho RT) \quad (2)$$

where C_r is the solubility of droplets of radius r, C_∞ the solubility of an infinitely large droplet, M is the molecular weight of the oil, γ is the interfacial tension, ρ the density of the oil, R the gas constant and T the absolute temperature. The increase in solubility with decreasing r will make the small droplets thermodynamically more unstable with respect to the larger ones.

The rate of dissolution or growth of a droplet can be expressed as

$$G = 4\pi Dr(C_S - C_o) \quad (3)$$

where C_S is the concentration of oil in the water phase in equilibrium with the droplet, C_o is the concentration at some distance from the droplet, large as compared to r, and D is the diffusion coefficient of the oil in water. Because the rate of growth or dissolution of a droplet must be equal to its rate of change of mass, one also may write

$$G = 4\pi r^2 \rho \, dr/dt \quad (4)$$

Higuchi and Misra consider a bimodal emulsion with n_2 droplets of radius r_2 and n_1 droplets of radius r_1. From the above equations and mass balance considerations the rate of change of the small sized droplets (r_1) was found to be:

$$\frac{dr_1}{dt} = -\frac{DC_\infty K}{\rho r_1^2}\left[\frac{n_2(r_2 - r_1)}{n_1 r_1 + n_2 r_2}\right] \quad (5)$$

where $K = (2\gamma M)/(\rho RT)$.

The rate of change will be directly proportional to the solubility of the oil. Moreover, it should be noted that if the radius of the droplet is decreased ten fold the rate of change will occur a thousand times faster.

Higuchi and Misra have also treated the more complex situation of a polydisperse emulsion. Davis and Smith find that the equation developed for degradation by the molecular diffusion route as expressed by Eq. (5) may explain the rate of degradation of different emulsions and the effect of the solubility of the internal phase on the stability of the emulsion. Higuchi and Misra also consider the effect of addition of a small quantity of an additive which is considerably less water soluble than the main constituent of the internal phase.

For a two component case one may write:

$$C_{xW} = k_x C_{xL} \exp(K_x/r) \quad (6)$$

$$C_{zW} = k_z C_{zL} \exp(K_z/r) \quad (7)$$

where C_{xW} is the equilibrium concentration of the main component x in water with a droplet size r, C_{xL} is the concentration in the droplet, C_{zW} and C_{zL} are the corresponding values for the minor component z,

$$K_x = 2\gamma\bar{V}_x/RT, \qquad K_z = 2\gamma\bar{V}_z/RT \quad (8)$$

k_x and k_z are the respective distribution coefficients, and $\bar{V}_x$ and $\bar{V}_z$ are the respective partial molal volumes of the components x and z in the droplet.

If $k_z << k_x$ the rate of degradation of the emulsion will be governed by the diffusion rate of z. Physically what happens is that the relatively slow rate of diffusion of z permits the x component to remain essentially equilibrated among the droplets. Therefore changes in the system occur only as fast as the change in distribution of the slowest diffusing component. Thus the degradation rate may be retarded by a factor $K_p = k_x/k_z$.

Further considerations lead to the conclusion that droplets in the micro range may be "completely" stabilized by addition of 1% of a "water insoluble" compound. For smaller size ranges the amount of the water insoluble compound must be proportionately greater.

In this treatment it is assumed that there is no restriction to diffusion by any interfacial film and there is no interaction between x and z or between the emulsifier and the compounds of the internal phase.

Hallworth (10) finds that in most cases the rate of degradation of the emulsions is considerably faster than may be estimated from the Higuchi-Misra theory. It should be noted that the droplet sizes in Hallworth's emulsions are considerably greater than those investigated by Davis and Smith. The importance of the two possible routes of degradation of the emulsions, coalescence or molecular diffusion, may be dependent upon the droplet size and size distribution. Also an interfacial coherent film may reduce the demulsification by either mechanism, i.e. by reducing the rate of coalescence or by presenting an interfacial barrier to molecular diffusion (13, 14).

Application of sodium hexadecyl sulphate and hexadecane give similar results as reported in the present paper with OPB and hexadecane. Subsequent polymerization using $K_2S_2O_8$ as initiator resulted in predominant initiation in the monomer droplets (15).

Summary and Conclusions

Emulsions of styrene with mixtures of octadecyl pyridinium (OPB) and long chain fatty alcohols were prepared in accordance with previous methods. Premixing of OPB with alcohol in water prior to addition of monomer leads to fine emulsions of styrene, obtained with moderate stirring. The stability of the emulsions increases markedly with increasing chain length of the alcohol.

A method for obtaining electron micrographs of the monomer emulsion has been developed employing an OsO_4-staining technique. This allows a direct comparison of the monomer droplet size distribution with that of the particles in the final latex. Polymerizations of the monomer emulsions were carried out with 2,2'-azobisisobutyronitrile (AIBN), benzoyl peroxide (BP) and cumene hydroperoxide (CHP). Depending upon the amount and ratio of emulsifier used for emulsification of monomer, the relative degree of initiation of polymerization in the droplets and in the aqueous phase could be varied. Emulsions were also prepared by means of a Manton Gaulin homogenizer. With hexadecanol as additive this gave emulsions which showed approximately the same stability as obtained with the ordinary preparative method. When hexadecanol was replaced by hexadecane as additive one obtained with the same small amount of additive a finely dispersed styrene emulsion which was far more stable than could be obtained with fatty alcohol of the same chain length.

Acknowledgements

Thanks are due to the Norwegian Agency for International Development (NORAD) for the award of a fellowship to A.R.M. Azad.

Literature cited

1. Ugelstad, J., El-Asser, M., and Vanderhoff, J., J.Pol.Sci., Polymer Letters (1973), 11, 503.
2. Ugelstad, J., Hansen, F.K., and Lange, S., Makromol.Chem. (1974), 175, 507.
3. Lange, S., Ugelstad, J., and Hansen, F.K., Proceedings of the 5th Scandinavian Symposium on Surface Chemistry (1973).
4. Azad, A.R.M., Fitch, R.M., and Ugelstad, J., ACS Symp.Series, No. 9, (1975), 135.
5. Lankveld, I.M.G., and Lyklema, J., J.Colloid Interface Sci. (1972), 41, 475.
6. Davies, J.T. and Haydon, D.A., Proc., 2nd Intern. Congress Surface Activity, London (1957), 1, 417.
7. Carless, J.E. and Hallworth, G.W., J. Colloid Int. Sci., (1968), 26, 75
8. Hallworth, G.W. and Carless, J.E., J.Pharm.Pharmacol. (1972), 24, Suppl. 71.
9. Hallworth, G.W. and Carless, J.E., Ibid (1973), 25, Suppl.87.
10. Hallworth, G.W. and Carless, J.E., In "Theory and Practice of Emulsion Technology", Symp. Brunel University, 10-18. September 1974, Preprint p. 265.
11. Davis, S.S. and Smith, A., Ibid p. 285.
12. Higuchi, W.J. and Misra, J., J.Pharm.Sci. (1962) 51, 459
13. Ghanem, A., Higuchi, W.J. and Simonelli, A.P., J.Pharm.Sci. (1969), 58, 165.
14. Goldberg, A.H., and Higuchi, W.J., Ibid (1969), 58, 1341.
15. Ugelstad, J. and Hansen, F.K., to be published.

2

The Effects of Polyvinyl Alcohols on the Polymerization of Vinyl Acetate

ALEXANDER S. DUNN, CHRISTOPHER J. TONGE, and SAMIR A. B. ANABTAWI

Department of Chemistry, The University of Manchester Institute of Science and Technology, Manchester M60 1QD, England

Polyvinyl alcohol is an important component in the emulsifying systems used in practical formulations for the emulsion polymerisation of vinyl acetate (1) but, although it is often used as a stabilising agent in suspension polymerisations, its use has not been found advantageous in the emulsion polymerisation of other monomers (2). Despite the unique importance of polyvinyl alcohol in the vinyl acetate polymerisation, few reports of its effects have been published. O'Donnell, Mesrobian, and Woodward (3) reported the effect of varying polyvinyl alcohol concentration to be in accordance with the Smith-Ewart theory (4) but it has subsequently been shown (2) that their results actually imply a first order dependence of rate on emulsifier concentration. By contrast, we found (5) that lower concentrations of the same grade of polyvinyl alcohol greatly reduced the rate of polymerisation of aqueous vinyl acetate solutions. These observations are reconciled by the recent work of Dimonie et al. (6) who find that the order in polyvinyl alcohol varies with vinyl acetate concentration from - 2.4 for a 1.5 mol dm^{-3} solution to 0.2 for a saturated solution. For vinyl acetate solutions at lower temperatures, Hayashi, Iwase, and Hojo find that the order in polyvinyl alcohol depends on the initiator used, being zero for a persulphate redox system and a low order (0.2 - 0.3) which increases with temperature for a peroxide redox system.(7).

Commercial grades of polyvinyl alcohol are customarily characterised by their residual acetyl content and the viscosity of their 4% aqueous solutions. Grades containing 10 - 20 mole per cent acetyl are preferred for use as emulsifying agents (8). These grades have a block copolymer structure: random copolymers can only be produced by complete hydrolysis and reacetylation and have comparatively low surface activity. The distribution of residual acetyl groups is critically affected by the hydrolysis procedure adopted: the 'blockiness' of the copolymer (and its surface activity) can be increased by adding a proportion of benzene to the alkaline hydrolysis solvent (9). Compatability of these polyvinyl acetate blocks with the latex particles explains the unique suitability of these grades of polyvinyl alcohol for the vinyl acetate emulsion polymerisation.

Experimental

Procedure. 2% v/v solutions of redistilled vinyl acetate were prepared in distilled water from which oxygen had been substantially expelled by boiling. The initiator was 0.020 % w/v potassium persulphate. Although the same rate of polymerisation is observed with this procedure as when vacuum degassing is used (5), the size of the latex particles formed is reduced slightly (10) presumably because of adsorption of surface-active oligomer formed during the inhibition period. Rates were determined dilatometrically at 60.0°C. At this temperature aqueous solutions contract by 15.7 % on complete polymerisation (5, 11) much less than in bulk polymerisation where the contraction is 26.8 %. This is attributable (10) to the effect of vinyl acetate on the hydrogen bonded structure of water: the contraction due to the polymerisation of the solute is partially compensated by the expansion of the solvent when the polymer precipitates. Stirred dilatometers were used for the emulsion polymerisations. The contraction observed after complete polymerisation of an 8.0 % v/v emulsion was 18% although a contraction of 21.6 % was expected on the basis of the figures given above for aqueous solution and bulk polymerisations: this may be a consequence of the increased solubility of vinyl acetate in aqueous solutions of polyvinyl alcohol(12).

Materials. The polyvinyl alcohols were commercial samples used as received. The DuPont 'Elvanol' was a sample used previously (5) obtained in 1959 as typical of the material in commercial use at that time: the rates of polymerisation observed with it were identical with those observed earlier (5). The 'Polyviol' grades were manufactured by Wacker Chemie in 1968, the 'Gohsenol' grades by Nippon Gohsei Kagaku Kogyo K.K., Osaka, Japan in 1969, and the 'Vinol' grades by Air Products & Chemicals Inc., Calvert City, Kentucky in 1974: these last were specially selected single batches, not blends of several batches adjusted to conform to specification.

Maximum Rates. Conversion-time curves are S-shaped but the rate is usually constant at its maximum value (R_{max}) between about 20% and 70% polymerisation: this value is used to characterise the rate of reaction.

Results

Effect of Polyvinyl Alcohol Concentration. Fig. 1 shows results with 2% v/v aqueous vinyl acetate solutions and 8% v/v emulsions together with the results of earlier work (3, 5). 'Gohsenol' GH-23 has a similar acetyl content to the 'Elvanol' used in the earlier work although its molecular weight is higher. Although it does reduce the rate of polymerisation of 2% solutions at higher concentrations, it has little effect on the rate at 0.36% w/v which was enough to halve the rate with the 'Elvanol'. On the other hand concentrations in the same range increase the rate of polymerisation of 8% emulsions as reported by O'Donnell et al. (3) although the actual rates are much higher at a similar initiator concentration despite the lower temperature used in the present experiments.

Fig. 2 shows the results for solutions in more detail. Although excellent repeatability was attained using a similar technique with 1% v/v solutions (13), it appears that the higher vapour pressure of 2% solutions (86 mmHg compared with 38 mmHg at 25°C (14)) can lead to some loss of monomer when filling dilatometers: the higher of the observed rates are thus likely to be the more reliable and are given more weight in drawing curves. It is clear that there is a considerable difference between the effects of different grades even those from the same manufacturer.

Effect of Acetyl Content of Polyvinyl Alcohol. Fig. 3 shows the effect of a low concentration (0.10% w/v) of 'Gohsenol' samples which have similar molecular weights but varying acetyl contents on the rate of polymerisation of vinyl acetate solutions. Increase of acetyl content decreases the rate of polymerisation. However this effect cannot account for the difference between the samples shown in Fig. 2 because these differ by only 0.1% in acetyl content. On the other hand any effect of acetyl content in the 'Vinol' series was small: rates of 5.8, 5.9, and 5.4 %/min were observed using 'Vinol' 325, 425, and 523 which have similar molecular weights and acetyl contents of 1.8, 4.0, and 12 mole % respectively.

Effect of Molecular Weight of Polyvinyl Alcohol. The table shows the rates observed with a standard concentration (0.36% w/v) of grades of similar acetyl content (12-13 mole %) for the polymerisation of 2% vinyl acetate solutions. The 4% solution viscosities are either the means of the range specified for the grade or, for the 'Gohsenol' series, actual batch values.

It is clear that the effects of other differences between samples far outweigh any effect that variation of molecular weight may have.

Possible Effect of Charged End-groups. A possible reason for differences between samples might be the process used for polymerising the original polyvinyl acetate. Emulsion polymerisation is likely to introduce a proportion of ionic (sulphate or carboxyl) end-groups which would not be expected if bulk polymerisation with benzoyl peroxide had been used. An Antweiler Microelectrophoresis apparatus was used to measure rates of electrophoresis of polyvinyl alcohols in solution in a pH 7.8 phosphate buffer. No significant difference was observed between

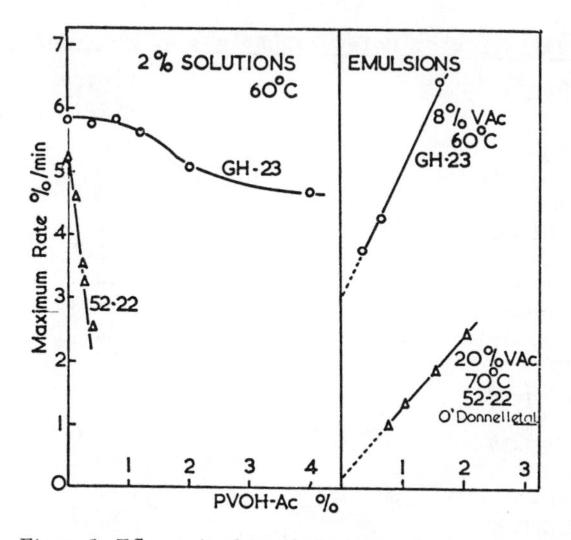

Figure 1. Effects of polyvinyl alcohol grades on rates of solution and emulsion polymerization of vinyl acetate

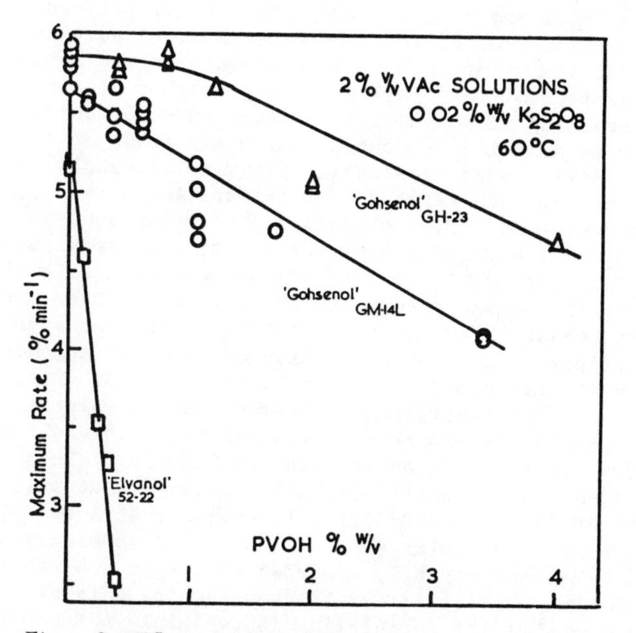

Figure 2. Effect of polyvinyl alcohol concentration on rate of vinyl acetate polymerization

Effect of Variation of Molecular Weight of Polyvinyl Alcohol

Grade	4% Viscosity (cp)	R_{max} (%/min)
Blank	-	5.8
Polyviol M05/140	5	3.1
Vinol 205	5	5.4
Gohsenol GL-05	5.2	4.5
Gohsenol GL-08	10.0	5.0
Polyviol M13/140	13	3.6
Gohsenol GM-14L	18.0	4.7
Elvanol 52-22	22	3.3
Vinol 523	22	5.1
Gohsenol GM-14	22.7	5.1
Polyviol W25/140	25	2.7
Gohsenol GH-17	30.0	4.5
Polyviol W40/140	40	3.3
Vinol 540	40	3.2
Gohsenol GH-20	41.7	5.8
Gohsenol GH-23	55.4	5.3

'Gohsenol' GM-14L and GH-23 (which were found to be weakly adsorbed) and 'Polyviol' W25/140 (which was strongly adsorbed by polyvinyl acetate latex).

Adsorption of Polyvinyl Alcohols by Polyvinyl Acetate Latex. It would be expected that the 'blockier' samples of polyvinyl alcohol would be the more strongly adsorbed on polyvinyl acetate latex particles. Adsorption isotherms (Fig. 4) were determined by a procedure similar to that used by Johnson and Lewis (15) to study adsorption of polyvinyl alcohol by carbon blacks. Polyvinyl alcohol remaining in solution is determined colorimetrically using a boric acid-iodine reagent which minimises the effect of structural differences between the polyvinyl alcohols on the colour developed: nevertheless the calibration curves for the different grades do differ slightly. The latex used was prepared by polymerising a 2.0% v/v solution of vinyl acetate for 2½ h at 60°C using 0.020% w/v potassium persulphate initiator. Residual initiator was decomposed by boiling for 5 h. The latex was monodisperse: the diameter of the particles was approximately 260 nm. It was found that adsorption of two grades ('Gohsenol' GM-14L, GH-23) was complete within 15 min at 20°C. Fig. 4 shows that there are large differences between the extents to which the different grades are adsorbed. Surprisingly 'Gohsenol' NH-17 with a very low residual acetyl content (1 mole %) is as strongly adsorbed as 'Gohsenol' GH-17 (12 mole % acetyl) whilst 'Gohsenol' GM-14L and GH-23 (also 12 mole % acetyl but with lower and higher molecular weights than GH-17) are only weakly adsorbed. The more strongly adsorbed grades do seem to be those which have the greatest effect in reducing the rate of polymerisation but the correlation is certainly not precise

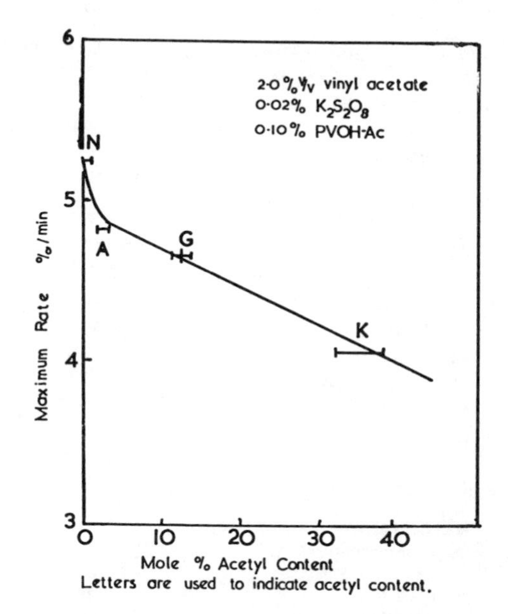

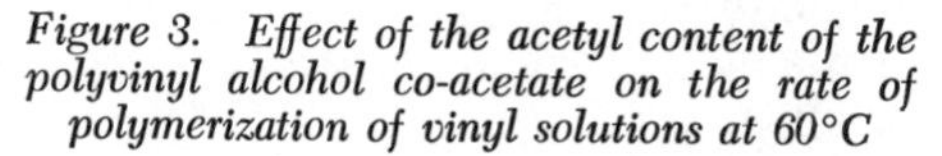

Figure 3. Effect of the acetyl content of the polyvinyl alcohol co-acetate on the rate of polymerization of vinyl solutions at 60°C

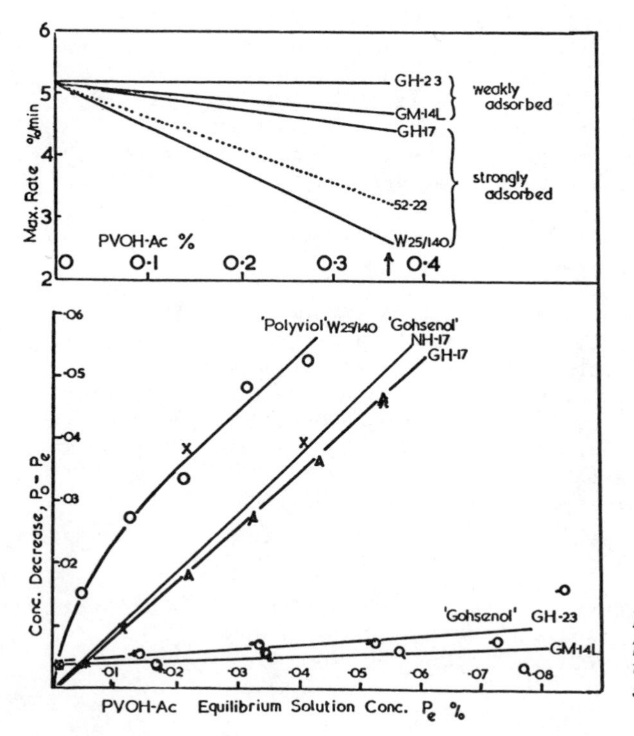

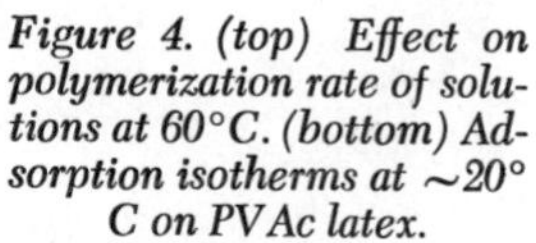

Figure 4. (top) Effect on polymerization rate of solutions at 60°C. (bottom) Adsorption isotherms at ~20° C on PVAc latex.

which implies that some other factor is also influencing one or other of these effects. It is possible that there are significant differences in adsorption enthalpies so that amounts adsorbed under polymerisation temperatures cannot be inferred from isotherms determined at room temperature.

Seeded Polymerisation after Polyvinyl Alcohol Adsorption.

To assess the effect of an adsorbed layer of polyvinyl alcohol on the rate of polymerisation, the rates of seeded polymerisation were measured after adsorption of various grades of polyvinyl alcohol on the seed latex. The seed latex was prepared in the same way as the latex used in the adsorption experiments. After adsorption of polyvinyl alcohol from a 0.10 % w/v solution, the concentration of the seed latex was 0.945 % w/v in a 2.0 % v/v vinyl acetate solution containing 7.4×10^{-6} mol dm^{-3} potassium persulphate. The rates of polymerisation (the means of two to four experiments) observed at 60°C are tabulated below.

Polyvinyl alcohol grade	R_{max}/% min^{-1}
Blank	3.53
Gohsenol NH-17	3.87
Gohsenol GH-17	3.61
Gohsenol GM-14 L	3.21
Gohsenol GH-23	3.18
Polyviol W25/140	2.68

Polyviol W25/140 which is the most strongly adsorbed grade (Fig. 4) has the greatest effect in reducing the rate of polymerisation but Gohsenol GM-14 L and GH-23 which are only weakly adsorbed (Fig. 4) are more effective than Gohsenol NH-17 and GH-17 which are more strongly adsorbed. Once again it seems that some other factor is having more effect on the rate of polymerisation than the extent of adsorption of the polyvinyl alcohol.

Discussion

We originally suggested[5] that transfer to the polyvinyl alcohol could explain the effect of 'Elvanol' 52-22 in reducing the rate of polymerisation of vinyl acetate solutions. The importance of this transfer reaction has recently been confirmed by Dimonie et al.(6) who use ferrous ion/hydrogen peroxide redox initiation. However Okamura, Yamashita, and Motoyama determined the transfer rate constant of vinyl acetate to polyvinyl acetate and polyvinyl alcohol in homogeneous solution: they found for

transfer to polyvinyl acetate $k_{f(Ac)} = 1.5 \times 10^{-4} mol^{-1} dm^3 s^{-1}$,
transfer to polyvinyl alcohol $k_{f(OH)} = 35 \times 10^{-4} mol^{-1} dm^3 s^{-1}$

so that the more extensively hydrolysed grades of polyvinyl alcohol might be expected to have most effect whereas the experiments with the 'Vinol' samples suggest that variation of acetyl content has little effect and those with 'Gohsenol' samples (Fig. 2) that the effect in decreasing the rate increases with acetyl content. However the hydrophilic blocks of the emulsifier may not be accessible to polymer radicals in the latex particles and the effect of increasing acetyl content might be to increase the concentration of accessible groups adsorbed on the surface of latex particles.

We have confirmed (10) that polyvinyl alcohol is readily oxidised by potassium persulphate (17) and induces a much enhanced rate of decomposition of the persulphate but we have also shown (18) that any effect under polymerisation conditions is small so that decreased rates cannot be attributed to reduction in initiator concentrations below their nominal value by wastage in side reactions.

Reynolds (19) has given a preliminary report of experiments which show that persulphate resembles periodate in oxidising the 1:2-glycol groups of which there are about 2% in polyvinyl alcohol although the reaction is not quantitative. Conjugated unsaturated ketone groups are also formed: these may retard polymerisation. Some samples of polyvinyl alcohol (e.g. a sample of 'Gelvatol' 20-30 used in other work) have undergone some oxidation during manufacture but no conjugated ketone groups could be detected in the 'Elvanol' 52-22 sample which had a large effect in retarding polymerisation.

However, in a recent publication, Shirinyan, Mnatsakanov, et al. (20) find that differences between the rates of vinyl acetate emulsion polymerisation observed with samples of similar polyvinyl alcohols manufactured by the same process in three different factories could be attributed to a condensation product of acetaldehyde derived from hydrolysis of residual vinyl acetate: this gave rise to a conjugated ketone type ultra-violet spectrum and could be extracted from the polyvinyl alcohol under suitable conditions. This could be the uncontrolled factor which appears to have confounded many of the experiments reported here. Even more recently the same laboratory (21) has reported that there is an optimum sequence length of hydroxyl groups in the polyvinyl alcohol-acetate block copolymer for polymerisation rate and dispersion stability.

The blocky structure of incompletely hydrolysed grades of polyvinyl alcohol probably results in micellisation in aqueous solution despite an early report to the contrary (3). Benzene has been found to be solubilised in polyvinyl alcohol solutions to an extent which increases with acetyl content (22). Enfiadzhyan et al. (12) show that the equilibrium solubility of vinyl esters is much increased at 60°C in aqueous solutions of polyvinyl alcohol (0.92 % acetyl): similar concentrations of methanol, ethanol etc. may well have a similar effect to this. The solubility of vinyl acetate increases linearly reaching 62% v/v in a 7% polyvinyl alcohol solution. However 11 hours had to be allowed to reach saturation so that a slow rate of solution may explain the persistence of monomer droplets under normal polymerisation conditions when the times involved would probably not exceed two or three hours. An increase in the solubility of the monomer in the aqueous phase would reduce its concentration in the latex particles and, consequently, the rate of emulsion polymerisation and variations in the amount of monomer dissolved in the aqueous phase may explain the comparatively poor repeatability of the rates observed in the presence of polyvinyl alcohol.

Tuzar and Kratochvil (23) have reported that styrene-butadiene block copolymers micellise in selective solvents for polystyrene and solubilise large amounts of polybutadiene homopolymer. Since the surface active grades of polyvinyl alcohol are polyvinyl alcohol-acetate block copolymers and water is a selective solvent for polyvinyl alcohol a similar effect may be expected which could affect the course of the vinyl acetate emulsion polymerisation.

Summary

The effects of commercial grades of polyvinyl alcohol obtained from several manufacturers on the rate of polymerisation of vinyl acetate aqueous solutions and emulsions initiated by potassium persulphate at 60°C have been investigated. Increasing concentrations of polyvinyl alcohol in the same range tend to reduce the rate of polymerisation of the solutions but increase the rate of polymerisation of emulsions. Considerable differences were noted between the effects of nominally similar grades from different sources. Attempts to correlate the effect on the rate of polymerisation of vinyl acetate solutions with the acetyl content and molecular weight of the polyvinyl alcohol appear to have been confounded by an uncontrolled variable. A subsequent publication from the U.S.S.R. indicates that this may be the concentration of a by-product formed during the hydrolysis of the polyvinyl acetate. Other factors which may account for the effects are also discussed.

Literature Cited

1. Warson, H., "Synthetic Resin Emulsions", Benn, London, 1972.
2. Reynolds, G.E.J., and Gulbekian, E.V., S.C.I. Monograph (1968) 30, 131.
3. O'Donnell, J.T., Mesrobian, R.B., and Woodward, A.E., J. Polym. Sci. (1958) 28, 171.
4. Smith, W.V., and Ewart, R.H., J. Chem. Phys. (1948) 16, 592.
5. Dunn, A.S., and Taylor, P.A., Makromol. Chem. (1965) 83, 207.
6. Dimonie, V., Donescu, D., Munteanu, M., Hagiopol, C., and Gavat, I., Rev. Roum. Chim. (1974) 19, 903.
7. Hayashi, S., Iwase, K., and Hojo, N., Polymer J. (1972) 3, 226.
8. Toyoshima, K., in Finch, C.A. ed. 'Polyvinyl Alcohol', Chapter 2, Wiley, London, 1973.

9. Hayashi, S., Nakano, C., and Motoyama, T., Kobunshi Kagaku, (1963) 20, 303.
10. Tonge, C.J., Ph.D. Thesis, Manchester, 1971.
11. Napper, D.H., and Parts, A.G., J. Polym. Sci., (1962) 61, 113.
12. Enfiadzhyan, M.A., Nazaryan, L.N., and Akopyan, A.E., Arm. Khim. Zhur. (1971) 24, 839.
13. Dunn, A.S., and Chong, L. C.-H., Br. Polym. J. (1970) 2, 49.
14. Coniglio, O.B., and Parts, A.G., Makromol. Chem. (1971) 150, 263.
15. Johnson, G.A., and Lewis K.E., Br. Polym. J. (1969) 1, 266.
16. Okamura, S., Yamashita, T., and Motoyama, T., Kobunshi Kagaku, (1958) 15, 170.
17. Beileryan, L.M., Samvelyan, A.I., Chaltkyan, O.A. and Vardanyan, Arm. Khim. Zhur. (1967) 20, 338.
18. Dunn, A.S., and Tonge, C.J., Polymer Preprints (1972) 13, 1261.
19. Reynolds, G.E.J., contribution to C.S. Colloid and Interface Science Group Discussion on 'Characterisation and Colloidal Properties of Polymer Latices', Bristol, 26/27 June 1974.
20. Shirinyan, V.T., Mnatsakanov, S.S., Shirikova, G.A., Poznyakova, F.O., Popova, G.S., and Khvostyntseva, T.V., Plast. Massy. (1974) 15 (8): English translation, Int. Polym. Sci. Tech. (1975) 2, T82 (1).
21. Shirinyan, V.T., Mnatsakanov, S.S., Gromov, V.V., Perlova, T.I. and Ivanchev, S.S., Vysokomol. Soed. A (1975) 17, 182 (1), cf. Chem. Abstr. (1975) 82, 140547b.
22. Shakhova, Ye.M., and Meyerson, S.I., Vysokomol. Soed. A, (1972) 14, 2097: English translation, Polym. Sci. USSR, (1972) 14, 2354.
23. Tuzar, Z., and Kratochvil, P., Makromol. Chem. (1972) 160, 301, (1973) 170, 177.

3

Micellar Size Effect in Emulsion Polymerization

IRJA PIIRMA and PAO-CHI WANG*

Institute of Polymer Science, The University of Akron, Akron, Ohio 44325

One of the methods for preparing monodispersed latices, i.e. latices of uniform particle size distribution, is to use mixed surfactants as the emulsifier in the emulsion polymerization process. The term mixed surfactants, in general, refers to mixtures of ionic and nonionic surfactants.

Besides giving latices of narrow particle size distribution, mixed surfactant systems have shown several other interesting characteristics which lighten some aspects concerning the mechanism of particle nucleation in emulsion polymerization process.

Woods, Dodge, Krieger, and Piece (1) obtained monodispersed polystyrene latices at about 50 percent conversion from a series of polymerization recipes with mixed surfactants of different ionic to nonionic ratios. They found that the size of the polystyrene particles in these latices decreased with increasing amount of the ionic component in the surfactant mixtures. To interpret their observations, they adopted the theory of mixed micelle formation by mixed surfactant in aqueous solutions. This concept had been previously confirmed by Nakagawa and Inoue (2) with electrophoresis experiments. Later, Kuriyama, Inoue and Nakagawa (3), showed that the size of the mixed micelles decreased with increasing amounts of ionic component in a mixed surfactant system. Thus, Woods, Dodge, Krieger and Piece recognized the important relationship between the size of the latex-particles

*Present address: Lucidol Division, Pennwalt Corp., Buffalo, N. Y.

and the size of the mixed micelles in their polymerization experiments and proposed that the larger polystyrene particles were associated with the larger micelles.

In the polymerization of styrene, using AIBN as initiator, Medvedev and co-workers (4) found that rates of polymerization showed a maximum and, the particle size, a minimum value as the concentrations of the ionic component in the mixed surfactant increased.

In the polymerization of styrene, using potassium persulfate as initiator, Roe (5) observed that the total number of particles in latices depended on the composition of the mixed surfactants and not on the total number of micelles. Therefore, he devaluated the micellar nucleation mechanism for emulsion polymerization as proposed by Harkins(6)-Smith-Ewart(7).

In a persulfate initiated styrene polymerization, using sodium lauryl sulfate, (SLS), and Emulphogene BC-840, as the mixed surfactants, Kamath(8) and Wang(9) found that the rate of polymerization increased rapidly with small increases in the ionic component, SLS. Their recipe of polymerization is reproduced in Table I, and Figure 1 shows the rate of polymerization, R_p, plotted against SLS concentration in the mixed surfactants. In that plot, the concentration of SLS was expressed in parts of SLS/5-parts BC-840/100 parts styrene. Based on this study, Kamath (8) proposed that particle-nucleation in his system was controlled by micellar nucleation mechanism following Harkins' and Smith-Ewart's theory.

The above cited information showed unanimously that, in a mixed-surfactant system of emulsion polymerization, the composition of the mixed surfactant affects the rate of polymerization. Since by Harkins-Smith-Ewart theory, rate of polymerization is proportional to the total number of particles in the system, composition of mixed surfactants seems to affect the efficiency of nucleation.

The kinetic studies of Kamath (8) is of particular interest. In emulsion polymerization of styrene, again according to Smith-Ewart's theory, a logarithmic plot of the number of particles formed, and, therefore, the rate of polymerization, R_p, against total surfactant concentrations, C_s, should generate a straight line.

Table I

Polymerization Recipe
(Ref. 8,9,15)

Ingredient	PHM
Water	180
Styrene	100
BC-840*	5
SLS**	Variable
K-persulfate	0.18
KOH	0.075

*BC-840: Emulphogene BC-840
Tridecyloxypoly(ethyleneoxy)-ethanol
Donated by GAF Company

**SLS: Sodium lauryl sulfate

However, by neglecting the variations in the composition of the mixed-surfactants, Kamath's data and the preliminary experimental results of the present study did not generate a straight line by such a plot, but a curve instead as illustrated by Figure 2. Therefore, it seemed that some other factor or factors was (were) affecting the rate of polymerization, or more precisely, the efficiency of nucleation. Since this system of polymerization was using mixed surfactant as emulsifier, mixed-micelles should have been formed, and, based on the findings of Kuriyana, Inoue and Nakagawa (3), the sizes of such mixed micelles should vary with varying composition of the mixed-surfactants. This variation was confirmed by light scattering measurements, applying a method used by other investigators (3,10,11, 12) for the same purpose. The results obtained are summarized in Figure 3, where it can be seen that the micellar weight of the mixed micelles dropped rapidly with increasing value of surfactant ratio, r, of the mixed-surfactants. Surfactant ratio, r, is defined as r = moles SLS/mole BC-840. Since it is reasonable to assume that the size of the mixed micelle is proportional to its weight, Figure 3 implies that each specific value r stands for a specific value of micellar size in this surfactant system.

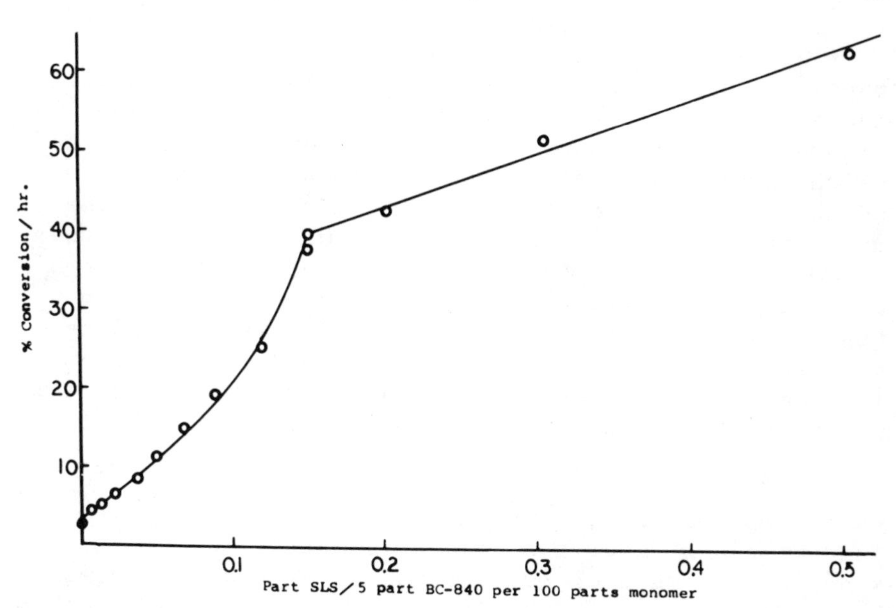

Figure 1. Effect of mixed surfactant composition on polymerization rate (SLS-BC 840 system; Ref. 8)

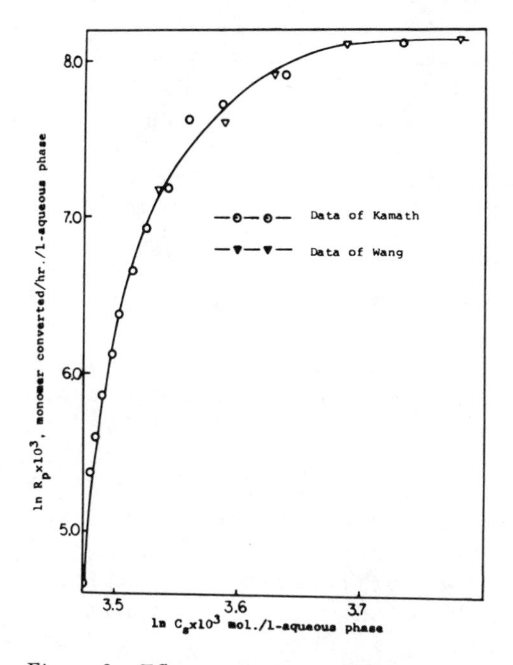

Figure 2. Effect of total surfactant concentration with varying surfactant ratio on polymerization rate

For a given total concentration of mixed surfactants, C_s, with known compositions, the value of surfactant ratio, r, can be calculated. Thus, data on Figures 2 and 3 can be combined to generate Figure 4 which shows a possible relationship between the size of the micelles and the rate of polymerization. Then, other things being equal, the rate of polymerization, R_p, of an emulsion polymerization process should be a function of two variables, namely, the total concentration of mixed surfactants, C_s, expressed in moles per liter aqueous phase, and the micellar weight, M_m, in grams per mole, which also can be expressed in terms of surfactant ratio, r, or by micellar size. Thus,

$$R_p = f_n (C_s, M_m) \quad (1)$$

which implies that, at a given concentration of initiator and under a specified temperature, the kinetics of emulsion polymerization should be expressed by a three parameter model, (R_p, C_s, M_m), rather than by the classical two parameter one, (R_p, C_s). This is illustrated schematically in Figure 5. For a single surfactant system of emulsion polymerization with rather narrow range of variations in surfactant concentration, the size and shape of the micelles should be constant, and thus the three parameter model of polymerization kinetics reduces itself to the classical two parameter one, and the logarithmic plot of rate against concentration yields a straight line with slope, x, to fit the relationship of

$$R_p \propto C_s^{\,x} \quad (2)$$

and x = 0.6 is one of the essentials of Smith-Ewart theory.

Therefore, the nonlinear relationship between rate of polymerization and the total surfactant concentration, as shown in Figure 2, was believed to be caused by a change in micellar size. Thus, the purpose of the present study was to verify the validity of the concept of micellar size effect in emulsion polymerization kinetics. Furthermore, although the Harkins-Smith-Ewart theory of micellar nucleation was proposed in 1948, and has found widespread application ever since, its validity is still challenged even for the case of polymerization of styrene (5). If micellar

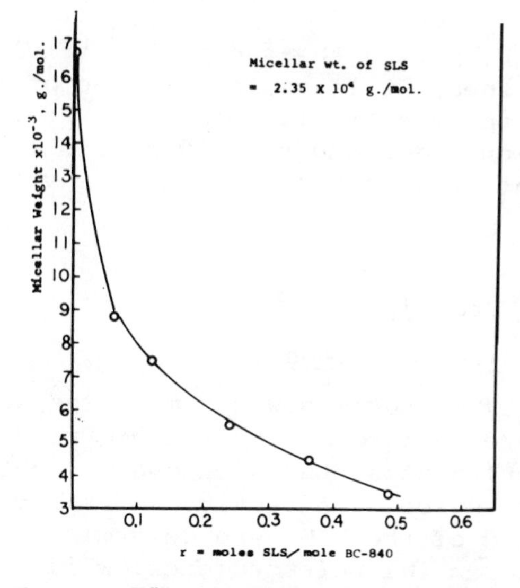

Figure 3. Effect of surfactant ration on micellar weight of mixed surfactant (SLS BC-840 system)

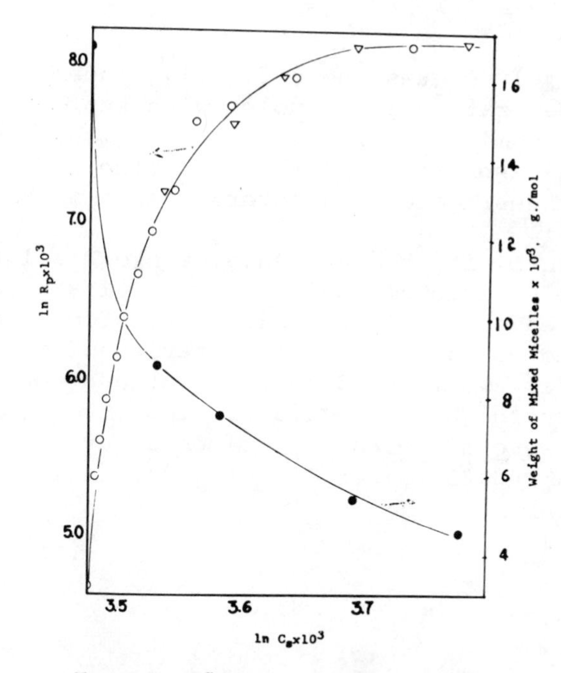

Figure 4. Effect of surfactant concentration with varying surfactant ratio on size of micelles and on polymerization rate

size should be proven to affect the efficiency of nucleation, then, at least in the case of styrene polymerization, the micellar size should be considered as an additional and important variable in the aqueous phase initiation as proposed by some investigators (5,13,14).

Experimental Procedures

I. Micellar Size Measurements.

Turbidity measurements for micellar weight calculations were carried out using the Price-Phoenix light scattering apparatus with green light of mercury as the light source. Refractive index increments of the BC-840 and of the SLS were determined by Brice-Phoenix differential refractometer, while that of the mixed surfactants were calculated using the following equation

$$(\partial n/\partial c)_m = \frac{C_s(\partial n/\partial c)_s + C_b(\partial n/\partial c)_b}{C_s + C_b} \qquad (3)$$

where $(\partial n/\partial c)$ denotes the refractive index increment, c the concentration in g./mole. with sub m, s, and b referring to mixed surfactant, sodium lauryl sulfate and BC-840 respectively. This equation has been used by other investigators in several mixed surfactant systems (3).

For turbidity measurements, a stock solution of desired mole ratio of sodium lauryl sulfate to BC-840 was prepared on a weight basis. The stock solution was diluted volumetrically to a series of different concentrations. These diluted solutions were allowed to stand overnight for equilibration of micelles. Solutions were filtered four times under pressure through HAWP 0.25 filter directly into scattering cells for measurements.

II. Polymerization Recipe.

Ingredients	Parts by Weight (grams)	Moles/liter aqueous phase
Styrene	100.0	5.342
Distilled water	180.0	-
Potassium persulfate	0.3	0.0062
Mixed surfactant*	variable	variable

*Nonionic component: Emulphogene BC-840, ave. mol. wt. 860, is a tridecyloxypoly (ethyleneoxy)ethanol was donated by GAF

Ionic component: Sodium lauryl sulfate, mol. wt. 288

The composition of mixed surfactant is designated by r, the surfactant ratio and expressed in moles SLS per mole of BC-840. This ratio can be adjusted as desired.

In the case where a series of polymerizations with identical r-values of mixed surfactant were needed, a stock solution with the desired r-value was made. This solution was then aged for 4 hours to have the surfactants dissolved completely before being used to prepare the polymerization emulsions. These polymerizations were then run in random order to minimize possible error caused by hydrolysis.

III. Polymerization.

Polymerizations were carried out in 8-ounce glass bottles with metal caps containing self-sealing butyl rubber gaskets. The capped bottles with their contents were rotated end-over-end at 45 rpm at 50°C. in a thermostatted water-bath. Samples for conversion and for particle size measurements were withdrawn at regular time intervals using hypodermic needle and syringe. Hydroquinone was used as a shortstop.

IV. Particle Size Analysis.

A JEM 120U electron microscope (Japan Electron Optics Co.) was used to obtain the photographic images of the particles. The final photographs were prepared using a photographic enlarger. The electron photographs were analyzed on a Carl-Zeiss TGZ-3 particle size analyzer. Three thousand particles were counted for each sample. Data from the analyzer were treated by a computer program to obtain the following quantities:

- Number average diameter of particles

$$\overline{D}_n = \Sigma\, n_i d_i / \Sigma\, n_i$$

with n_i particles of diameter d_i

- Weight average diameter of particles

$$\overline{D}_w = \Sigma\, n_i d_i^4 / \Sigma\, n_i d_i^3$$

- Volume average diameter of particles

$$\overline{D}_v = [\Sigma\, n_i d_i^3 / \Sigma\, n_i]^{1/3}$$

- Number of particles/ml-aqueous phase

$$N_m = [M] \times \%\ \text{conversion}/V \cdot \rho$$

where V = volume of particles $= \frac{1}{6} \pi \overline{D}_v^3$

ρ = density of polystyrene = 1.05 g/ml.

- Particle size distribution, PSD, expressed by

i) $\overline{D}_w / \overline{D}_n$

ii) Standard deviation of diameter

$$\text{S.D.} = [(D_i - \overline{D}_n)^2 / (f-1)]^{\frac{1}{2}}$$

where f = Number of particles counted.

Results and Discussion

I. Rates of Polymerization with Recipes of Identical Size of Micelles.

The basic concept of the present study was to show, other things being equal, that the rate of polymerization is affected by the size of the micelles and not by the total surfactant concentration as expressed by Equation (1). This micellar size effect was believed to be the reason why a nonlinear, i.e., a convex curve, relationship between ln Rp and ln C_s was obtained with emulsion polymerization systems of changing surfactant compositions as illustrated by Figure 2.

To check the validity of this concept, the simplest and most straight-forward approach seemed to be to carry out rate studies of emulsion polymerizations with recipes of identical micellar sizes. Since, as mentioned previously, each specific value of surfactant ratio, r, of the mixed surfactant stands for a specific size of the mixed micelles, the experimental approach boils down to run several series of kinetic studies with different surfactant ratios between series, but with varying surfactant concentrations within each series. The standard recipe for such experiments was described in the Experimental Section. This standard recipe is essentially identical with the one used by Kamath (8), Wang (9) and Letchford (15) with the exception of eliminating KOH.

The purpose of adding KOH to recipes of emulsion polymerization is mainly to control the pH-value of the reaction medium. However, being an electrolyte, KOH affects the size of the changed micelles (12,16-18). For mixed surfactant systems, this effect may be different for different values of surfactant ratios of the mixed surfactants. Furthermore, it was observed, during the earlier period of the present study, that the dilute solutions of sodium lauryl sulfate became turbid in the presence of KOH, presumably due to alkaline hydrolysis. Therefore, KOH was not used in this study.

Using the standard recipe mentioned above, eight different micellar sizes, i.e., different r-values were used, and for each r-value six concentrations, thus resulting in 48 rate curves. The conversion-time plot for r = 0.051 is shown in Figure 6 where it can be seen that there is a linear portion in every curve of such plots in this series of experiments. This agrees with the typical behavior of the so-called constant-rate period of emulsion polymerization systems as proposed by Harkins in 1948 and has been confirmed by many other investigators ever since. The slopes of the linear portions of these curves were taken as the rate of polymerization at corresponding surfactant concentration. The logarithmic plot of these rates and corresponding surfactant concentrations are shown in Figure 7. A perfect straight line is obtained which

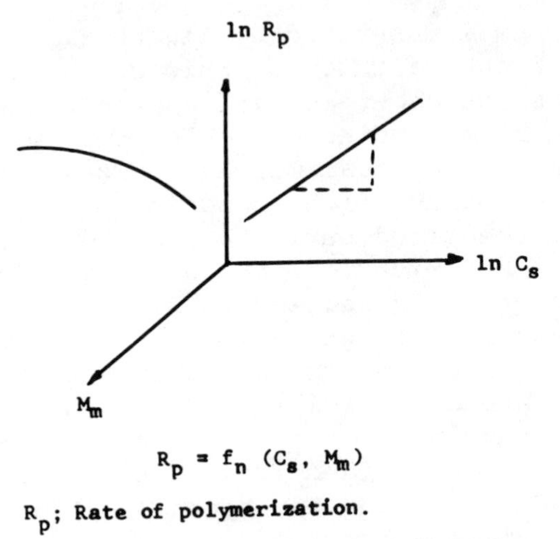

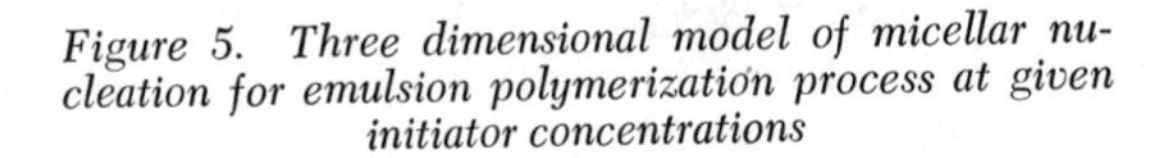
Figure 5. Three dimensional model of micellar nucleation for emulsion polymerization process at given initiator concentrations

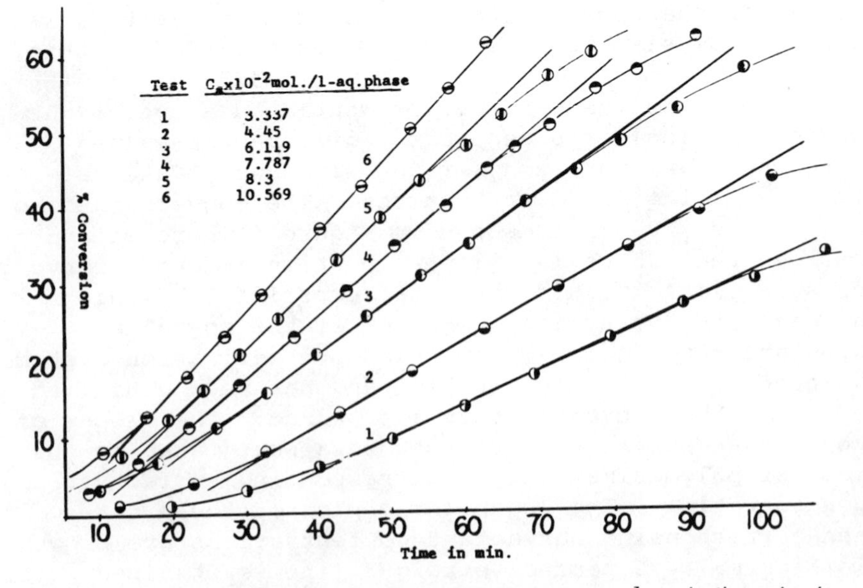

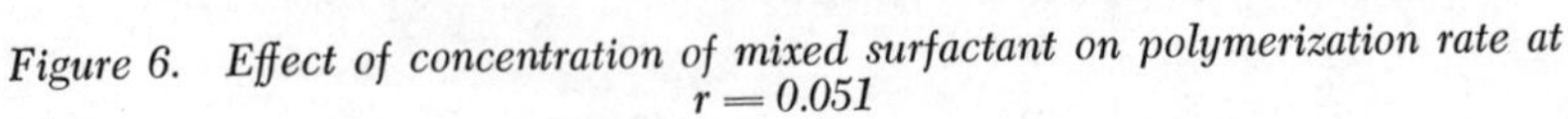
Figure 6. Effect of concentration of mixed surfactant on polymerization rate at r = 0.051

is usually the case with single emulsifier, i.e., with emulsifiers of constant micellar size.

By following identical experimental approach, seven other straight lines of such logarithmic plots were obtained with different surfactant ratios. Regression equation of the form

$$Y = xZ + B \qquad (4)$$

of these straight lines were obtained by repressional analysis and are listed in Table II. In the above regression equation Y corresponds to $\ln R_p \times 10^3$, Z to $\ln C_s \times 10^3$, and x is the slope of the line.

Since each of these 8 straight lines of the logarithmic plot of R_p vs C_s were obtained from a constant value of surfactant ratio, r, and thus a constant value of the size of the micelles, this series of experiments indicates that:

1. micellar size does play a role in affecting the rate of polymerization, and
2. the nonlinear relationship between R_p and C_s in a logarithmic plot, as shown in Figure 2, was caused by the variation of size of micelles as the concentration of the total surfactant varied.

It is interesting and important to note from Table II that the eight straight lines are of different slopes. This indicates that the x-value in the relationship of Equation (2)

$$R_p \propto C_s^{x}$$

should also be dependent on size of the mixed micelles. In the particular system of the present study, the relationship between x-value and micellar size, as expressed in surfactant ratio, r, is

$$x > 0.6 \text{ when } r < 0.2$$

$$x = 0.6 \text{ when } r > 0.2$$

This is shown graphically in Figure 8.

Strictly speaking, if the size of the micelle is to play a role in emulsion polymerization, it should be the size of the monomer-swollen surfactant micelle , and not the monomer free one . However, these two sets of "sizes" should be proportional in the present case

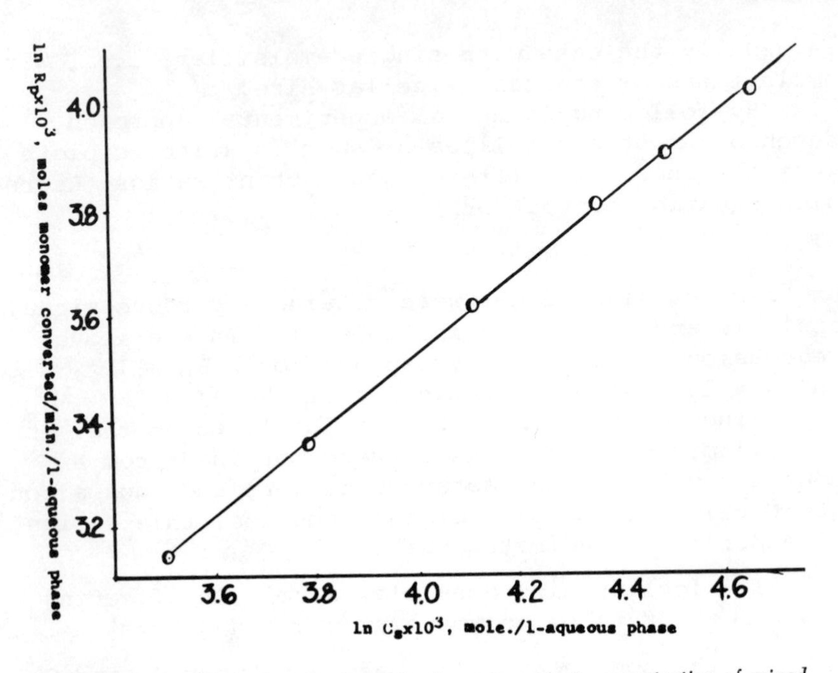

Figure 7. Logarithmic plot of polymerization rate against concentration of mixed surfactants at $r = 0.051$

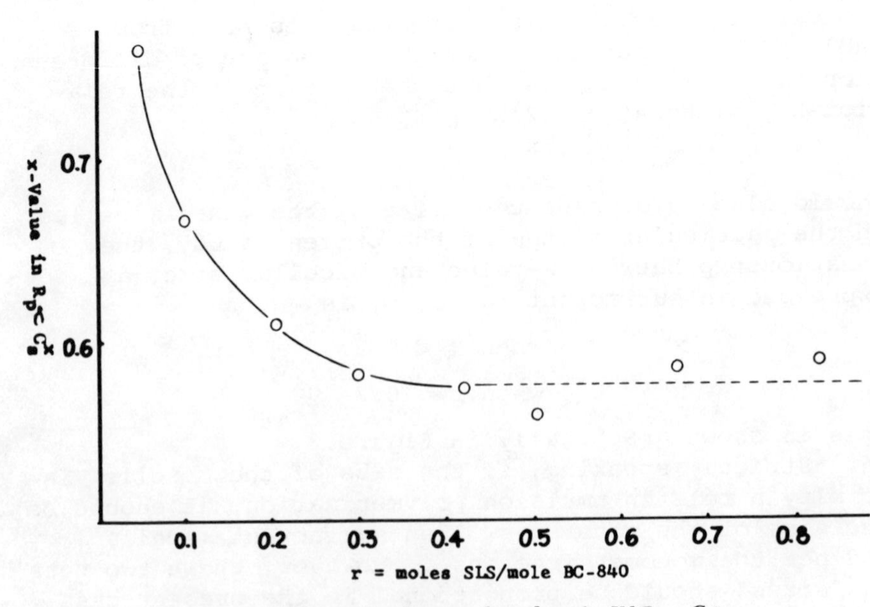

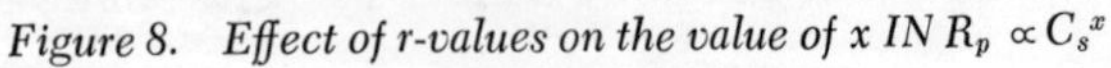
Figure 8. Effect of r-values on the value of x IN $R_p \propto C_s^x$

of study. The argument is that at low values of surfactant ratio, r, the monomer free micelles are large in size but low in charge density. Therefore, they can be considered low in polarity. This type of mixed micelles should have higher solubilizing power towards hydrophobic monomers such as styrene. Thus, the resulting monomer-swollen micelles are also large in size. Similar argument leads to the conception that smaller mixed micelles generate monomer-swollen micelles of smaller size. Accordingly, either set of these mixed micelles can be used for the purpose of interpretation of the polymerization behaviors in this mixed surfactant system.

Since the x-value in the relationship of $R_p \propto C_s^x$ depends on the size of monomer-swollen micelles and the latter is, in turn, related to the solubilizing power of the monomer-free micelles and the hydrophobic properties of the monomers, the "micellar size effect" should predict the following:

1. For a given surfactant, the x-value should depend on water-solubility of the monomer Monomers with solubility less than that of styrene should give x-values which are greater than 0.6.

2. For a given monomer, the x-values should be different with different surfactants if these surfactants give micelles of different sizes under similar reaction conditions.

Therefore, the "micellar size effect" offers, at least, an explanation for the variations observed in the x-values in emulsion polymerization with various monomer-surfactant combinations. Verification of this argument in detail is under investigation, however, and discussion of these results is beyond the scope of the present paper.

II. Rates of Polymerization with Recipes of Constant Surfactant Concentrations but Varying Micellar Size.

What will happen to rates of polymerization in a series of polymerizations where total surfactant concentrations are constant, but the size of the micelles, as expressed by surfactant ratio, r, are varying from test to test? This information can be derived from the linear regression equations shown in Table II. Since these equations are of the form

$$\ln R_p \quad 10^3 = x \ln C_s \cdot 10^3 + B \qquad (4b)$$

where x and B are constants but different for different r-values. The rate of polymerization can be calculated if both the values of r and C_s are known. Thus, the values of rate of polymerization, expressed as $\ln R_p$ 10^3, so obtained are plotted against the corresponding r-values at fixed values of total surfactant concentration, C_s. These plots are summarized in Figure 9. The curves show clearly that, at any given value of total surfactant concentration, the rate of polymerization increases rapidly with increasing values of surfactant ratio in region where r-values are small, however, it levels off when r-values become larger. These curves have a great resemblance with the one shown in Figure 2, where both the total surfactant-concentration and the surfactant ratio were changing.

If the nucleation process in this polymerization system is controlled primarily by the Harkins-Smith-Ewart mechanism, the following equation should apply.

$$R_p = k_p \, [M] \, N_p/2 \qquad (5)$$

or, simply

$$R_p \propto N_p \qquad (5b)$$

and, since not every one of the original micelles in an emulsion polymerization system becomes a polymer particle (6,7), the following equation should be true

$$R_p \propto N_p = P(N) \, N_m \qquad (6)$$

where, in the above equations:

N_p = Number of polymer particles per unit volume of aqueous phase

N_m = Number of starting mixed micelles per unit volume of aqueous phase

$P(N)$ Percentage of the total micelles to become polymer particles, or, the probability of a monomer-swollen micelle to become a polymer particle, or, the probability of nucleation, and

R_p, k_p, [M] are rate, rate constant and monomer concentration in particle respectively.

Equation (5) or (5b) is the highly important deduction of Harkins-Smith-Ewart theory. Its validity has been fully confirmed for many cases of polymerization (19). Furthermore, although it is difficult to determine the number of particles, N_p, accurately (19) this simple relationship has been used to determine the absolute value of the rate constant, k_p, satisfactorily for the polymerization of butadiene and isoprene by Smith (20) and by Morton et al.(21). Conditions where the rate of polymerization is not proportional to the number of particles are where Trommsdorff's effect (22-24) or Gordon's unsteady state (25) principles apply. However, the existence of linear portions of the conversion-time plots proves the absence of these principles in this system.

Accepting that the simple relationship of Equation (5) and (5b) are valid in the present case of polymerization, Equation (6) shows some characteristics of the probability of nucleation, P(N), of the present system of polymerization.

Since the sizes of the mixed micelles in the present mixed surfactant system are known to decrease with increasing surfactant ratio, r, the total number of mixed micelles must increase in a series of recipes with the same amount of mixed surfactant but of increasing surfactant ratios. Therefore, the curves in Figure 9 indicate that:

1. both P(N) and N_m were increasing with increasing r-values in region of $r < 0.2$. This is shown by the rapid increase in the total number of polymer particles, N_p, in that region as represented by the rapid increase in rates of polymerization, R_p, however,

2. the total number of particles, N_p, becomes constant and independent at surfactant ratios in the region of $r > 0.2$. Since the total number of mixed micelles, N_m, is still increasing with increasing r-values in that region, the probability of nucleation, P(N), must decrease correspondingly to keep the product $P(N) \cdot N_m = N_p$ constant.

Since different surfactant ratios stand for different values of the size of the mixed micelles, the above reasoning leads to the following possibilities.

1. The probability, P(N), of a monomer-swollen surfactant micelle to become a polymer particle is a function of its size.
2. In a series of polymerizations, the value of P(N) seems to pass through a maximum with increasing size of the surfactant micelles.

III. Rates of Polymerization with Recipes of Constant Number of Micelles but Varying Micellar Size.

Since it seemed to be true that the probability of nucleation, P(N), is maximum with micelles of a certain suitable size, it would be interesting and necessary to find just at what surfactant ratio, i.e., micellar size, this maximum P(N) is located in the present system of polymerization. This information can again be obtained from those linear regression equations in Table II. Since for a given surfactant ratio, r, the micellar size or micellar weight, is known from the results of light scattering studies (Figure 3), the total surfactant concentration, expressed in $\ln C_s \times 10^3$, can be calculated to give the desired number of mixed micelles. Frcm these calculated values of $\ln C_s \times 10^3$, the rate of polymerization is then obtained from the proper regression equation with specified surfactant ratio, r. This calculation is illustrated for r = 0.207 as an example below.

Let: M_m = weight of mixed micelles, g./mol.

m_w = calculated molecular weight of the mixed surfactant with surfactant ratio, r.

$$m_w = (288\ r + 860)/(1 + r)$$

Where: 288 = molecular wt. of SLS
860 = ave. mol. wt. of BC-840.

N_m = number of mixed micelles/l-aqueous phase

C_s = moles of mixed surfactants/l-aqueous phase

A = Avogadro's number.

Table II

REGRESSION EQUATIONS OF RATE OF POLYMERIZATION ON CONCENTRATION OF MIXED SURFACTANT AT GIVEN VALUES OF SURFACTANT RATIOS

Surfactant Ratio r	Regression Equation $y = bx + a$	Standard Deviation Of Slope, S_b	Correlation Coeff. r_c
0.0509	$y = 0.759x + 0.480$	0.0083	0.9997
0.1032	$y = 0.669x + 1.094$	0.0083	0.9996
0.2073	$y = 0.609x + 1.502$	0.0146	0.9985
0.3015	$y = 0.582x + 1.674$	0.0077	0.9995
0.4274	$y = 0.573x + 1.571$	0.0132	0.9986
0.5041	$y = 0.558x + 1.802$	0.0249	0.9960
0.6642	$Y = 0.587x + 1.707$	0.0154	0.9987
0.8205	$Y = 0.589x + 1.734$	0.0297	0.9954

$x = \ln C_s x10^3$, mol./l-aqueous; $y = \ln R_p x10^3$, moles monomer converted/min./l-aqueous phase.

Then

$$C_s = N_m/m_w \;/\; A/M_m = N_m\, M_m/A\, m_w$$

for $r = 0.207$ and $N_m = 4.5 \times 10^{21}$/l-aqueous phase

$$m_w = (288 \times 0.207 + 860)/(1 + 0.207) = 761.9$$

$$m_w = 6.0 \times 10^3 \text{ g./mol.}$$

$$C_s = 4.5 \times 10^{21} \times 6.0 \times 10^3/6.02 \times 10^{23} \times 762$$

$$= 5.884 \times 10^{-2} \text{ mol./l-aqueous phase}$$

$$\ln C_s \times 10^3 = 4.074$$

From the regression equation for $r = 0.207$, we obtain

$$\ln R_p \times 10^3 = 0.609 \times 4.074 + 1.502 = 3.983$$

Using this illustrated procedure, values for rates of polymerization were obtained at several levels of total number of starting mixed micelles, N_m, with varying surfactant ratios. Figure 10 shows a graph of the calculated $\ln R_p$ values plotted against r-values, at several levels of total number of starting micelles. These curves show that the rate of polymerization gives a maximum with increasing values of surfactant ratio, r. This maximum value is located at approximately $r = 0.2$. A maximum in rate was also observed by Medvedev and co-workers (4) with increasing ionic component in surfactant-mixture without fixing the total number of micelles in each individual test of the polymerization.

To explain the shape and to explore the meaning of the curves shown in Figure 10, Equation (6), has again to be considered. In the present case of discussion, the total number of mixed micelles, N_m, is a constant for all surfactant ratios. Therefore, the total number of polymer particles, N_p, in this system is related to P(N) only. The fact that the value of N_p, as represented by rate of polymerization R_p, goes through a maximum with increasing value of surfactant ratio, indicates that the probability of nucleation, P(N), must correspondingly also have a maximum value. This argument leads directly to the conclusion that:

> The maximum value of the probability of nucleation, P(N) is associated with a suitable size

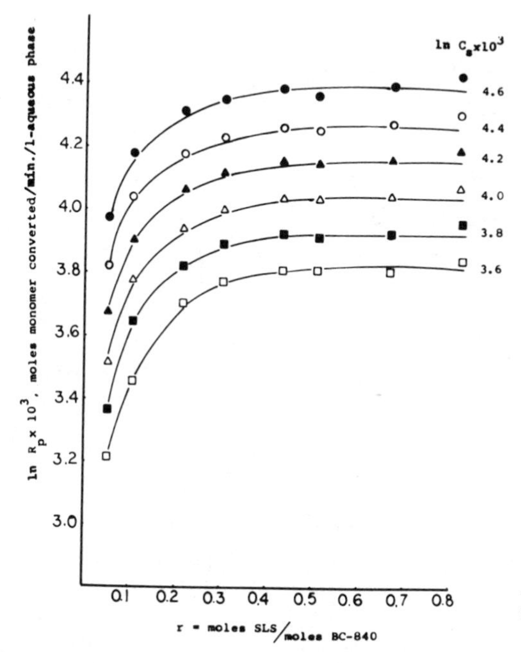

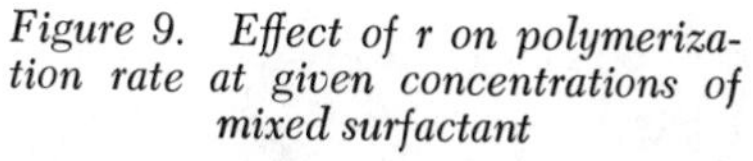
Figure 9. Effect of r on polymerization rate at given concentrations of mixed surfactant

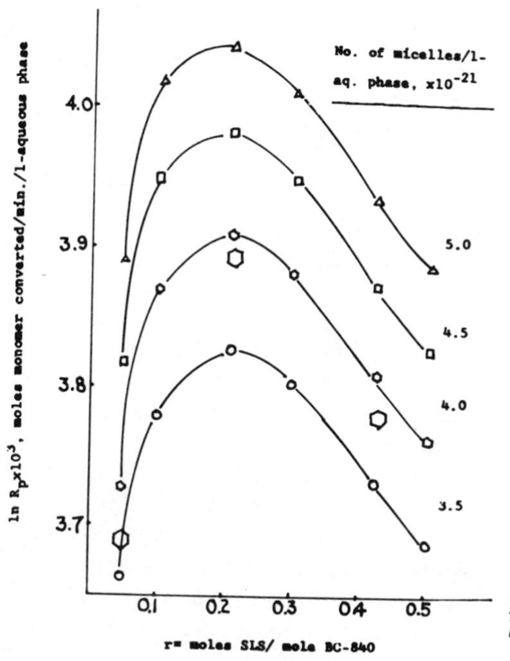

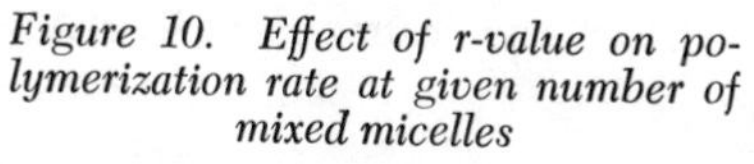
Figure 10. Effect of r-value on polymerization rate at given number of mixed micelles

of the mixed micelles.

In the present system of polymerization, this optimum occurs at a size of the mixed micelles which corresponds to a surfactant ratio of about 0.2.

IV. Hypothesis of Micellar Size Effect on Nucleation. Confirmation of the Hypothesis.

Experimental results and interpretations so far presented lead to the justification of proposing a hypothesis concerning the micellar size effect on particle nucleation in mixed surfactant systems of emulsion polymerization. Essentials of this hypothesis are as follows:

1. The probability, P(N), of a monomer-swollen surfactant micelle to become a nucleus for particle growth is a function of its size.
2. High P(N) associates with a suitable size of micelle.
3. For a given number of micelles, N_m, the number of particules being nucleated should be $N_p = P(N) \cdot N_m$.
4. High P(N) results in
   - fast rate of polymerization due to larger number of particles, N_p, being nucleated, and
   - narrow particle size distribution, due to shorter time period of nucleation.
5. The x-value in the Smith-Ewart relationship of $R_p \propto C_s^x$ is also being affected by P(N).

Therefore, this hypothesis claims that the size of surfactant micelles plays an important role in an emulsion polymerization process. Since, in the present system of study, the maximum value of P(N) happened at a surfactant ratio of about 0.2, all the above predictions should be true with polymerization recipe at that specified value of surfactant ratio.

To confirm these predictions, however, three polymerization experiments were carried out with recipes of surfactant ratios of 0.051, 0.207 and 0.427. These three recipes were made to have the same total starting number of mixed micelles which was 4.0×10^{21}

micelles per liter of aqueous phase. From these three polymerizations, latex samples were taken at conversions falling in the beginning of the region of constant rate period of polymerization. These samples were used for the particle number and the particle size distribution, PSD, determinations.

The experimental programs for the above three confirmation tests were made on the basis of the following concepts:

1. From mixed surfactant systems of emulsion polymerization, monodispersed latices were usually obtained at fairly low conversions with rather wide variations in emulsifier compositions (1). Therefore, samples for the determination of the particle size distribution in this system should be taken at relatively low conversions, otherwise, monodispersed latices will be obtained due to competitive growth from all samples regardless of the surfactant ratios in the recipe of polymerization. These particles will be different in size, but not in size distribution.

2. To obtain significantly different values in the particle size distribution and in the number of particles formed, comparative tests should be carried out with r-values considerably different one from another. Otherwise, significantly different results might not be obtainable due to the inherently high error in the determination of total number of particles, N_p (19). Since the micellar weight determinations in this series of experiments were limited to the range of r = 0.05 to r = 0.5, which was also the range of r-values within which the maximum P(N) happened, the selected three values of surfactant ratios for the confirmation tests were considered proper and sufficient.

From these three polymerizations, the conversion-time plots are shown in Figure 11. Obviously, the rates of polymerization in these three test runs were not the same, the one at r = 0.207 is higher than the other two. This agrees with the predictions (points represented by hexagonals in Figure 10).

Three thousand particles were counted for each sample. The frequency of occurrence of particles of various sizes is expressed graphically in Figure 12 as frequency polygons. These polygons showed these three latex samples to have particles of different average sizes even though they had been grown to almost equal level of conversions. It is interesting to note that the particle sizes obtained from recipes with surfactant ratios of r = 0.051 and r = 0.427 were both larger than the ones obtained from r = 0.207. Since the size of the mixed micelles at r = 0.207 is in between the other two, the size of polymer particles does not correlate linearly with the size of mixed micelles, but was controlled by the probability of nucleation, P(N).

From the observed values of the particle size analysis of each of the latex samples, the number and weight average diameters of the particles, the size distributions, and the standard deviations, S.D., of the average diameters were calculated. Knowing the rate of polymerization and the total number of particles in a system, the rate of polymerization per particle, R_{pp}, can also be calculated. The values are listed on Table III, where it can be seen that, in comparison with the other two recipes, the one with surfactant ratio of r = 0.207 has the following features.

1. Highest R_p.

   i.e., faster rate than the other two although the total number of the micelles was the same for all three.

2. largest value of N_p.

   i.e., more particles formed from the same number of micelles.

3. Narrowest PSD

   i.e., $\overline{D}_w/\overline{D}_n$ value of 1.02 vs 1.04 for the other two.

The significant differences in N_p and in PSD, in terms of S.D., were confirmed by the Student t-Test and the F-test, using the procedures described by Li (27).

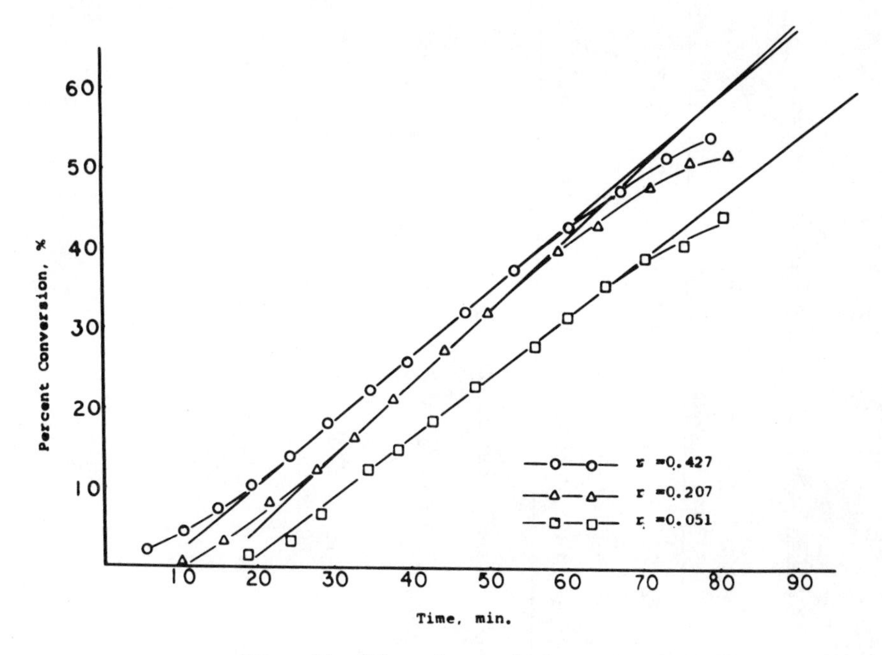

Figure 11. Effect of r on polymerization rate

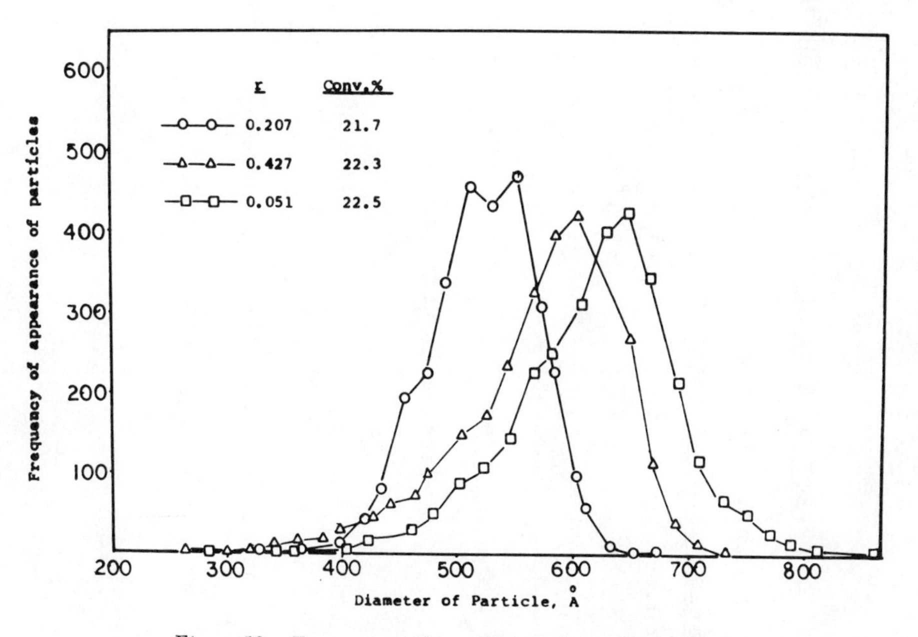

Figure 12. Frequency polygon of particle size distribution

Table III

EFFECT OF SURFACTANT RATIO ON NUMBER AND SIZE DISTRIBUTION OF LATEX PARTICLES

Surfactant Ratio, r	Percent Conversion	$\overline{D}_w$	$\overline{D}_n$	$\overline{D}_w/\overline{D}_n$	S.D.	$\overline{D}_v$	N_p $x10^{-15}$	R_p	R_{pp} $x10^{20}$
0.051	22.5	647.61	617.89	1.038	72.79	626.14	0.93	0.0400	4.32
0.207	21.7	538.08	521.94	1.024	47.42	526.17	1.51	0.0489	3.24
0.427	22.3	600.67	574.38	1.037	68.48	582.12	1.14	0.0438	3.84

$\overline{D}_w$, $\overline{D}_n$, $\overline{D}_v$; Weight, Number, and Volume average of diameters of the particles.

S.D.; Standard Deviation particle diameter $= [(D_i-D_n)^2/(f-1)]^{\frac{1}{2}}$

f = Number of particles counted, 3000 for r=0.051
3001 for r=0.207
3003 for r=0.427

R_p; Rate of polymerization, moles monomer converted/min./l-aqueous phase.

R_{pp}; Rate per particle.

Since the probablity of nucleation for a given micelle can be calculated by

$$P(N) = N_p/N_m$$

the values of P(N) for the three confirmation tests are as follows:

surfactant ratio r	$P(N) \times 10^4$
0.051	2.33
0.207	3.78
0.427	2.85

Thus the results of our calculations do confirm our predictions.

Conclusions

We have shown that with the use of a mixed surfactant system in styrene emulsion polymerization, the composition of the mixed surfactant has an effect on the rate of polymerization, the number of particles formed and the particle size distribution. We have also shown that a change in the ratio, r of the two surfactants in the mixture results in a considerable change in the micellar weight of the resultant mixed micelles. We have thus proposed and proven that the efficiency of nucleation of particles (even when the same number of micelles is used in the experiment) is dependent on the size of the mixed micelle, and that there is an optimum size at which the polymerization rate is the fastest and the particle size distribution is the narrowest.

The linear relationship between the rate of polymerization, R_p, and the total surfactant concentration, C_s, in a logarithmic plot is valid only for systems where the micellar size, M_m, is constant. For systems where micellar size is a variable, other things being equal, the kinetics of emulsion polymerization should be expressed by a three parameter model, involving R_p, M_m, and C_s, rather than by the classical two parameter model which involves the rate dependence on the total emulsifier concentration only. The two parameter model gives a nonlinear relationship between $\ln R_p$ and $\ln C_s$ if M_m is a variable. This variation in the micellar weight can be brought about also by the addition of electrolytes to the emulsion system, and, of course, by the change in the chain length of the surfactant (18).

The x-value in the relationship of $R_p \propto C_s^x$ is

also affected by the micellar size of the surfactant, and larger x-values are expected for highly water-insoluble monomers. This explains, at least, some of the abnormal behavior of some emulsion polymerization systems as compared with the so-called normal requirements of the Smith-Ewart's theory.

The present study thus supports a theory of particle nucleation where micelles do play a vital part, it does not necessarily support or negate micellar initiation.

Literature Cited

1. M. E. Woods, J. S. Dodge, J. M. Krieger and P. E. Piece; J. Paint Tech., 40, 57, 543 (1968).
2. T. Nakagawa and H. Inoue; Nippon Kagaku Zasshi, 78, 636 (1957).
3. K. Kuriyama, H. Inoue and T. Nakagawa; Kolloid Z. Z. Polym., 183(1), 68 (1962).
4. S. S. Medvedev, J. A. Griskova, A. V. Zurikov, L. I. Sedakova and G. B. Boreino; J. Macromol. Sci.,-Chem. A-7(3), 715 (1973).
5. C. P. Roe; Ind. Eng. Chem., 60, 9, 20 Sept. 1968.
6. W. D. Harkins; J. Am. Chem. Soc., 69, 1428 (1947).
7. W. K. Smith and R. H. Ewart; J. Chem. Phys., 16 592 (1948).
8. V. Kamath; Ph.D. Thesis, University of Akron, Akron Ohio (1973).
9. A. Wang; M.S. Thesis, University of Akron, Akron, Ohio (1973).
10. J. M. Corkill, J. F. Gordman and R. H. Ottewill; Trans. Faraday Soc., 57, 1627 (1961).
11. P. Debye; J. Phys. Chem., 53, 1 (1949).
12. J. N. Phillips and K. J. Mysel; J. Phys. Chem., 59, 325 (1955).
13. I. D. Robb; J. Poly. Sci., A-1, 7, 417 (1969).
14. R. M. Fitch; Br. Polym. J. 5, 467 (1973).

15. R. J. Letchford; M.S. Thesis, University of Akron, Akron, Ohio (1973).
16. J. L. Mateo and I. Cohen, J. Polymer Sci., A2, 711 (1964).
17. P. H. Elworthy, A. T. Florence and C. B. Macfarline; "Solubilization by Surface-Active Agent and its Applications in Chemistry and the Biological Science", Chapman and Hall, L.T.D. London, 1968.
18. K. Shinoda, B. Tamashumi, T. Nakagawa and T. I. Semura;"Colloidal Surfactants", Academic Press, New York (1963).
19. P. J. Flory, "Principles of Polymer Chemistry"; Cornell University Press, 1969.
20. W. V. Smith; J. Am. Chem. Soc., 70, 3695 (1948).
21. M. Morton, P. P. Salatiello, H. Landfield; J. Polym. Sci., 8, 111 (1952).
22. H. Gerrens, Z. Elektrochem., 60, 400 (1956).
23. B. M. E. van der Hoff; J. Polym. Sci., 33, 487 (1958).
24. B. M. E. van der Hoff; J. Polym. Sci., 44, 241 (1960).
25. J. L. Gardon; J. Polym. Sci., A-1, 6, 623, 643, 665, 687, 2853, 2859 (1968).
26. J. W. Vanderhoff, J. F. Vitkuske, E. B. Bradford and T. Alfrey, Jr., J. Polym. Sci., 20, 225 (1956).
27. Jerome, C. R. Li; "Statistical Inference", 105, 185, Adward Brothers, Inc., 1964.

4

Interfacial Phenomena in Emulsion Polymerization of Polar Monomers

V. I. YELISEYEVA and A. V. ZUIKOV

The Institute of Physical Chemistry, Academy of Sciences of the USSR, Moscow, USSR

A substantial amount of recent experimental data demonstrate that the model of styrene emulsion polymerization (1,2) on which the quantitative theory is based (3), is not capable of adequate interpretation of polymerization in many real systems. An attempt to use the theoretical relationships to describe polymerization of such industrially important monomers as vinyl acetate, vinyl chloride, alkylacrylates, as well as copolymerization of common monomers with functionally substituted ones, leads to a conclusion that the theory disregards some of the essential factors of the process. Therefore, this theory cannot be a foundation for polymerization technology of the above monomers to be modernized and automatized.

The model proposed by Harkins and Yurzhenko doesn't take into account, the intermolecular interactions on the interface, the most important factor determining behaviour of the colloidal system. Thus, this model assumes that the molecule area of an emulsifier in a micelle and in an adsorption layer of a polymer-monomer particle have identical values, and the newly formed surface is stabilized immediately after its formation. As a result, the surface of particles per unit volume is defined as the surface occupied by the unit weight of the emulsifier multiplied by the latters concentration. The surface calculated in this manner is used to derive the number of particles which remaines unaltered and is the fundamental parameter in the polymerization rate equation.

Medvedev (4), deriving rate equation, take into account the possibility of particles' coalescence which does not alter the total surface; but the reaction order with respect to emulsifier (0.5) in his equation has a constant value and is not related to emulsifier activity in a given system.

However, it was shown long ago, that the surface activity, i.e. the substance capability to adsorption on the interface is the lower the smaller is the difference between polarities of the phases (5), that is in the case of emulsion polymerization the higher is the monomer polarity.

The assumption of the significance of emulsifier adsorption capability to the kinetics of emulsion polymerization was suggested in our paper (6), for polar monomers, and by Roe (7) and Robb (8) for styrene. Paxton (9) demonstrated that the adsorption area occupied by a molecule of a given emulsifier (Na-dodecyl benzyl sulphonate) on the surface of polymethylmethacrylate latex particles, is 1.31 $(nm)^2$, and so exceeds by a factor of 2.5 a similar area on the surface of polystyrene latex, equal to 0.53$(nm)^2$.

We have demonstrated that the coverage of particles by the anion-active emulsifier (sodium alkyl sulphonate) in the polymethylacrylate latex does not exceed 30%, and reaches 43% in the polybutylacrylate latex (10).

The data obtained by Brodnyan and Kelly (11) on adsorption of water-soluble macromolecules on latex particles indicate that both the nature of macromolecules and that of the substrate are essential for adsorption.

Breitenbach et al. (12) established that in the case of vinyl acetate emulsion polymerization the dispertion medium of the resultant latex contains the amount of emulsifier exceeding by a factor of two the critical micelle concentration (CMC), i.e. the reaction system is not depleted of the emulsifier. We obtained similar results in the case of methylacrylate polymerizing with alkyl- and alkylarylsulphonates.

Comparison of published data on vinylacetate polymerization kinetics, for which a steady-state period is typical (13), to the kinetics of variation of the number of particles which decreases during the process up to a factor of 40 (14), permits us to conclude that there is no correlation between the rate and the number of particles. This conclusion was supported by Medvedev et al. (15) in the case of emulsion polymerization of methylmethacrylate. We deduce from the above data that the emulsifier concentration itself does not determine either the total surface of the disperse phase or the number of particles during polymerization of polar monomers.

The present paper is devoted to the investigation of emulsion polymerization of lower alkyl acrylates in terms of the interface properties: interface tension, adsorption capability of emulsifiers of different molecular structures, and stability of the particles formed. The data obtained, together with those discussed above, made it possible to conclude that the rate dependence on emulsifier concentration in the processes involving flocculation of particles, must be more complicated than is predicted by well known equations (3,4). The reason for particle flocculation is a low emulsifier adsorption energy on the interface, decreasing as monomer solubility in water increases and depending, for a given monomer, on the nature of the emulsifier.

Experimental procedure

Monomers. The following lower esters of acrylic acid were used: methylacrylate (MA), ethylacrylate (EA), butylacrylate (BA), and also styrene (St). These were technical-grade products, purified by conventional methods.

Emulsifiers. SLS, or sodium lauryl sulphate, manufactured by "Serva" (DBR) was used without additional purification.

TX-100(S), or isooctyl phenylpolyglycolsulphate. The product used was obtained by sulphating Triton X - 100 emulsifier manufactured by Rohm and Haas Comp. Sulphating was performed by means of the sulphaminic acid. Product was redeposited from butyl alcohol.

C-30-polyglycol ether of cetyl alcohol. Laboratory product; the degree of oxyethylation: 30.

APGS-alkylpolyglycolsulphate. Laboratory product; the degree of oxyethylation: 10.

Initiators. Ammonium persulphate (APS), analytical grade, was redeposited twice from bidistilled water. Benzoyl peroxide (BP). Technical product, twice redeposited from acetone.

Molecular weight regulator. tert-Dodecyl mercaptan (t-DDMC), purified by double distillation in vacuum; used in polymerization experiments accompanied by measurements of electric conductivity.

Dispersion medium. Bidistillated water.

The degree of adsorption saturation and the area occupied by an emulsifier molecule in the saturated adsorption layer on the surface of latex particles were determined by means of the adsorption titration method (16). Sizes of latex particles were determined by electron microscopy. Since the glass transition temperature of the investigated polyacrylates is below room temperature, the particles of latexes were preliminary solidified under a γ-ray source to eliminate errors due to deformation of polymer globules on the substrate.

The synthesis of latexes for adsorption titration was conducted in a thermostated glass reactor provided with a stirrer and an attachment for feeding in inert gas. The polymerization temperature is 328°K in the case of MA, and increases to 333°K, when EA, BA and St are used. Phase ratio was 1:4 (by weight). Compositions used for latex synthesis and latex characteristics are listed in Table I.

TABLE I

Characteristics of Latexes Used for Adsorption Titration

Monomer	SLS mole per monomer mole, x10^{-3}	APS mole per monomer mole, x10^{-3}	Conversion, %	γ, (dyne/cm)
MA	2.96	0.188	99.5	49.5
EA	6.93	0.307	98.7	47.5
BA	13.44	0.588	∼100.0	53.6

The kinetics of emulsifier adsorption in the course of polymerization was investigated by means of a dilatometer with sealed-in platinum electrodes for measuring electrical conductivity of the system. (17)

The investigation of the kinetics and of the number of particles in the course of the process was conducted in a special dilatometer which enables one to extract the minimum volume samples (0.03 ml) and requires no compensation for the diminution of the reactive mixture (18). The extracted samples were diluted with the required amount of the emulsifier solution, containing the inhibitor (p-benzoquinon), and were then used to determine the number of particles by flow ultramicroscopy (19) and electron microscopy.

The interface surface tension was calculated from the data on mutual solubility (20) of monomer and water determined by means of refractometer or interferometer.

The emulsifier adsorption energy ($\mathcal{E}$) was calculated from the isotherm of the interface surface tension, obtained by means of a stalagmometer with automatic recording of the number of drops. The Langmuir equation was used for calculations:

$$\frac{\mathcal{E}}{RT} = -\ln \frac{-\frac{\partial \gamma_{1,2}}{\partial c}}{RT\Delta}$$

where Δ (the surface layer thickness) was taken to be 1.2 nm.

Results and Discussion

The results obtained from adsorption of SLS on the water-monomer interface (Figure1) demonstrate that the $\gamma_{1,2}$ vs C isotherm for methyl-, ethyl- and butylacrylate (T=298°K) are essentially different even for nearby members of the alkylacrylate homologous series. The calculated values of adsorption energy for methyl-, ethyl- and butylacrylate for $c \to 0$ were found equal to 5.1, 5.6 and 6.5 Kcal/mole, respectively.

Formally the monomer polarity is not nearly sufficient to explain the differences in the adsorption energies of a given emulsifier on the water-alkylacrylates interfaces (5): according to (21) the dipole moments of lower alkylacrylates have practically identical values. It appears possible that polarity of the interface itself*) is significant here; this polarity increases as the monomer solubility in water increases from BA to MA; the decrease of the free energy of the interface, caused by it, results in its turn in decreased adsorption energy of emulsifier.

Figure 2 shows similar isotherms for ethylacrylate, obtained from the adsorption data for various emulsifiers. It is seen that these isotherms considerably differ. The calculated values of the adsorption energy for different emulsifiers on the ethylacrylate -water interface (T=293°K) are listed in Table II.

The investigation of kinetics of consumption of emulsifiers of different structure, viz SLS and TX-100(S), in the course of alkylacrylates polymerization has demonstrated that the rate and degree of adsorption depend in both cases on the alkyl

*) That is the degree of polarization of monomer polar groups (-CO) orientated to water phase and the absence of clear cut boundary between phases.

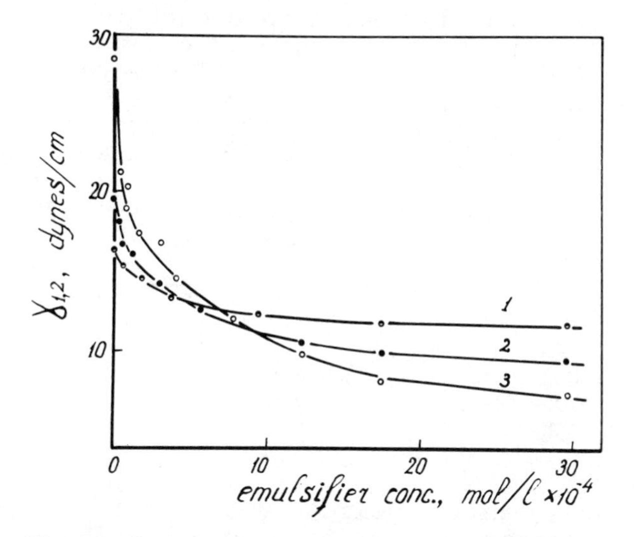

Figure 1. Surface tension isotherms of SLS on the MA-(1), EA-(2), and BA-(3) water interfaces. T = 298°K.

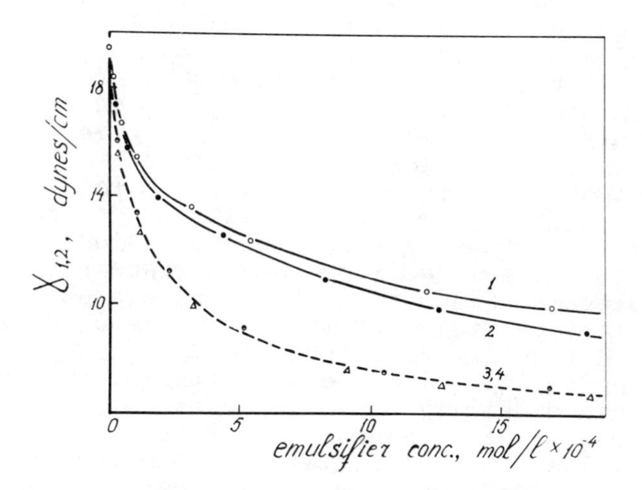

Figure 2. Surface tension isotherms on the EA-water interface: 1—SLS, 2—APGS, 3—C-30 (◓), 4—TX-100(S) (△). T = 298°K.

length in a given homologous series of monomers. This is seen from the adsorption vs conversion curves in Figures 3, a and b: both the rate of adsorption and its equilibrium value significantly decrease with a decrease in an alkyl length, i.e. when monomer water solubility is lessened. This difference is pronounced more sharply in the case of SLS as compared to TX-100(S).

TABLE II

Emulsifier	Energy of adsorption, Kcal/mole
SLS	5.6
TX-100(S)	6.5
APGS	5.9
C-30	6.4

In order to investigate emulsifier adsorption by the particle surface in final latexes, the latters were specially synthesized by polymerization of MA, EA, BA and St in the presence of SLS and APS.

The characteristics of obtained latexes are listed in Table III.

The results of adsorption titration of latexes, presented in Figure 4, show that the equilibrium adsorption is significantly and regularly decreased with an increase of monomer solubility in water. The area occupied by an emulsifier molecule in the saturated adsorption layer is correspondingly increased; this is seen in Figure 5 showing this area to be the function of the surface tension $\gamma_{1,2}$ on the monomer-water interface, which we assumed to give a measure of the interface polarity. The data of Table III indicate that increase of solubility in water, from styrene to methylacrylate, results in enlarged particle size, increased adsorption saturation of these particles, as well as increased surface per emulsifier molecule. Since all latexes were stable, we concluded that particles of the polystyrene latex were protected from flocculation already at 50% saturation, while the particles of the polymethyl acrylate latex reach this stability only at almost complete saturation. However, the adsorption layers of methylacrylate particles in the latex are, even at nearly maximum saturation, two times less dense than that on styrene particles with 50% saturation. Since the rate of MA polymerization is higher than that of styrene (K_p of these monomers at 333°K are equal to 1260 and 190

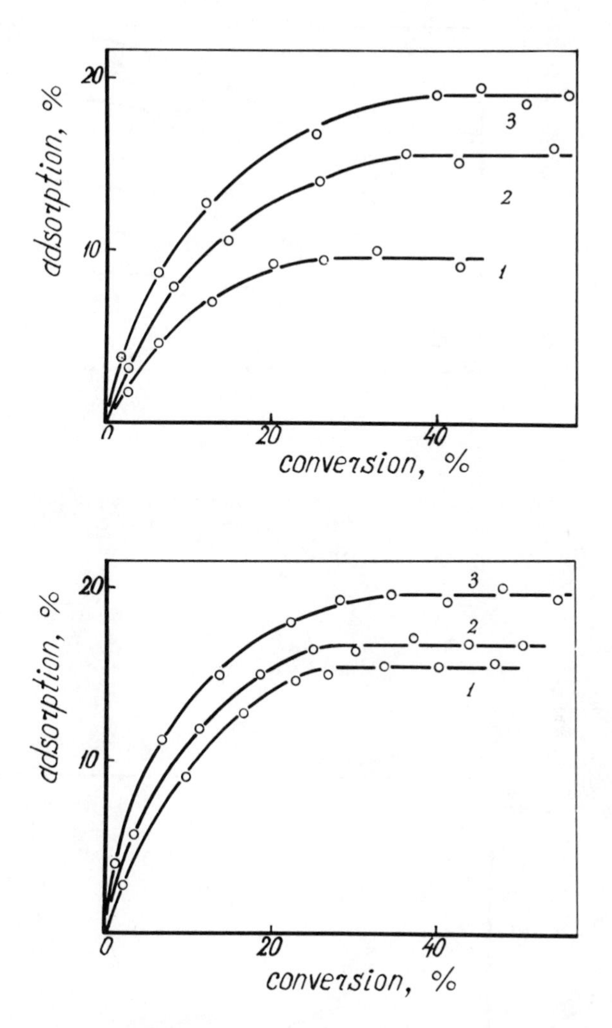

Figure 3. Emulsifier adsorption during polymerization. a) (top) SLS; b) (bottom) TX-100(S); 1—MA, 2—EA, 3—BA. [SLS] and [TX-100(S)]—2%, [tDDM]—1% of the monomer weight. The monomer-to-water phase ratio is 1:9 (by weight). T = 318°K. [BP]—0.2% was used as an initiator.

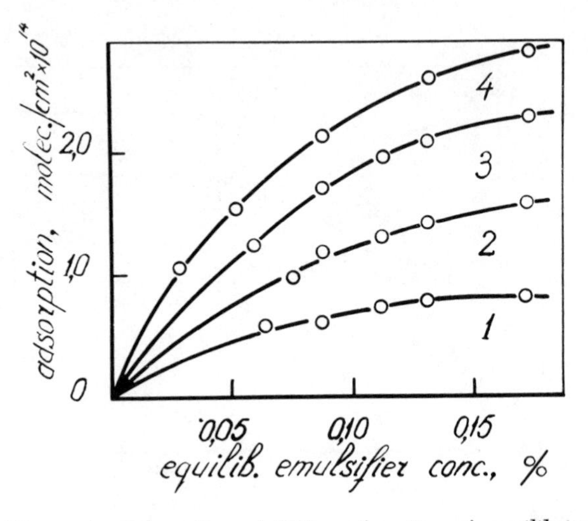

Figure 4. Adsorption of SLS as function of equilibrium concentration of emulsifier in the aqueous phase: 1—MA, 2—EA, 3—BA, 4—St

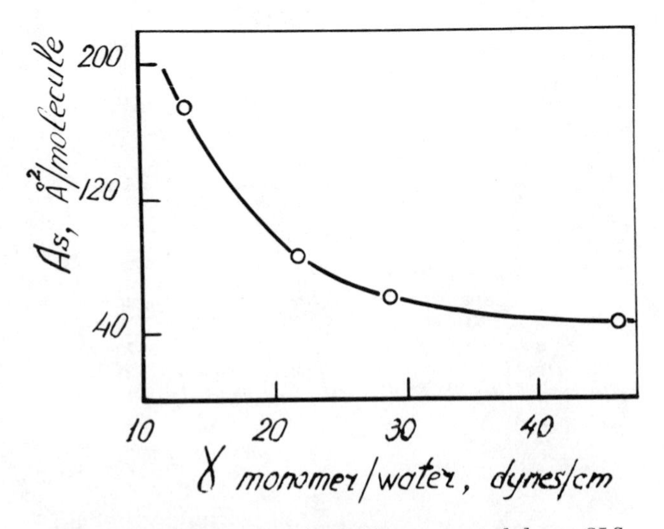

Figure 5. Variation of the surface occupied by a SLS molecule in the saturated layer as a function of surface tension at the monomer–water interface

TABLE III

Physico-chemical Properties of Latexes Obtained by Polymerization of Different Monomers

Monomer	Monomer solubility in water, wt.%, 298°K	$\gamma_{1,2}$, dynes/cm	Particle diameter, nm	Latex concentration, %	Total surface per 1ml, m^2	Surface per emulsifier molecule, $(nm)^2$		Degree of adsorption saturation, %
						in latex, calculated	in the saturation adsorption layer	
MA	5.69	13.0	125	20.5	8.28	1.920	1.753	90.6
EA	1.84	21.5	96.0	19.3	10.67	1.238	0.862	78.2
BA	0.34	28.5	88.0	20.6	13.00	1.057	0.615	63.8
St	0.03	46.5	71.0	19.7	16.35	0.945	0.493	51.7

litre/mole/s, respectively), and the emulsifier adsorption rate at MA polymerization can only be lower (for lower value of $\chi_{1,2}$ see Figure 5), the data of Table III can be interpreted as the evidence of domination of flocculation mechanism of particle formation in the polymerization of a polar monomer.

Summerizing the results obtained we come to a conclusion that the adsorption capability of the interface or emulsifier adsorption energy, strongly depending on the nature of monomer and emulsifier are of considerable significance for the colloidal behaviour of the polymerizing system. This, in turn, must influence the polymerization process itself.

Figure 6 presents a plot of the rate (R) and particles number (N) (N being determined by the flow ultramicroscopy technique) as functions of conversion during polymerization of EA (phase ratio is 1:9). The Figure shows that identical behaviour patterns are observed at different concentrations of SLS: R undergoes a steep rise, then reaches a constant level, and at 40-45% conversion decreases until the process is completed; N first goes through a maximum, drops down to a certain level until a deep conversion is reached (similar data are reported by Fitch (22) for a very dilute (1%) methylmethacrylate latex on the basis of electron-microscopic observations; the number of particles was not compared to rate in these experiments). No correlation, characteristic of styrene polymerization, is observed between N and R. The maximum on the curve of particle number indicates the flocculation mechanism of their formation.

The equilibrium ethylacrylate concentration in particles was determined by measuring vapor pressure (23) in order to explain characteristic points on the kinetic curves (Figure 6). This concentration varied in experiments with variable diameter of particles in the range of 3.58 - 4.15 volume fractions of monomer in polymer* (T=318°K) which corresponded to disappearance of monomer drops from the reaction system on the average at conversion ~ 20%.

Figure 6 shows that constancy of N, determined by the flow ultramicroscopy method, is achieved at conversion of 20-25%, which would be easily attributed to the stopped formation of a new surface, since monomer is completely absorbed by particles at this stage, and it is at the expense of this monomer that

* According to Gardon (24), this figure for ethylacrilate is 4.15.

further polymerization proceeds. However, electron micrograph (Figure 7) demonsrated the presence in the Experiment 2 (Figure 6) of particles with diameters below 10 nm up to the conversion of 51%. These particles are gradually disappearing due to their flocculation with larger particles; they are not recorded by flow ultramicroscopy which detects only the particles with diameters larger than 40 nm. Therefore, the moment when the total number of particles reaches constancy is in fact shifted towards high conversions. Figures 8a,b,c represent electron micrographs of ethylacrylate latex samples (phase ratio 1:2) extracted during a steady-state period (see the caption to the Figure). The micrografphs demonstrate that intensive flocculation takes place during this period of polymerization. For the same experiment Figure 9 gives the curve of variation of N obtained from the flow microscopy data corrected by the data of electron microscopy*. These curves also indicate that during polymerization from 15.6% up to 45.5% conversion N decreases by 1-1.5 orders of magnitude; (this corresponds to the data we have published earlier on the content of microglobules in acrylate latex particles (25)(Figure 10). However the approximative calculation shows that decreasing of particle number by factor 15 in this interval of conversion (see Figures 8a and 8c) does not alter the total surface significantly. It can be seen from Figures 8 and 9 that a boundary exists between primary globules, which can be explained by hydrophilysation of their surface.

The results of the present investigation of particle formation mechanism by means of electron microscopy correspond to the data of Dunn and Chong (26) who proved, by using the DLVO theory of stability (27), that very small particles (with diameter less than 20 nm) formed during polymerization of vinyl acetate are less stable, so that their flocculation with larger particles is possible immediately after they

---

*) The correction was carried out by determining from electron micrographs the ratio of the general number of particles to that of particles larger then d=40 nm and by multiplying the number of particles, found by flow microscopy, by a corresponding ratio. We assumed that this approach led to smaller errors in the number of particles as compared to finding their average diameter from electron micrographs and dividing the volume concentration of the latex by the calculated mean volume of the particles.

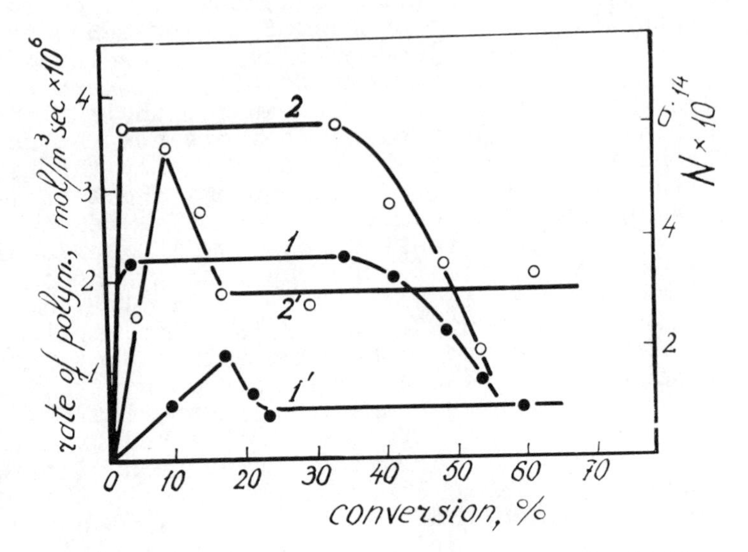

Figure 6. Polymerization rate (1, 2) and particle number (1′ 2′) as function of conversion for EA polymerization in the presence of SLS: 1—0.3%; 2—8% of the monomer weight. Phase ratio: 1:9. T = 318°K. [APS]—0.5% of monomer weight.

Figure 7. Electron microphotograph of the EA-latex particles at 51% conversion. [SLS]—8%, [APS]—0.5%. Phase ratio: 1:9. T = 318°K. Magnification ×18.700.

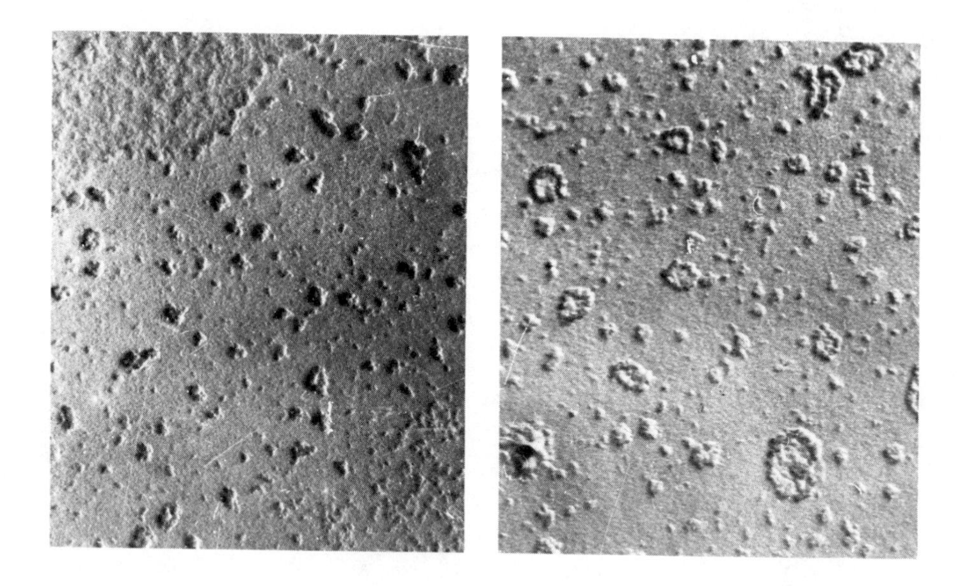

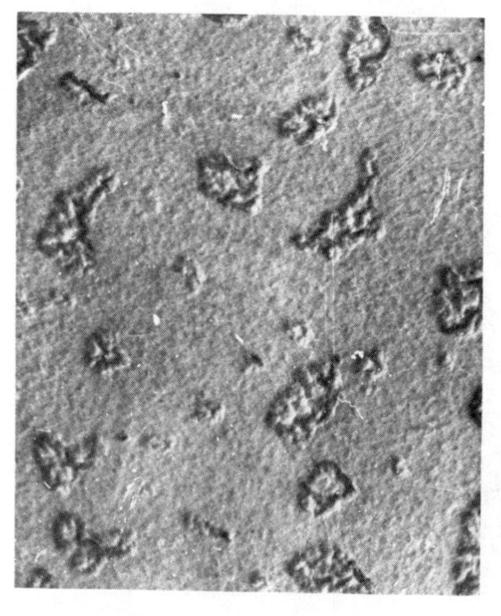

Figure 8. Electron microphotographs of the EA-latex particles for conversions: a) (top left) 15.6%, b) (top right) 38%, c) (left) 45.5%. $C_{15}H_{31}SO_3Na$ concentration 2%, [APS]—0.3% of monomer weight. Phase ratio 1:2. $T = 333°K$. Magnification ×18.700.

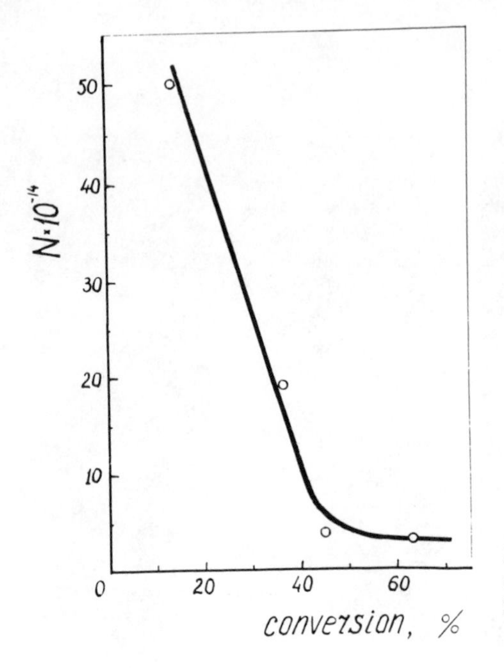

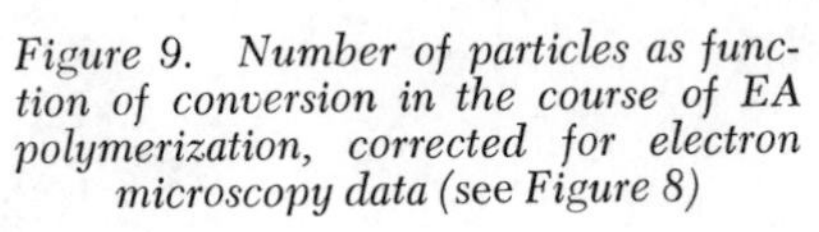

Figure 9. Number of particles as function of conversion in the course of EA polymerization, corrected for electron microscopy data (see *Figure 8*)

Figure 10. Electron microphotograph of particles of acrylate latexes after oxygen etching. Magnification ×127.500

are formed. Figure 8 demonstrates that flocculation of small particles develops gradually.

It is noteworthy that the maximum of the particle number curve was observed in our experiments also at such concentrations of emulsifier at which the polymerizing system is not depleted of emulsifier ($\gamma_{lat} \leqslant \gamma_{cmc}$) which was the case for the Experiment 2 (Figure 6). Therefore, flocculation of particles is not the consequence of emulsifier lack in the system, as was suggested by Brooks (28), but is due to the limited capability of its adsorption.

As for the characteristic point on rate-conversion curves corresponding to the end of the steady-state interval, it was observed at conversion exceeding the one at which monomer drops disappeared, when APS were used as initiator (Figure 6). This fact can be accounted for by the flocculation mechanism of particle formation and the complexity of latex particle structure. We assume that in this case olygomer radicals formed in water phase initiate the polymerization in particles. These radicals can't reach the inner primary globules due to diffusional impediments and polymerization proceeds only in the outer zone after flocculation takes place. The rate constancy after monomer drops disappearance is conditioned due to monomer redistribution between inner "dead" and outer "living" primary globules. The assumption is confirmed by the fact that the prolongation of a steady state period takes place only when a water soluble initiator is used: such a phenomena was not observed in the case of an oil soluble initiator. This can be seen from the kinetic curves of alkyl acrylate polymerization in the presence of benzoyl peroxide (Fig. 2) The end of steady state periods correspond to 15.8; 26.6; 32.4% conversion for MA, EA and BA respectively, which roughly coincide with the equilibrium swelling of latex particles (see (24) and this work's data).

The investigation of polymer rate dependence on initiator concentration C_i (with ionic strength of the solution equalized) and emulsifier concentration C_{em} (for various molecular structures of emulsifier) permitted us to establish that it can be described by the following equation:

$$R = K \left[C_{em}\right]^{X} \cdot \left[C_i\right]^{0.5}$$

where X is in accordance with an emulsifier type. The correlation is observed between the value of X and the energy of emulsifier adsorption (Table II) on the

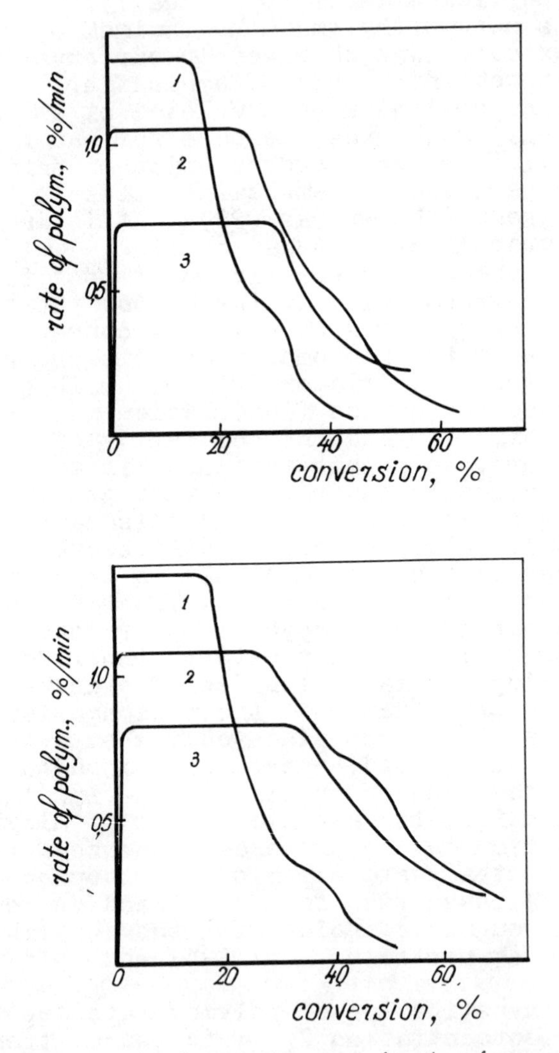

Figure 11. Polymerization rate as function of conversion: a) (top) SLS, b) (bottom) TX-100(S). 1—MA, 2—EA, 3—BA. Experimental conditions are indicated in Figure 3.

monomer-water interface; this is illustrated by the following comparison:

TABLE IV

Emulsifier	X	Energy of adsorption Kcal/mole
SLS	0.21	5.6
APGS	0.25	5.9
C-30	0.38	6.4

The observed correlation can be explained by a stronger dependence of particle surface on emulsifier concentration when the energy of its adsorption increases.

Conclusions

Investigation of emulsion polymerization of lower alkyl acrylates in terms of the properties of the interface made it possible to establish its significance for the colloidal behaviour of systems as well as for process kinetics. The study of emulsifier adsorption on water-monomer, water-monomer-polymer particles and water-latex particles interfaces shows that a correlation exists between their values. Emulsifier adsorption during polymerization as well as a degree of defence of the surface formed depending on the emulsifier's nature decrease with increasing monomer solubility in water. It leads to particles flocculation, which is typical of emulsion polymerization of polar monomers. The polymerization is characterized with a rather long steady-state period in spite of poor adsorption of emulsifier and a continuous flocculation of particles. This fact as well as correlation of a rate order with respect to emulsifier with its adsorption energy and continuation of the steady state period after the disappearance of monomer drops point to the surface zone of particles as the main site of polymerization. The existance of the constant rate interval after free monomer disappears (at persulphate initiation) can be explained by the proceeding of polymerization in the outer zone of particles and by the redistribution of monomer between inner and outer primary globules.

Abstract

Emulsion polymerization of lower alkyl acrylates was studied in terms of interfacial phenomena. Adsorption of emulsifier depending on its nature reduces during polymerization water solubility of an alkyl acrylate increases. Molecular area of sodium lauryl sulphate in saturated adsorption layers increases from 0.5 $(nm)^2$ to 1.75 $(nm)^2$ in styrene, butyl-, ethyl- and methylacrylate latexes.

The energy of emulsifier adsorption on monomer - water interface depends on the emulsifier nature and is due to monomer water solubility. The reaction order with respect to the emulsifier depends on the chemical nature of the latter and seemingly correlates with the emulsifier adsorption energy on monomer -water interface. At the persulphate initiation the steady-state interval continues up to the conversion exeeding one at which monomer drops disappear. This deviation from "classical" kinetics was explained by the flocculation mechanism of particle formation, and by the redistribution of monomer between "dead" and "living" primary globules.

Literature Cited

1. Yurzhenko A.I., Mints S.M., DAN SSSR, (1947), 47, 2, 106-108.
2. Harkins W.D., J.Amer.Chem.Soc., (1947), 69, 1428-1444.
3. Smith W.V., Ewart R.H., J.Chem.Phys., (1948), 16, 5, 529-599.
4. Medvedev S.S., in: "Kinetics and Mechanism of Formation and Transformation of Macromolecules", Nauka Publs, Kiev, 1968, pp..5-24.
5. Rhebinder P.A., Zeitshrift fur Phys. Chemie, (1927), 129, 11.
6. Yeliseyeva V.I., Acta Chem. Acad.Sci.Hung., (1972), 71, 465-481.
7. Roe Ch.P., Industr.Eng.Chem., (1968), 60, 9, 20-34.
8. Robb Y.D., J.Polym.Sci., (1969), A-1, 2, 417-427.
9. Paxton Th., J.Coll.Interface Sci., (1969), 31, 1, 19-30.
10. Yeliseyeva V.I., Bakayeva T.V., Visokomol.Soyed., (1969), 1A, 2186-2190.
11. Brodnyan Y.G., Kelly E.T., J.Polym.Sci., (1969), C-27, 263-273.
12. Breitenbach J.W., Kuchner K., Fritzl H., Tarnowiecki H., Br. Polym.J., (1970), 1/2, 13-17.
13. Friis N., Singer K., Kops Y., (1971), I.U.P.A.C. Intern.Sympos.on Macromol., Helsinki, Preprints.
14. Romanova O.S., (1972), Thesis, Leningrad.
15. Krishan T., Margaritova M.F., Medvedev S.S., Visokomol. Soyed., (1969), 5, 585.

16. Maron S.H., Elder M.E., Ulevitch I.N., J.Coll. Sci., (1954), 9, 263.
17. Petrova S.A., (1973), Thesis, Moscow.
18. Yeliseyeva V.I., Zuikov A.V., Mamadaliyev A., Zhurnal Fiz. Khimii, (1975) to be published.
19. Kudryavtseva N.M., Deryagin B.V., Koll. Zh., (1963), 25, 739-741.
20. Donahue D.J., Bartel F.E., J.Physic.Chem., (1952) 56, 480-485.
21. Granzhan V.A., Zhurnal V.Ch.O. im.Mendeleyeva, (1969), 14, 2, 223.
22. Fitch R.M., Brit. Polym.J., (1973), 5, 467-483.
23. Vanzo E., Marchesault R.H., Stannet V., J.Coll.Sci., (1965), 20, 62-71.
24. Gardon J.L., J.Polym. Sci., (1968), A-1, 6, 2859-2879.
25. Yeliseyeva V.I., Zharkova N.G., Lukyanovich V.M., Visokomol.Soed., (1967), 9A, 11, 2478.
26. Dunn A.S., Chong Z.C-H., Brit. Polym. J., (1970), 2, 1/2, 49-59.
27. Deryagin B.V., Landau L.D., Zh.Eksperim. i Teoret. Fiziki, (1945), 15, 11, 663-682.
28. Brooks B.W., Brit.Polym.J., (1970), 2, 197-201.

5

Gel-Effect in Emulsion Polymerization of Vinyl Monomers

N. FRIIS and A. E. HAMIELEC

Department of Chemical Engineering, McMaster University, Hamilton, Ontario, Canada

In the bulk polymerization of vinyl monomers at ordinary polymerization temperatures (50-100°C), the termination reactions involving macroradicals become diffusion controlled and the termination rate constant decreases by three or in some cases even by four orders of magnitude in the conversion interval 0-100%. This decrease in termination rate constant, which will be referred to as gel-effect, always causes a significant increase in rate of polymerization and can also shift the molecular weight distribution to higher molecular weights, but the magnitude of the shift depends upon which reactions control molecular weight development. By increase in polymerization rate we mean the increase over the rate which would have been observed had gel-effect been absent. In the absence of gel-effect the rate of bulk polymerization would be first order with respect to monomer concentration and hence decrease with reaction time. Therefore, if for instance a constant polymerization rate is observed experimentally, this means that gel-effect has increased the polymerization rate and is therefore significant.

In bulk polymerization at a given temperature the increase in rate due to gel-effect is almost independent of experimental conditions, as for instance initiator concentration.

In emulsion polymerization the situation is entirely different in that the increase in rate due to gel-effect depends on experimental conditions such as initiator concentration, particle size and particle number. Therefore, accounting for gel-effect in emulsion polymerization is considerably more complex than in bulk polymerization. However, we have recently shown that the increase in rate due to gel-effect in emulsion polymerization of various monomers can be accounted for quantitatively by means of data from bulk polymerization (1, 2, 3, 4).

It is the purpose of this paper to outline a general technique of treating gel-effect in emulsion polymerization and to discuss the role of gel-effect in emulsion polymerization of various monomers.

We will furthermore present data showing the effect of polymerization temperature on limiting conversion. With polymerizations below the glass transition point of the polymer, the monomer-polymer solution reaches its glass transition point at a conversion <100%. At this point reactions involving small molecules, such as propagation, become diffusion controlled. This causes the rate of polymerization to fall to virtually zero in the normal polymerization time scale, i.e. in practice the reaction rate reaches a limiting conversion <100%.

Calculation of Gel-Effect Data

In Figure 1 are shown experimental conversion histories from bulk polymerization of styrene, methyl methacrylate, and vinyl acetate. It appears that with any of these monomers the rate of polymerization increases substantially during reaction, i.e. gel-effect is important in bulk polymerization of these monomers. The effect is particularly pronounced with methyl methacrylate.

By application of the simple rate expression for homogeneous bulk polymerization

$$R_p = k_p[M] \left(\frac{fk_i[I]}{k_{tp}}\right)^{1/2} \qquad (1)$$

we have from these and similar data "extracted" the gel-effect, i.e. by fitting Equation (1) to experimental data we have measured the decrease in termination rate constant as function of conversion. Figure 2 shows the relative change in k_{tp} as function of conversion thus obtained for the three monomers at 50°C, and it appears that in any case k_{tp} decreases by at least three orders of magnitude during the polymerization. In Table I are summarized in mathematical form gel-effect data of the three monomers. We have, of course, assumed that for an isothermal polymerization all rate constants other than k_{tp} are independent of conversion.

Emulsion Polymerization Kinetics

In emulsion polymerization a single polymer particle can be regarded as a locus of bulk polymerization with intermittent initiation. A decrease in termination rate constant, which is observed in bulk polymerization, should, therefore, also occur in a single polymer particle and must be taken into account in model simulation of emulsion polymerization. However, since the polymer particles from the very beginning of the reaction contain a certain percentage of polymer, typically 20-40%, then the termination rate constant is always smaller in the beginning of emulsion polymerization than in the similar homogeneous bulk

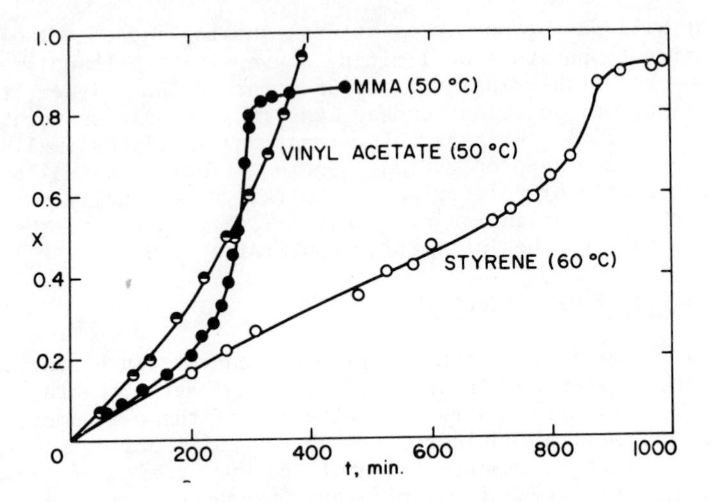

Figure 1. Conversion histories in bulk polymerization of methyl methacrylate, styrene, and vinyl acetate

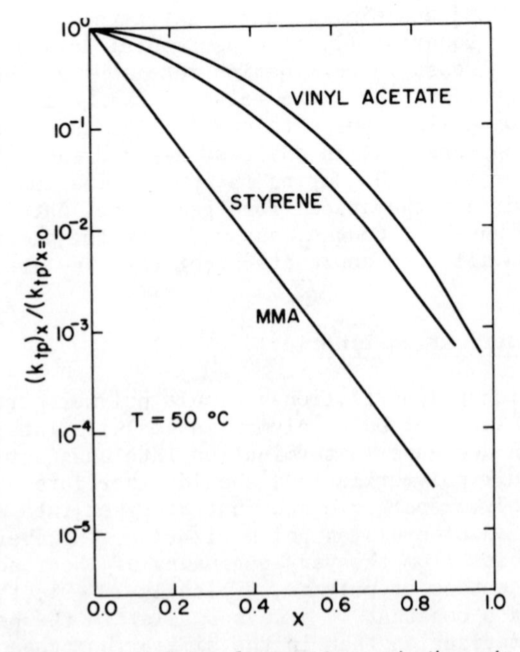

Figure 2. Relative decrease in termination rate constants as function of conversion at 50°C

TABLE I

GEL-EFFECT DATA

Methyl Methacrylate (Valid Temperature Range, 40-90°C):

$$(k_{tp})_x/(k_{tp})_{x=0} \quad [\frac{1}{1-x} \mathrm{Exp}(Bx + Cx^2)]^2$$

$$B = -41.54 + 0.1082 \cdot (T\ °K)$$

$$C = 23.46 - 0.0785 \cdot (T\ °K)$$

Styrene (Valid Temperature Range, 50-200°C):

$$(k_{tp})_x/(k_{tp})_{x=0} = [\mathrm{Exp}\ (-(Bx + Cx^2 + Dx^3)]^2$$

$$B = 2.57 - 5.05 \cdot 10^{-3} \cdot (T\ °K)$$

$$C = 9.56 - 1.76 \cdot 10^{-2} \cdot (T\ °K)$$

$$D = -3.03 + 7.85 \cdot 10^{-3} \cdot (T\ °K)$$

Vinyl Acetate (50°C):

$$(k_{tp})_x/(k_{tp})_{x=0} = \mathrm{Exp}(Bx + Cx^2 + Dx^3)$$

$$B = -0.4407$$

$$C = -6.7530$$

$$D = -0.3495$$

process, and therefore the relative change in termination rate constant is always somewhat smaller in emulsion than in bulk.

After Smith and Ewart (5), we define the kinetics of emulsion polymerization in terms of the recurrence formula,

$$N_{n-1}(\rho'/N) + N_{n+1}k_d(n+1) + N_{n+2}[(n+2)(n+1)/v]$$

$$= N_n[(\rho'/N) + k_d n + k_{tp}n(n-1)/v] \qquad (2)$$

N_n denotes number of particles with n free radicals, ρ' is the total rate by which radicals enter all N particles, k_d and k_{tp} denote desorption rate constant and termination rate constant respectively, and v is the average volume of a polymer particle. The general solution of Equation (2) is represented graphically in Figure 3 after Ugelstad (6), where the average number of radicals per particle, $\tilde{n}$, is plotted as function of the parameters α' and m defined in Figure 3. Given the parameters α' and m, $\tilde{n}$ can be obtained from Figure 3, and from $\tilde{n}$ the rate of polymerization is calculated as

$$R_p = k_p\ [M_p]\ N\tilde{n} \qquad (3)$$

k_p and $[M_p]$ are propagation rate constant and monomer concentration within polymer particles respectively.

It is generally believed that desorption of radicals from polymer particles is negligible in emulsion polymerization of styrene and methyl methacrylate, i.e. with these monomers $k_d \simeq 0$, and therefore $m \simeq 0$. Let us now in this case consider the effect of a decrease in k_{tp} by three to four orders of magnitude on $\tilde{n}$ and thus on the rate of polymerization. Assume that the number of polymer particles, N, becomes constant relatively early in the polymerization, for instance at 10-15% conversion. If at this stage of the reaction α' is less than 10^{-4}, then a decrease in k_{tp} by three to four orders of magnitude has no effect on $\tilde{n}$, which will remain constantly equal to 0.5. Such a small value of α' is obtained either if the initiation rate ρ is very low or the number of particles is very large, and in this case gel-effect will not affect the polymerization rate, which will be first order with respect to $[M_p]$. If α' in the early stages of polymerization lies between 10^{-4} and 10^{-2} then the same decrease in k_{tp} will cause a certain increase in $\tilde{n}$ during polymerization. The polymerization rate will no longer be first order with respect to monomer concentration throughout the reaction. It may increase slightly, remain constant, or decrease during polymerization, depending on the initial value of α' and the magnitude of gel-effect. Finally, for values of α' larger than 10^{-2} gel-effect will cause a considerable acceleration in polymerization rate. This is the case when N is relatively small or

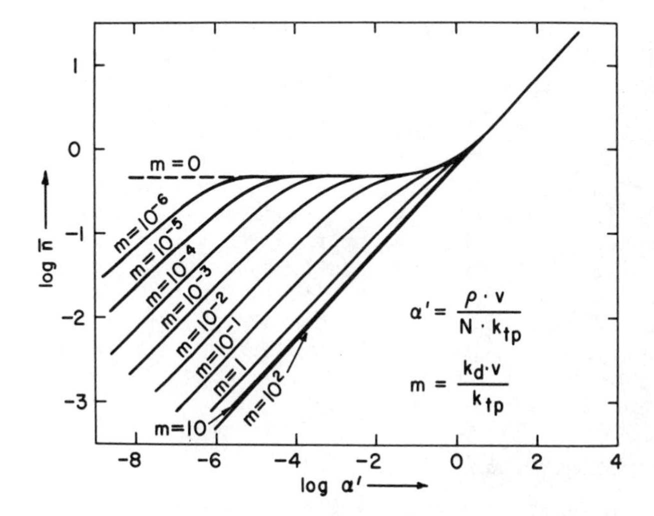

Figure 3. Average number of radicals per particle, ñ, as a function of the parameters α′ and m

the initiation rate is relatively large.

Since the polymerization conditions for styrene and methyl methacrylate often are chosen such that α' in the early stages of polymerization (after N has become constant) lies between 10^{-2} and 1, it is understandable that gel-effect usually is important in emulsion polymerization of these monomers and therefore must be accounted for since it completely dominates the shape of the conversion history.

In vinyl acetate emulsion polymerization radical desorption is important, i.e. with this monomer $k_d \neq 0$ and therefore $m \neq 0$. Typical values of m and α' lie in the intervals 10^{-1} - 10^{-3} and 10^{-3} - 10^{-6} respectively in the early stages of polymerization. A decrease in k_{tp} during polymerization will cause α' and m to increase. A decrease in m will tend to reduce $\tilde{n}$ while an increase in α' will cause an increase in $\tilde{n}$. It follows from Figure 3 that the effect of a decrease in m on $\tilde{n}$ is only slightly smaller than the effect of a corresponding increase in α', i.e. the net-effect on $\tilde{n}$ is very small and therefore gel-effect has a much smaller influence on polymerization rate in vinyl acetate than in typical styrene and methyl methacrylate emulsion polymerization. However, since the desorption rate constant k_d decreases somewhat during vinyl acetate polymerization, $\tilde{n}$ will in fact increase and therefore the rate of polymerization will not be first order with respect to $[M_p]$. But, this is not due to gel-effect as defined in this paper.

Simulation of Emulsion Polymerization

By application of the gel-effect data given in Table 1, it is now possible to account for gel-effect in the simulation of emulsion polymerization. The value of k_{tp} as calculated from Table 1 is substituted into Equation (2) which is solved for $\tilde{n}$. R_p is then calculated from Equation (3). The conversion history is obtained by repeating this calculation stepwise at successively higher conversions and simultaneous integration of Equation (3).

Systems Where Radical Desorption is Negligible. Styrene and methyl methacrylate emulsion polymerization are examples of systems where radical desorption can be neglected. In Figures 4 and 5 are shown comparisons between experimental and theoretical conversion histories in methyl methacrylate and styrene polymerization. The solid curves represent the model, and it appears that there is excellent agreement between theory and experiment. The values of the rate constants used for the theoretical simulations are reported in previous publications (2, 3). The dashed curves represent the corresponding theoretical curves in the calculation of which gel-effect has been neglected, that is, k_{tp} is kept constant at a value for low viscosity solutions. It appears that neglecting gel-effect in the simulation of styrene and methyl methacrylate emulsion polymerization produces large deviations from the experimental curves. In emulsion polymerization of these systems particle number and size play an important role in the influence of gel-effect on rate of polymerization through the change of α' with conversion. From the definition of α', it is clear that a decrease in the number of polymer particles and an increase in particle size makes the reduction in termination rate constant more significant in increasing the average number of radicals per particle and therefore the rate of polymerization. In other words, gel-effect will cause a stronger acceleration in the rate of polymerization the larger the particle volume and the smaller the number of polymer particles. This effect is clearly illustrated in Figure 4.

Systems Where Radical Desorption is Important. Vinyl chloride and vinyl acetate emulsion polymerization are examples of systems where radical desorption is important. Typical data for vinyl acetate emulsion polymerization together with model predictions are shown in Figure 6. The dashed curve represents the model where gel-effect is not accounted for and the solid curve includes corrections for gel-effect. As expected, the difference in the model predictions is small. The decrease in the termination rate constant for vinyl acetate is about the same as that for styrene (refer to Figure 2), however, the effect of radical desorption in the case of vinyl acetate counteracts the gel effect. While this is generally true for both vinyl chloride and vinyl acetate emulsion polymerizations there are special situations, such as very high initiation rates with large polymer particles where the rate of radical desorption can no longer counteract the gel effect and therefore in such cases an autoacceleration in the polymerization rate will be experienced. Referring to Figure 3, it is clear that this occurs with either values of $m > 1$ or values of $\alpha' > 0.1$.

Limiting Conversions

We have so far discussed mainly the effects of diffusion controlled termination reactions. However, for polymerizations which occur near the glass transition point of the monomer-polymer mixture reactions involving small molecules can become diffusion controlled. When the propagation reactions become diffusion controlled, the rate of polymerization rapidly falls to a value close to zero. This is clearly shown for the polymerization of styrene and methyl methacrylate in Figure 1. The glass transition temperatures for pure PMMA and PS are about 110°C and 85°C, respectively. However, with vinyl acetate where the glass transition point of pure PVAc (~28°C) is below the polymerization temperature a limiting conversion near 100% can be reached. A PMMA-MMA solution containing 15% monomer has a glass

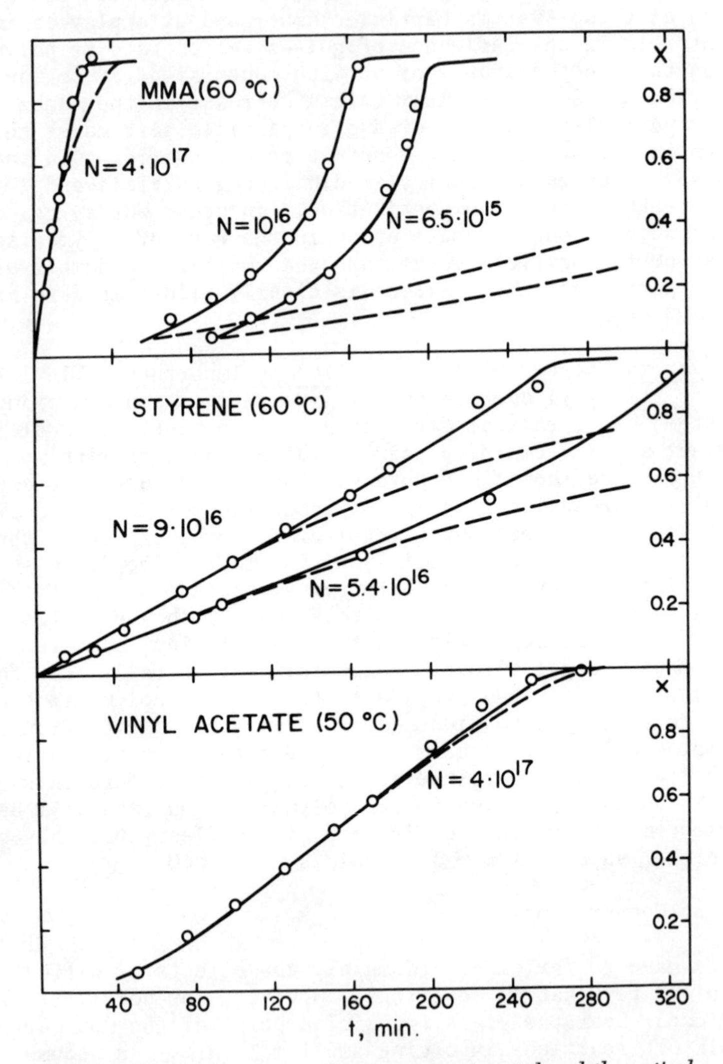

Figures 4, 5, 6. Comparison between experimental and theoretical conversion histories in emulsion polymerization of methyl methacrylate, styrene, and vinyl acetate. (—), model with gel-effect; (– –), model without gel-effect.

transition temperature of 50°C. A PS-S solution containing 3.5% monomer has a glass transition temperature of about 60°C. Data showing the effect of polymerization temperature on limiting conversion for PMMA, PVC and PS are shown in Figure 7. An extrapolation of these data to a limiting conversion of 100% gives T_g of the pure polymer which is in reasonable agreement with T_g values measured by differential scanning calorimetry.

In conclusion we may say that a conversion near 100% can only be reached within reasonable time if the polymerization temperature is above the glass transition point of the polymer. This conclusion holds for bulk, emulsion, and suspension polymerization but, of course, not for solution polymerization where the solvent-polymer mixture usually has a glass transition point which is well below ordinary polymerization temperatures.

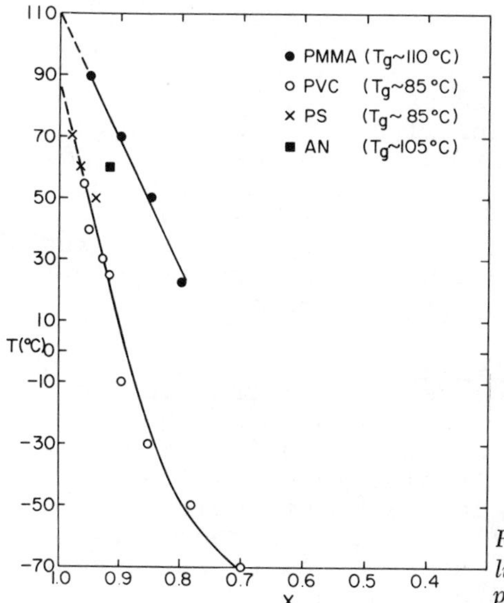

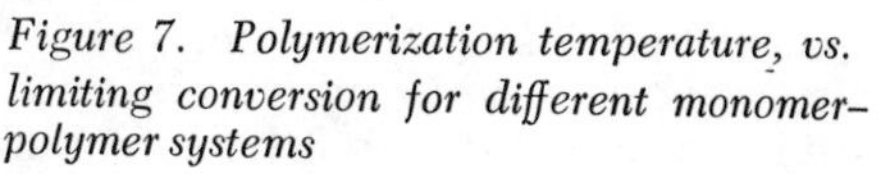
Figure 7. Polymerization temperature, vs. limiting conversion for different monomer-polymer systems

Literature Cited

(1) Friis, N. and Nyhagen, L., J. Appl. Poly. Sci., (1973), 17, p. 2311.

(2) Friis, N. and Hamielec, A.E., J. Poly. Sci., (1974), 12, p. 251.

(3) Friis, N. and Hamielec, A.E., J. Poly. Sci., (1973), 11, p. 3341.

(4) Friis, N. and Hamielec, A.E., "Kinetics of Vinyl Chloride and Vinyl Acetate Emulsion Polymerization", J. Appl. Poly. Sci., (1975), 19, p.97.

(5) Smith, W.V. and Ewart, R.H., J. Chem. Phys., (1948), 16, p. 592.

(6) Ugelstad, F., Mork, P.C. and Aasen, J.E., J. Poly. Sci., (1967), 5, p. 2281.

6

Emulsion Polymerization of Multifunctional Monomers—Preparation of Reactive Microgels

W. OBRECHT, U. SEITZ, and W. FUNKE

Institut für Technische Chemie der Universität, Stuttgart, D 7000 Stuttgart-80, Pfaffenwaldring 55, West Germany

1. Introduction

Reactive microgels are crosslinked polymer particles which have a diameter of some 100 Å and contain pendant reactive groups like vinyl groups. Such microgels may be copolymerized with bifunctional monomers to obtain crosslinked polymers in which the microgel particles act as multifunctional crosslinking sites (1). In some respects such crosslinked polymers are comparable with other multiphase polymer systems like block or graft polymers.

2. Preparation of reactive microgels

For the preparation of reactive microgels two methods have been applied, one of which is the polymerization of multifunctional monomers like divinylbenzene (DVB) or ethyleneglykol-dimethacrylate (EDMA) in highly dilute solution in which intramolecular crosslinking is favored (2). By this method pure microgels may be prepared, however the reaction conditions must be chosen thoroughly in order to avoid macrogelation. A more convenient method makes use of polymerizing the above mentioned monomers in emulsion. In this case, crosslinking is limited to colloidal particles, however it introduces the problem of separating the particles from the adsorbed emulsifier before they can be used for further reactions.

In both cases the isolation of the reactive microgels from the reaction batch requires careful and elaborate series of procedures at low temperatures in the absence of oxygen (1,3). Otherwise the reactive particles will agglomerate irreversibly.

The following procedure has proved to be most effective: at first, the residual monomer is removed by freeze-drying of the reaction mixture. Then the emulsifier and the salts are separated by dialysis.

After washing several times with methanol, the microgel fraction is freeze-dried from benzene and residual solvents are stripped off under reduced pressure (10^{-5} mbar). The microgels thus obtained are white, dust-like powders with a sedimentation volume of 20-50 ml/g. They can be redispersed to form colloidal organic solutions.

3. Characterization of reactive microgels

The numbers of pendant vinyl groups can be determined quantitatively by addition reactions or, more conveniently, by IR-spectroscopy. Depending on the reaction conditions of the polymerization, from 20 to 90% of the maximum possible amount of pendant vinyl groups (one per each monomer unit) have been found unreacted. Depending on the monomers used for microgel preparation the conversion versus time plots are significantly different. Microgels made of pure 1,4-DVB at first show a rapidly increasing conversion which rather abruptly changes after one hour of reaction time to a very slow further increase "Figure 1". The characteristic break observed in the conversion may be explained by the change of the very rapid reaction of vinyl groups attached to the particle surface to the slow, diffusion-controlled reaction of the unsaturated groups within the particles. However, this break is not found with microgels which have a lower density of crosslinks (e.g. those made from monomers like technical DVB or mixtures of 1,4-DVB/styrene). In these cases the reaction at the surface is already superposed by the reaction of vinyl groups located in the interior of the microgel particles.

Shapes and diameters of microgel particles have been determined by electron microscopy and soap titration. Knowing both mean diameter and number of pendant vinyl groups at the particle's surface, it is possible to estimate the concentration of the vinyl groups as well as their average distances on the surface of the microgel particles. For example in the case of microgels from 1,4-DVB which have a mean diameter of 96 Å and 20% vinyl groups at the surface, the concentration of the vinyl groups is 1,59 vinyl groups per 100 $Å^2$ and the mean distance between neighbored vinyl groups is 8,5 Å (3).

4. Emulsion polymerization of 1,4-DVB

It is well known that in the emulsion polymerization of styrene, particle size and size distribution can be varied by changing the amount of emulsifier, the initiator concentration, the ratio of monomer to water and the temperature. In contrast to the emulsion polymerization of styrene however, the size and size distribution of the microgel particles prepared from 1,4-DVB are hardly influenced by these experimental parameters. Furthermore, some distinct differences are observed even when the same experimental conditions have been applied, which will be discussed below:

4.1 Number of particles. Emulsion polymerization of 1,4-DVB yields significantly more and smaller polymer particles than that of styrene (table I).

Table I shows that the diameters of the polymers from styrene are approximately twice as large as those from 1,4-DVB, and in all experiments with 1,4-DVB at least 6 times more particles have been formed than with styrene.* The maximum diameters of spherical particles from 1,4-DVB which can be obtained, are about 500 Å. Larger particles are mostly irregularly shaped (see 4.2) whereas in the case of styrene, latex particles of 2000 Å and more may be prepared easily. These differences may be explained by a discussion of the SMITH-EWART theory (4). According to this theory the number N of latex particles formed in the emulsion polymerization of styrene is given by

$$N = \text{const.} \left(\frac{\varsigma}{\mu}\right)^{2/5} \left(a_s [S]\right)^{3/5}$$

where ς is the rate of formation of free radicals, μ is the rate of volume increase of one particle, a_s is the interfacial area occupied by one surfactant

*) Comparing the diameters of the microgel particles the amount of coagulum formed in the emulsion polymerization of 1,4-DVB has not been taken into account. Supposed the coagulum would be distributed among the number N of counted microgel particles the mass of each microgel particle as well as its diameter would grow. In case of exp. no. 27 where 21,7% coagulum were formed the mean diameter would increase from $\bar{d}_o$ = 250 Å to $\bar{d}_o$ = 270 Å.
This difference can be neglected since it does not affect in any way our experimental findings the mean diameters of styrene particles being approximately twice as high as those of the microgels.

Table I. Diameters and Numbers of various Particles prepared by Emulsion Polymerization of 1,4-DVB and Styrene

Exp. No.	30		30/1		27	
Monomers	1,4-DVB	St	1,4-DVB	St	1,4-DVB	St
$\bar{d}_o$ by soap titr. [Å]	304	645	355	810	250	565
$\bar{d}_o$ by EM [Å]	315	643	-	-	277	570
N by soap titr. $[cm_w^{-3}]$	$5{,}9 \cdot 10^{15}$	$9{,}9 \cdot 10^{14}$	$4{,}0 \cdot 10^{15}$	$5 \cdot 10^{14}$	$1{,}3 \cdot 10^{16}$	$1{,}5 \cdot 10^{15}$
$N_{1,4\text{-}DVB}/N_{st}$	6,0		8,1		8,9	

Experimental conditions:

in all cases: temperature: 50^oC; emulsifier: Sodium-Laurysulfate (=S); ratio monomer to water (v/v): 1:6,25; initiator: $K_2S_2O_8$ (=J); $[NaHCO_3] = 2 \cdot 10^{-2}$ mole·dm_w^{-3}

in exp. 30 : $[J] = 1 \cdot 10^{-2}$ mole/l_w; $[S] = 2 \cdot 10^{-2}$ mole/l_w

in exp. 30/1: $[J] = 0{,}5 \cdot 10^{-2}$ mole/l_w; $[S] = 2 \cdot 10^{-2}$ mole/l_w

in exp. 27 : $[J] = 1 \cdot 10^{-2}$ mole/l_w; $[S] = 4 \cdot 10^{-2}$ mole/l_w

amount of coagulum (with 1,4-DVB, see 4.3): in exp. no. 30 : 37,3%
in exp. no. 30/1: 32,89%
in exp. no. 27 : 21,7%

molecule and [S] the concentration of surfactant.

In the case of the comparative experiments compiled in table I a_s and [S] have been kept constant. If different numbers of particles are formed, either ρ or μ (or both) have changed. If ρ may be expected to remain constant (the $K_2S_2O_8$ concentration is the same) μ must be about 100 times smaller in the case of 1,4-DVB than in the case of styrene to produce 6 times more particles. Though it can be expected that a latex particle of a multifunctional monomer (1,4-DVB) which produces a crosslinked polymer has not the same rate of volume increase as a monomer-swollen polystyrene latex particle, such a great difference in μ is not very likely. More likely the determining factor is ρ which is not the same in both cases though the same amount of $K_2S_2O_8$ has been used.

If styrene is polymerized at 50°C in emulsion, the initiating radicals are essentially produced by the decomposition of the persulfate to radical anions. In case of monomers like 1,4-DVB, however, additional radicals in substantial amounts are formed by thermal initiation. Therefore the radical formation is significantly higher, more micelles are initiated and consequently more but smaller particles are formed.

The possibility of a thermal initiation respectively polymerization of 1,4-DVB at 50°C in emulsion without $K_2S_2O_8$ or other initiators has been established.

4.2 Agglomeration processes. In contrast to the emulsion polymerization of styrene, agglomeration processes take place in the emulsion polymerization of 1,4-DVB. They can be observed in electron micrographs as inhomogeneities regarding structure and shape of the particles.

Concerning this the following effects are of great influence: it may be conceived that contrary to the normal termination mechanism, microgels from 1,4-DVB contain more than one radical at average time; this is due to diffusional and steric hindrance of radical combination. Therefore these particles are highly reactive both on behalf of remaining vinyl groups and of radical chain ends attached to the surfaces of the particles.

The second important effect is the number of micelles initiated which is very much higher in the case of 1,4-DVB than in the case of styrene (see 4.1). Thus the total surface area increases faster in the case of 1,4-DVB than in the case of styrene and a deficiency of emulsifier molecules occurs at a very early stage.

In order to compensate this, the particles agglomerate. Because of the high reactivity of the particles (pendant vinyl groups and pendant radicals) these agglomeration processes are irreversible.

There are two significant cases of agglomerated particles differing in size as well as in contour.

4.2.1 Primary agglomerates. These particles are spherical and have a maximum diameter of 450 Å. They are formed by agglomeration of smaller (below 50 Å) homogeneous primary particles (initiated micelles); the space remaining between these primary particles has obviously been filled up by polymerization of further monomer. Therefore these particles may only be identified as agglomerates after a careful inspection with the electron microscope "Figure 2"

4.2.2 Secondary agglomerates. If the polymerization is continued to a very high conversion, further agglomeration of the primary agglomerates takes place. In general these particles have a diameter of 1000-3000 Å and are irregularly shaped. When not enough monomer is left, these agglomerates can not grow further to attain a spherical form again "Figure 3". In figure 4 the agglomeration processes are described schematically.

4.3 Appearance of coagulum. Towards the end of the reaction besides a colloidal solution of microgel particles some flocculate may be found. With the electron microscope two kinds of flocculates can be distinguished. The first type looks like coagulated secondary agglomerates and the second like polymerized monomer droplets (with diameters of about 2000-5000 Å). (These polymerized monomer droplets are also found in the case of 1,3,5-trivinylbenzene.)

The amount of flocculated polymer formed in the emulsion polymerization of 1,4-DVB at approximately 100 % conversion varies with the amount of emulsifier (Na-laurysulfate), the ratio of monomer to water, the initiator concentration ($K_2S_2O_8$) and temperature "Figure 5".

Formation of flocculated polymer is most pronounced at low soap concentration and vanishes with a higher soap content.

The authors gratefully acknowledge the financial support granted by the Deutsche Forschungsgemeinschaft, and one of us (W.O.) thanks for a fellowship sponsored by the University of Stuttgart.

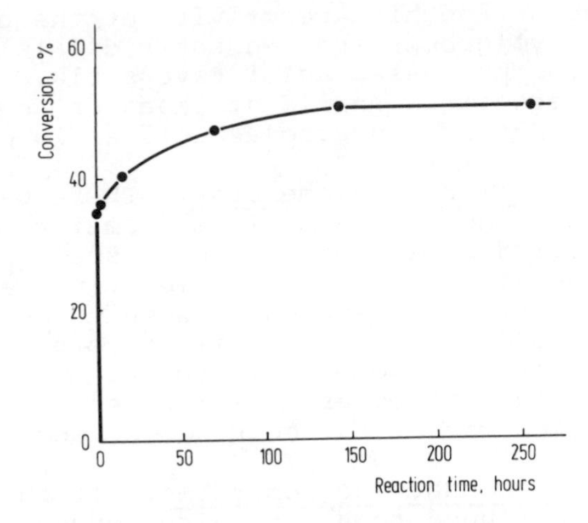

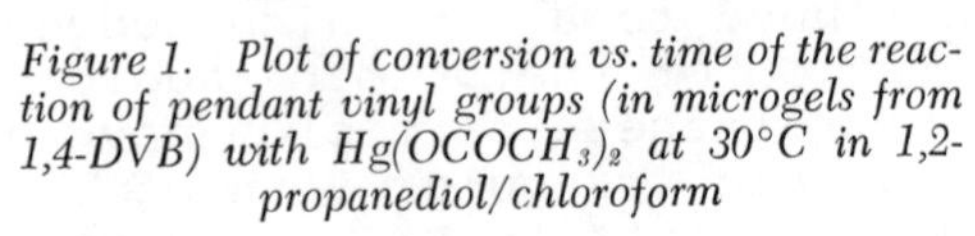

Figure 1. Plot of conversion vs. time of the reaction of pendant vinyl groups (in microgels from 1,4-DVB) with $Hg(OCOCH_3)_2$ at 30°C in 1,2-propanediol/chloroform

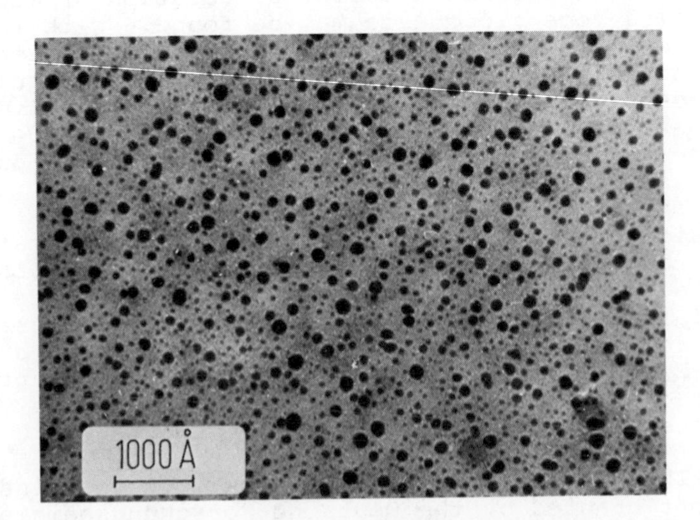

Figure 2. Primary agglomerates from 1,4-DVB

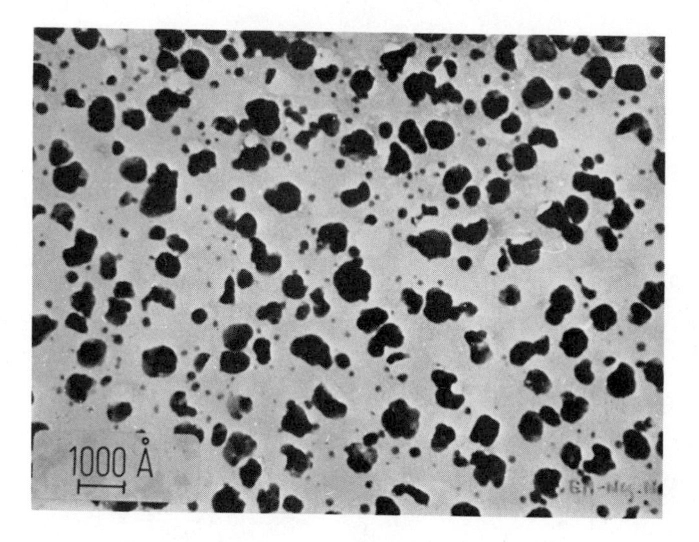

Figure 3. Secondary agglomerates from 1,4-DVB

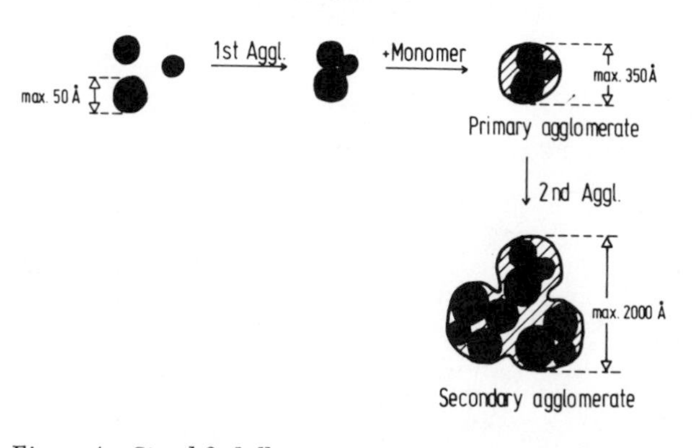

Figure 4. Simplified illustration of the agglomeration processes during the emulsion polymerization of 1,4-DVB

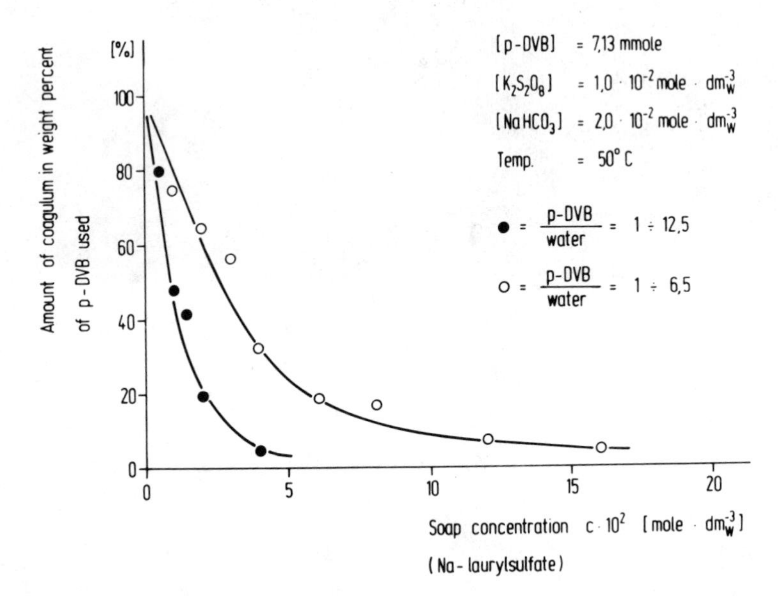

Figure 5. Plot of the amount of coagulum vs. soap concentration in the emulsion polymerization of 1,4-DVB at two different volume ratios monomer/water

Literature cited

(1) FUNKE, W., BEER, W. and SEITZ, U. Fortschrittsber. Kolloide, Polymere 57 (1975) 48

(2) FUNKE, W., being published

(3) OBRECHT, W., SEITZ, U. and FUNKE, W. Amer.Chem. Soc., Div.Polym.Chem., Prepr. 16(1)(1975) 149

(4) SMITH, W.V. and EWART, R.H. J. Chem.Physics 16 (1948) 592

7

Kinetics and Mechanism of the Emulsion Polymerization of Vinyl Acetate

M. NOMURA,[**] M. HARADA,[*] W. EGUCHI,[*] and S. NAGATA

Department of Chemical Engineering, Kyoto University, Kyoto, Japan

To date it has been a well-known fact that the Smith and Ewart theory does not apply to the emulsion polymerization of highly water-soluble monomers such as vinyl acetate and vinyl chloride. Therefore, many experimental and theoretical investigations have been performed during the past decade to understand and explain the mechanism of the emulsion polymerization of such monomers. Recently, Ugelstad et al.(1)(2) have succeeded in explaining the kinetics of the emulsion polymerization of vinyl chloride introducing the mechanism of rapid radical desorption and reabsorption in the polymer particles, while the authors (3)(4) have developed a theoretical expression for the rate coefficient of radical escape process from the polymer particles assuming that the majority of escaping radicals is a monomeric one, and successfully explained the role of polymer particles in the emulsion polymerization of a wide variety of monomers including, of course, vinyl acetate and vinyl chloride. Recently, several investigations have been carried out to reexamine the kinetics and mechanism of the emulsion polymerization of vinyl acetate and vinyl chloride(5)(6), and clarified that the kinetics of the emulsion polymerization of these monomers are essentially identical. However, the mechanism of particle nucleation still remain equivocal and hence one cannot estimate the conversion-time relation in regard to monomer consumption theoretically.

The purpose of this paper is first to clarify the detailed characteristics of the emulsion polymerization of vinyl acetate using sodium lauryl sulfate as emulsifier and potassium persulfate as initiator, and second to propose a new reaction model, based on our theory relating to the role of polymer particles, which enable us to predict the number of polymer particles produced and consequently , the progress of vinyl acetate emulsion polymerization.

Experimental

A schematic diagram of the experimental apparatus and the di-

Present address

*) Institute of Atomic Energy, Kyoto Univ. Uji, Japan

**) Dept. of Industrial Chem., Fukui National Univ. Fukui, Japan

mensions of the reactor vessel and the impeller are shown in Fig.1. The reactor is a cylindorical glass vessel with a dished bottom, equipped with four-bladed paddle type impeller and four baffle plates located on the vessel wall at 90° intervals. Vinyl acetate monomer was distilled twice under vacuum, stored at -20°C and redistilled before use. Sodium lauryl sulfate and potassium persulfate of extra pure grade were used without further purification.

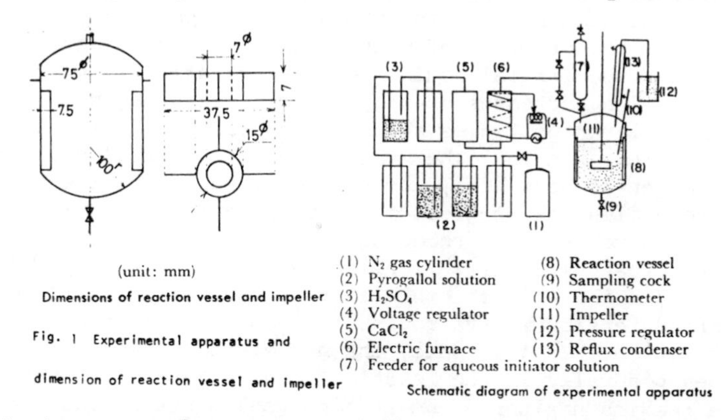

Dimensions of reaction vessel and impeller

Schematic diagram of experimental apparatus

Fig. 1 Experimental apparatus and dimension of reaction vessel and impeller

The start-up procedures are as follows. The reaction vessel was charged with the desired amounts of purified water, emulsifier and monomer, and the dissolved oxygen was removed by bubbling purified nitrogen gas through the reaction mixture for at least half an hour. The aqueous initiator solution previously deoxygenated with the nitrogen gas was then fed to the reaction vessel and the polymerization reaction was started. In all cases the reaction temperature was maintained at 50 ± 0.5°C by means of a thermostatted water bath and impeller speed at 400 r.p.m.. Monomer conversion was determined gravimetrically and the degree of polymerization by the viscosity in benzen solution method employing Nakajima's equation given below(7).

$$[\eta] = 5.36 \times 10^{-4} \, [\, M_\mu \,]^{0.62} \qquad (1)$$

The number of polymer particles was determined from the monomer conversion X_M and the volume average diameter of the polymer particles d_p measured with an electron microscope.

$$d_p^3 = \Sigma n_i \, d_{pi}^3 / \Sigma n_i \qquad (2) \qquad N_T = 6\, M_o \, X_M / \pi \, d_p^3 \, \rho_p \qquad (3)$$

where M_o = initial monomer concentration(g/cc-water), N_T = the number of polymer particles(particles/cc-water) and ρ_p = density of polyvinyl acetate.

The monomer concentration in the monomer-swollen polymer particles was determined by chemical analysis after separating the remaining monomer droplets in the sample with a centrifuge. The details are given in the previous paper(4).

Results and discussion

Characteristics of vinyl acetate emulsion polymerization.

(1) Relationship between the variation of surface tension of aqueous phase of reaction mixture and the number of polymer particles produced:

Fig.2 shows a typical representation of the variation in the surface tension of the reaction mixture with reaction time. Fig.3 gives the relationship between the number of polymer particles produced and the progress of polymerization, corresponding to Fig.2. When the initial emulsifier concentration is very low, the surface tension increases sharply from the very beginning and the number of polymer particles seems to be constant from the start of polymerization. On the other hand, when the initial emulsifier concentration is very high, the surface tension remains almost constant regardless of progress of polymerization and polymer particles appears to generate throughout the polymerization process. This means that emulsifier micelles exist to the end of polymerization. When the initial emulsifier concentration is in between, the surface tension starts to increase abruptly in the course of polymerization. In this case, the number of polymer particles increases gradually in the reaction interval where the surface tension remains unaltered and attains to a constant value at a certain conversion of monomer where the surface tension starts to increase abruptly. Considering these characteristics which are very closely resemble to those observed in styrene emulsion polymerization(8), we may deduce that polymer particles generate from emulsifier micelles.

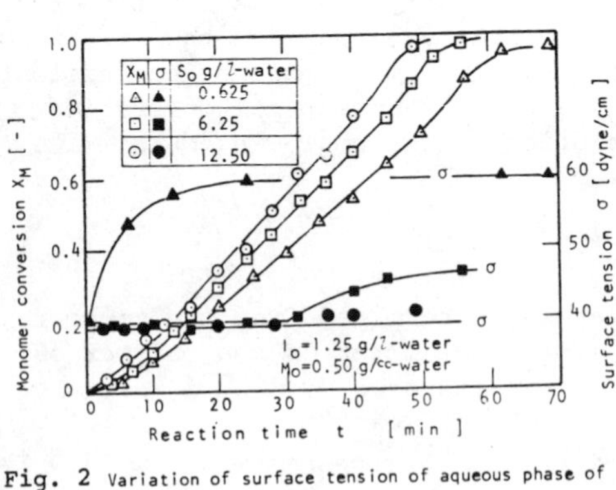

Fig. 2 Variation of surface tension of aqueous phase of reaction mixture with progress of polymerization.

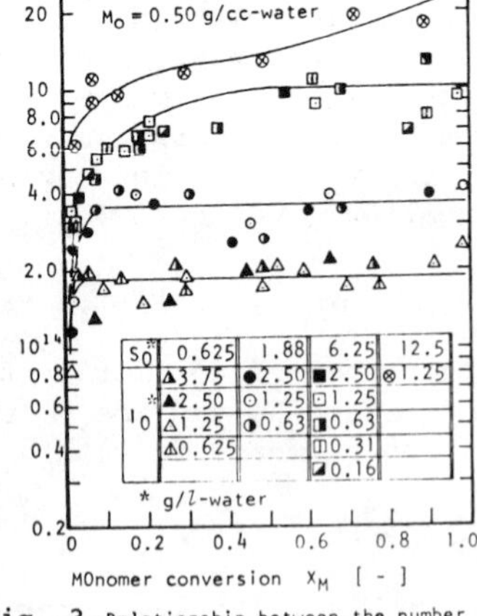

Fig. 3 Relationship between the number of polymer particles and monomer conversion

(2) Effect of emulsifier concentration upon the number of polymer particles and the progress of polymerization:

Fig. 4 shows the effect of initial emulsifier concentration on the number of polymer particles produced. From the log-log plot of N_T versus s_o, it can be seen that at lower range of emulsifier concentration the emulsifier dependence exponent for particle number is 2, but at higher range of emulsifier concentration, this exponent changes to the 0.92 power. This will be a clear evidence that the mechanism of particle nucleation in a lower range of the initial emulsifier concentration must be different from that in a higher range. Therefore, it can be deduced that in the lower range of emulsifier concentration the nucleation of polymer particles occurs possibly by precipitation of polymer formed by homogeneous polymerization in the water phase and in the higher range of emulsifier concentration particle nucleation from micelles will become dominant, though the change of particle nucleation mechanism occurs at the emulsifier concentration somewhat lower than CMC of sodium lauryl sulfate in pure water. Fig. 5 shows monomer conversion versus time plots at different emulsifier concentrations with initiator and monomer concentrations fixed, corresponding to Fig. 4 . It is seen that polymerization rate is almost linear from 15 to over 80 % of monomer conversion, increa-

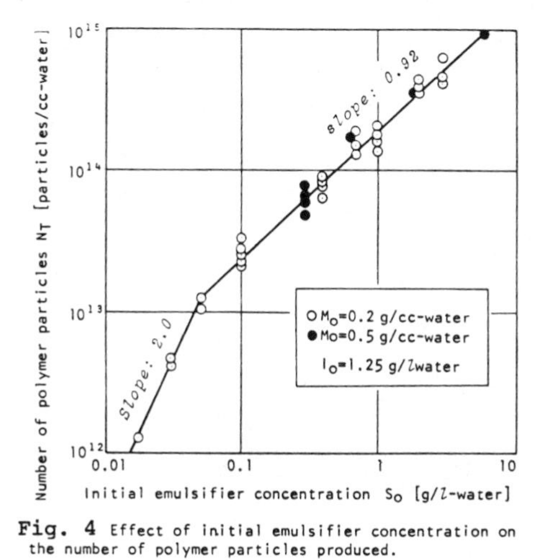

Fig. 4 Effect of initial emulsifier concentration on the number of polymer particles produced.

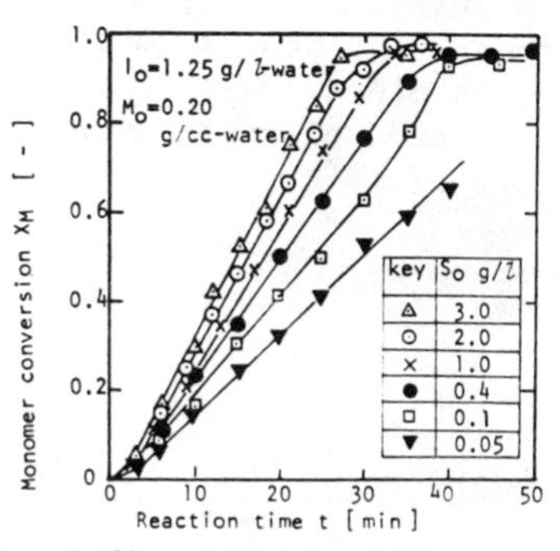

Fig. 5 Effect of initial emulsifier concentration on the course of polymerization.

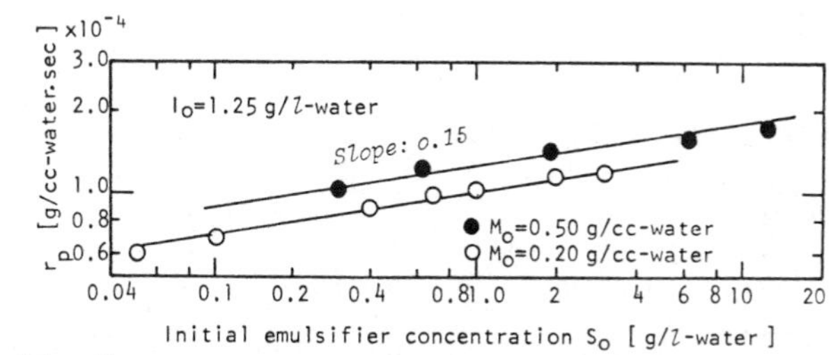

Fig.6 Effect of initial emulsifier concentration on the rate of polymerization.

sing slightly with increasing emulsifier concentrations. The slope of a linear portion of conversion-time curve varies approximately in proportion to the 0.15 power of the initial emulsifier concentration, as shown in Fig. 6 . From the relationship between emulsifier concentration and the number of polymer particles and polymerization rate, respectively, it can be found that the rate of polymerization is proportional to the 0.16 power of the number of polymer particles.

(3) Effect of initiator concentration upon the number of polymer particles and progress of polymedization:

Figs.7 and 8 respectively show the effect of initial initiator concentration on the number of polymer particles and the progress of polymerization at fixed initial emulsifier and monomer concentrations. It can be concluded that the number of polymer particles is independent of initial initiator concentration, as shown in Fig.7. Fig.9 shows log-log plots of polymerization rate r_p(g/cc-H_2O·sec) versus initiator concentration. r_p is calculated from the slope of the linear portion of the monomer conversion versus time plot shown in Fig.8. The order of reaction with respect to the initiator concentration is found to be approximately 0.5 from Fig.9. The same result was obtained by recent investigation of Friis et al. (5).

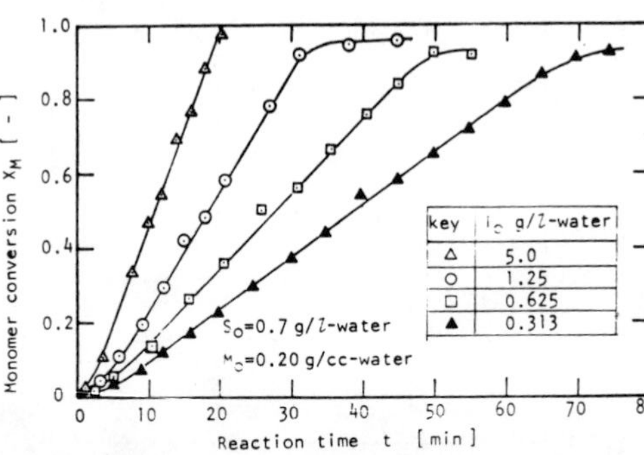

Fig.8 Effect of Initiator concentration on the course of polymerization.

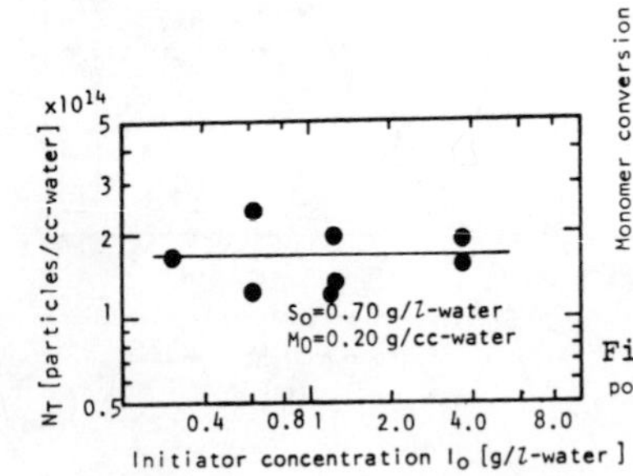

Fig.7 Effect of initial initiator concentration on the number of polymer particles produced.

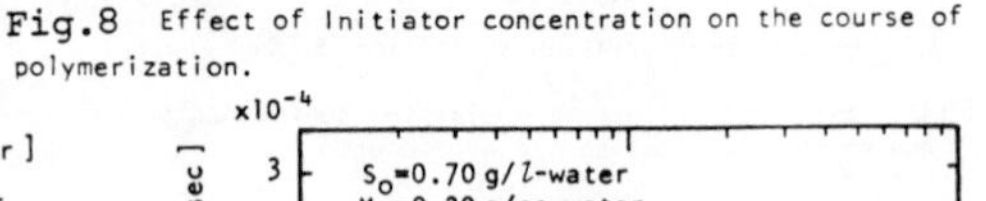

Fig.9 Effect of initial initiator concentration on the rate of polymerization.

(4) Effect of monomer concentration upon the number of polymer particles and progress of polymerization:

Fig.10 shows log-log plots of N_T versus initial monomer concentration M_o(g/cc-water). No effect of monomer concentration is observed on the number of polymer particles, though the data points shows some scatter. Fig.11 gives the relationship between monomer conversion and reaction time, corresponding to Fig.10. Fig.12 shows the relationship between polymerization rate r_p(g/cc-water·sec) calculated from the slope of the linear portion of monomer conversion versus time plots and monomer concentration M_O. It is found that the rate of polymerization is proportional to the 0.38 power of monomer concentration. This result is quite different from styrene emulsion polymerization where monomer concentration does not affect the rate of polymerization.

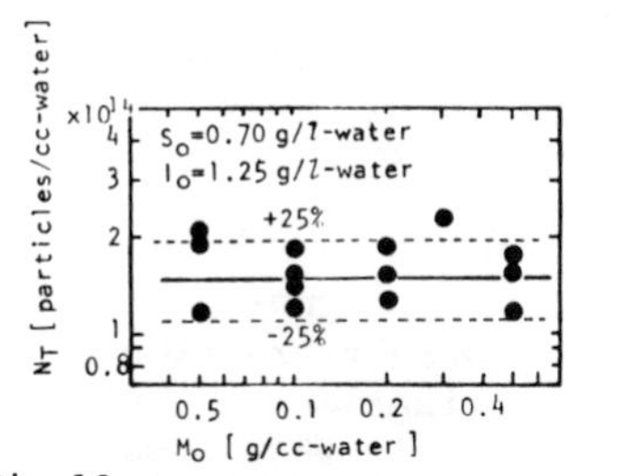

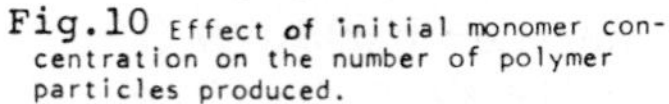
Fig.10 Effect of initial monomer concentration on the number of polymer particles produced.

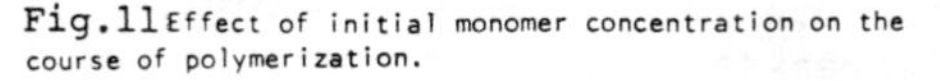
Fig.11 Effect of initial monomer concentration on the course of polymerization.

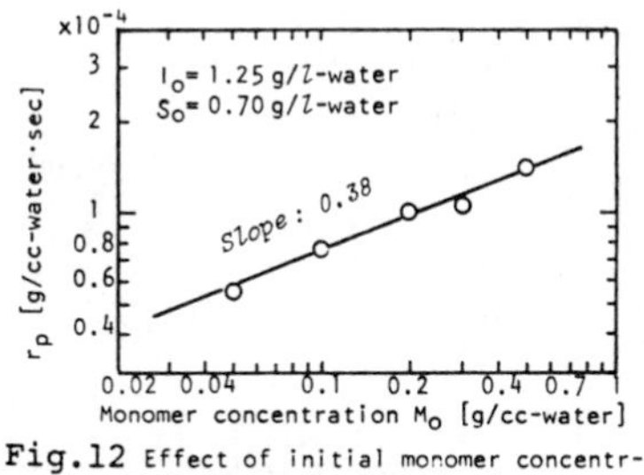

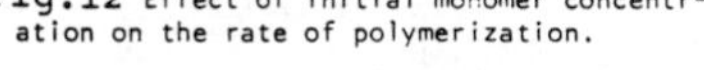
Fig.12 Effect of initial monomer concentration on the rate of polymerization.

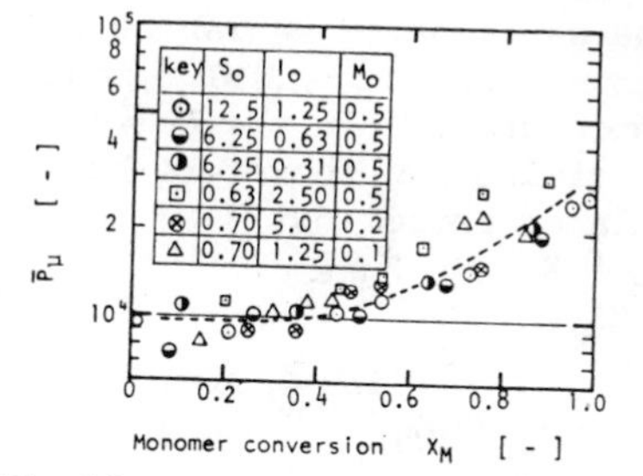

Fig.13 Effects of emulsifier, initiator and monomer concentration on the viscosity-average degree of polymerization.

(5) Effect of initiator, emulsifier and monomer concentrations on molecular weight development:

Fig.13 shows the effect of initiator, emulsifier and monomer concentrations on the viscosity-average degree of polymerization. the variation is identical with each other independently of initiator,emulsifier and monomer concentrations. This leads to the conclusion, as a matter of course, that particle number and its volume do not affect the molecular weight development in vinyl acetate emulsion polymerization. The constant value of $\bar{P}_\mu$ in the lower range of monomer conversion can be reasonably explained by assuming that chain transfer reaction to monomer molecules is a dominant factor determining the degree of polymerization. Based on this supposition, the viscosity-average degree of polymerizat ion can be calculated by the following simple equations as long as the mathematical form of molecular weight distribution coincides with that of the"most probable" distribution.

$$\bar{P}_N = k_p / k_{mf} \qquad (4) \qquad \bar{P}_\mu = \{\Gamma(a+2)\}^{1/a} \cdot \bar{P}_N = 1.82\,\bar{P}_N \qquad (5)$$

where k_p = propagation rate constant, k_{mf} = transfer rate constant to monomer, a = the power number in Mark-Houwink's equation shown by Eq.(1) and $\bar{P}_N$ = number-average degree of polymerization. When the value $k_{mf}/k_p = 1.98 \times 10^{-4}$ obtained in homogeneous polymerization (9)is applied to Eq.(4) and then to Eq.(5), $\bar{P}_\mu$ is apploximately calculated to be 9.2×10^3. This value is a good agreement with observed values. Recently, Friis and Hamielec(6)have dealt with molecular weight development in vinyl acetate emulsion polymerization and proposed a mathematical model.

(6) Variation of monomer weight fraction in polymer particles:

Fig.14 shows the variation of monomer weight fraction in the polymer particles ψ in the course of polymerization only with emulsifier concentration changed(4). From this figure, it can be deduced that monomer concentration in the polymer particles is constant as long as monomer droplets exist in the water phase and then decreases linearly to zero as the polymerization proceeds. The separate monomer droplets in the water phase vanish at 23% conversion(= X_{MC2}). The monomer concentration in the polymer particles can therefore be calculated as follows.

(i) $X_{MC2} \geq X_M \geq 0$

$$M_p = M_{pc} = 8.95 \text{ mole}/l \qquad (6)$$

(ii) $1.0 > X_{MC} \geq X_{MC2}$

$$M_p = M_{pc}(1-X_M)/(1-X_{MC2}) \qquad (7)$$

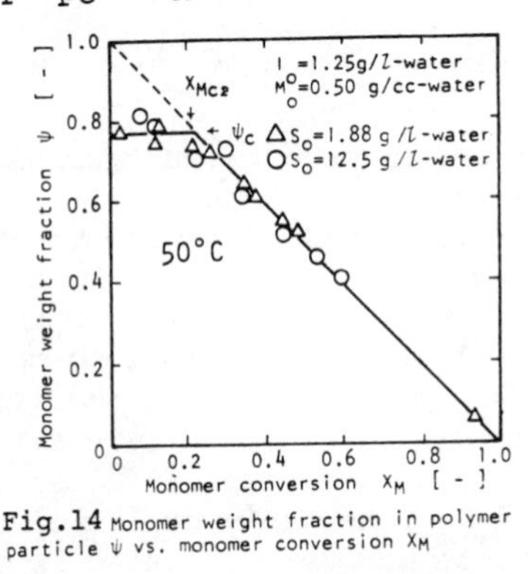

Fig.14 Monomer weight fraction in polymer particle ψ vs. monomer conversion X_M

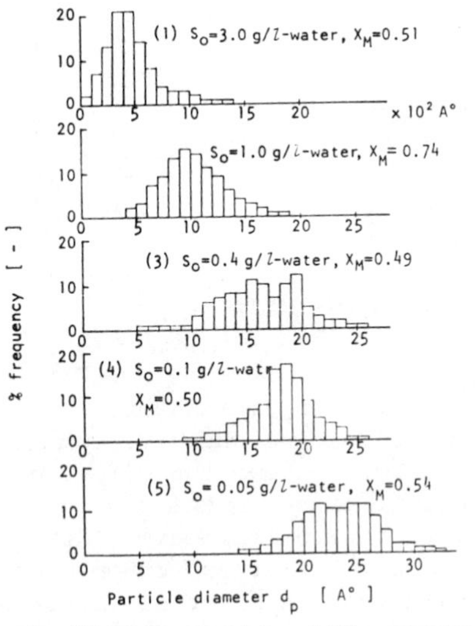

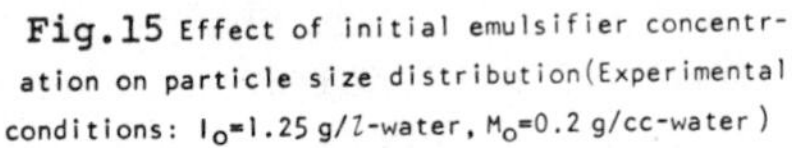

Fig.15 Effect of initial emulsifier concentration on particle size distribution(Experimental conditions: I_o=1.25 g/l-water, M_o=0.2 g/cc-water)

(7) Effect of emulsifier concentration upon particle size distribution:

Fig.15 shows particle size distribution of latex particles produced at various emulsifier concentrations and at about 50 %

monomer conversion. As can be seen in Fig.15, particle size distribution changes from normal to log-normal as emulsifier concentration increases, and particle nucleation seems to continue even at 51 % conversion when S_o= 3.0 g/l-water, because polymer particles the diameter of which is near zero can be seen at this conversion.

(8) Summary:

The following are summerized for kinetic behavior of vinyl acetate emulsion polymerization from the data presented here.

$$r_p \propto S_o^{0.15} I_o^{0.50} M_o^{0.38} \quad (8) \qquad N_T \propto S_o^{0.92} I_o^{0.0} M_o^{0.0} \quad (9)$$

The average degree of polymerization is independent of the reaction variables such as the emulsifier, initiator, monomer concentrations and the number of polymer particles.

The role of polymer particles.

From recent investigations(4)(5), it is generally concluded that polymerization takes place exclusively in the polymer particles in vinyl acetate emulsion polymerization. As a detailed description of the role of polymer particles was presented in the previous paper(4), a brief explanation will therefore be reviewed here.

The rate of emulsion polymerization can conventionally be expressed by:

$$r_p = k_p M_p M_g \bar{n} N_T / N_A \qquad (10)$$

where r_p= the rate of polymerization(g/cc-H_2O·sec), k_p = propagation rate constant(l/mole·sec), Mg = molecular weight of monomer, $\bar{n}$ = the average number of radicals per particle(molecules/particle), N_T = the number of polymer particles(particles/cc-H_2O) and N_A = Avogadro's number.

In Fig.16 is shown, as an example, the variation of value of $\bar{n}$ with the progress of polymerization, calculated with the use of Eqs.(10),(6) and (7), the observed values of N_T shown in Fig.4, and r_p, obtained from graphical differentiation of monomer conversion versus time curve in Fig.5. In this case, it is assumed that k_p = 3300 l/mole·sec(4).

Recently, it has been proved by Ugelstad et al.(1) for vinyl chloride system and by the authors for vinyl acetate system that the average number of radicals per particle $\bar{n}$ can be predicted by the following equations derived by assuming that only particles containing 0, 1 and 2 radicals need be considered(4).

$$\bar{n} = [\, r_i/2k_f N_T + r_i v_p/2k_{tp} N_T \,]^{1/2} \qquad (11)$$

$$r_i = 2\,k_d f\, I_o \qquad (12)$$

where r_i= the production rate of initiator radicals in the water phase(molecules/cc-H_2O·sec), k_d= the rate constant of initiator decomposition(1/sec), f= initiator efficiency, v_p = particle volume (cm^3/particle), k_{tp} = the rate constant of termination in the polymer particles(cm^3/molecule·sec) and K_f = the rate coefficient of radical escape from the polymer particles(1/sec).

A quantitative discussion is given in the authors' literatures(3)(4) for the theoretical derivation and the application of k_f. In vinyl acetate emulsion polymerization, the value of k_f may be calculated by;

$$k_f = (12 D_w \delta / m d_p^2)(k_{mf}/k_p)^{*)} \quad {}^{**)} \tag{13}$$

where D_w = diffusion coefficient of monomeric radicals in the water phase(cm^2/sec), m = partition coefficient of monomeric radicals between water and particle phases(-), d_p = particle diameter(cm), $\delta = (1 + D_w/D_p m)^{-1}$, D_p = diffusion coefficient of monomeric radicals in the polymer particles(cm^2/sec) and k_{mf} = chain transfer rate constant to monomer(l/mole·sec).

As to the particle diameter, the following equations hold. In the interval where monomer droplets exist($X_{Mc2} > X_M \geq 0$),

$$d_p = [6 M_o / \pi \rho_p (1 - \psi_c)]^{1/3} (X_M/N_T)^{1/3}. \tag{14}$$

In the interval where monomer droplets disappear($X_{Mc2} \leq X_M$),

$$d_p = [6 M_o / \pi \rho_p (1 - \psi_c)]^{1/3} (X_{Mc2}/N_T)^{1/3}. \tag{15}$$

where ρ_p means the density of monomer-swollen polymer particles and can be approximated as unity.

Rewriting Eq.(11), we have:

$$\bar{n} = (r_i / 2k_f N_T)^{1/2} [1 + (k_f v_p / k_{tp})]^{1/2}. \tag{16}$$

The order of the value of $k_f v_p / k_{tp}$ is calculated to be in the range 0.001 - 0.01, corresponding to the diametr of polymer particles of 0.01μm to 0.1μm produced in this system. The values of the constants used in this calculation are as follows.

$D_w = 1.86 \times 10^{-5}$, $k_{mf}/k_p = 1.98 \times 10^{-4}$. $k_{tp} = 11.68 \times 10^7$, $\delta = \delta_c = 1.0$

m = 27.1

where, the value of D_w is estimated by Wilke's equation(10). The value of δ is uncertain in the present stage but seems to be constant and rather equal to unity because the value of D_p appears to be constant and not so smaller than that of D_w in the early stage of polymerization where the polymer particles are saturated with monomer. The value of m can be evaluated to be 27.1 for monomeric

---

*) derived assuming that the majority of radicals escaping paricles are monomeric radicals.

**) Another derivation is presented in Appendix I

radicals with the assumption that the properties of monomeric radicals are approximately equal to those of the monomer, from the equation given below using $M_p = 8.95$ mole/l and the saturation concentration of the monomer in the water phase, $M_{wc} = 0.33$ mole/l at 50 °C (11).

$$M_p = m\, M_w \tag{17}$$

It appears, therefore, that the value of $k_f v_p / k_{tp}$ in Eq.(16) may be completely neglected compared to unity especially in the early stage of polymerization (such as in the interval I) where particle diameter is very small. So, in place of Eq.(16) in such an early stage of polymerization we have:

$$\bar{n} = (r_i / 2k_f N_T)^{1/2} \tag{18}$$

Inserting Eqs.(12),(13) and (14) into Eq.(18), we get:

$$\bar{n} = (k_p m k_d f / 3 D_w k_{mf} \delta_c)^{1/2} (3/4\pi\rho_p (1-\psi_c))^{1/3} (I_o/N_T)^{1/2} (M_o X_M / N_T)^{1/3} \tag{19}$$

Raising each side of Eq.(19) to the 2nd power and rearranging it lead to:

$$3\left(\frac{\bar{n}^2 N_T}{m I_o}\right)\left(\frac{k_{mf}}{k_p}\right) = \left(\frac{k_d f}{D_w \delta_c}\right)\left(\frac{3}{4\pi\rho_p (1-\psi_c)}\right)^{2/3}\left(\frac{M_o X_M}{N_T}\right)^{2/3} \tag{20}$$

The values of the terms $(\bar{n}^2 N_T / m I_o)(k_{mf}/k_p)$ and $(M_o X_M / N_T)^{2/3}$ can be calculated using the literature value $k_{mf}/k_p = 1.98 \times 10^{-4}$ at 50°C, the observed values of $\bar{n}$, X_M, N_T, m and the constants previously given. In Fig.17 is plotted the observed values of $\bar{n}$ in accordance with Eq.(20). As can be seen in Fig.17, linear relationship is obtained between these. This appears to mean that Eq.(18) is valid for the prediction of the value of $\bar{n}$ in the early stage of vinyl acetate emulsion polymerization system. The value of $k_d f$ at 50°C can therefore be estimated from the slope of the best straight line through the data points using the values $D_w = 1.86 \times 10^{-5}$ and $\delta_c = 1.0$ and other known constants.

$$k_d f = 15.0 \times 10^{-6} \quad 1/\text{sec} \tag{21}$$

This value is about 20 times greater than that obtained in the emulsion polymerization of styrene ($k_d f = 6.7 \times 10^{-7}$ 1/sec) (8) and about 10 times greater than the value $k_d = 1.5 \times 10^{-6}$ 1/sec obtained by Kolthoff et al. (12). Breitenbach et al. reported that the decomposition rate of potassium persulfate in the water phase is about 10 times higher than that obtained by Kolthoff et al. in the presence of vinyl acetate. Morris et al. also reported that the decomposition of potassium persulfate is accelerated by vinyl acetate (13). Considering these suggestions, the value of $k_d f$ obtained here appears to be reasonable.

From the analysis given above, it proves that Eq.(18) is completely applicable in the earlier stage of polymerization such as

Interval I and II. Therefore, Eq.(18) may be applicable to the higher range of monomer conversion. If it is assumed that Eq.(18) is applicable to the whole range of monomer conversion for estimating the value of $\bar{n}$, the following two equations can be obtained: In the range of $0 < X_M < X_{Mc2}$,

$$\bar{n} = \left(\frac{k_d f m k_p}{3D_w k_{mf} \delta_c}\right)^{1/2} \left(\frac{3}{4\pi\rho_p(1-\psi_c)}\right)^{1/3} M_o^{1/3} I_o^{1/2} N_T^{-5/6} X_M^{1/3} \quad (22)$$

In the range of $X_{Mc2} < X_M < 1.0$ where monomer droplets disappear,

$$\bar{n} = \left(\frac{k_d f m k_p}{3D_w k_{mf} \delta}\right)^{1/2} \left(\frac{3}{4\pi\rho_p(1-\psi_c)}\right)^{1/3} M_o^{1/3} I_o^{1/2} N_T^{-5/6} X_{Mc2}^{1/3} \quad (23)$$

In Fig.18 is shown a comparison between the observed and calculated values of $\bar{n}$ where $\bar{n}_{cal.}$ means the value calculated from Eqs. (22) and (23) and the value of δ in Eq.(23) is assumed to be equal to $\delta_c (=1.0)$ for simplicity because the variation of the value of δ cannot be followed theoretically in the present stage. As can be seen in Fig.18, a good agreement is obtained between $\bar{n}_{obs.}$ and $\bar{n}_{cal.}$ up to about 35 % conversion. However, the values of $\bar{n}_{obs.}$ become gradually greater than those of $\bar{n}_{cal.}$ with progress of polymerization. It seems reasonable to consider that this discrepancy between $\bar{n}_{obs.}$ and $\bar{n}_{cal.}$ arises from neglecting the term $k_f v_p/k_{tp}$ in Eq.(16), that is, the term $r_i v_p/2k_{tp}N_T$ in Eq.(11) and the increase in the value of $\bar{n}$ is due to the increase in the value of the term $r_i v_p/2k_{tp}N_T$. But the values of r_i, v_p and N_T are approximately constant in the range of $X_M > X_{Mc2}$. It seems therefore reasonable to assume that the increase in the value of the term $r_i v_p/2k_{tp}N_T$ should be mainly due to the decrease in the value of k_{tp} with the progress of polymerization. Based on this supposition, the variation of k_{tp} with the progress of polymerization can be calculated with the use of Eqs.(11) to (15) and the constants previously given. The results of calculation are shown in Fig.19. The experimentally obtained values of k_{tp} gather around a straight line independently of experimental conditions, and can be found to be expressed as a function of only monomer conversion or monomer weight fraction of polymer in the polymer particles ω. Thus we have;

$$k_{tp} = 7.8 \times 10^{-13} \exp.[-8.3\,\omega] \ \mathrm{cm^3/molecule \cdot sec}$$

$$= 4.7 \times 10^{8} \exp.[-8.3\,\omega] \ l/\mathrm{mole \cdot sec} \quad (24)$$

where ω is the weight fraction of polymer in the polymer particles and exactly equal to X_M when $X_M > X_{Mc2} (=0.23)$ and to a constant value 0.23 when $X_M < 0.23$. In the lower range of the value of ω, the experimental value for k_{tp} found here is in good agreement with the literature data by Matheson(14), Dixson-Lewis(15) and Burnet et al.(16). Recently, Friis and Nyhagen(5) have investigated the termination rate constant in the bulk polymerization of vinyl acetate at 50 °C with α,α'-azoisobutyronitrile as initiator and shown that k_t is experimentally expressed as an exponential function of only monomer conversion which is very similar to Eq.(24). Moreover, their data show good agreement with our data obtained here. This fact appears to mean that the analysis given above is reasonable, and Eqs.(11) to (15) and (24) are valid for prediction of $\bar{n}$. The solid lines shown in Fig.16 represent the

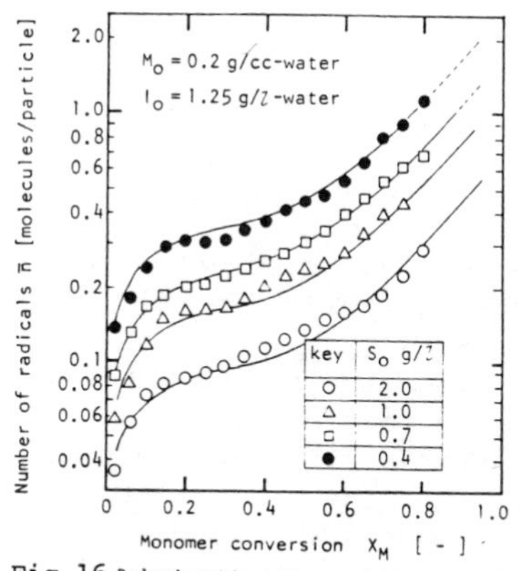

Fig.16 Relationship between the number of radicals per particle and the monomer conversion.

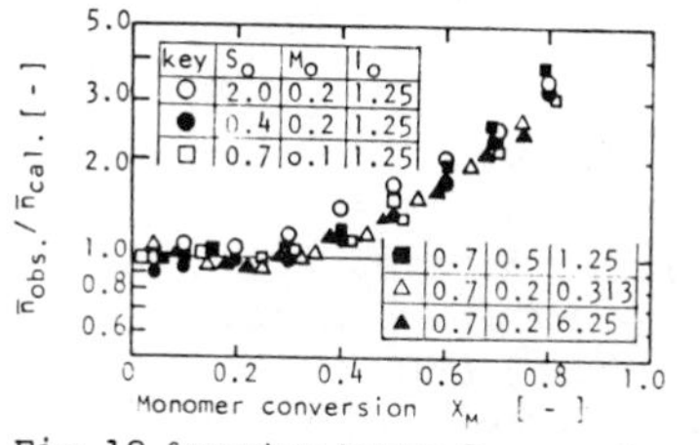

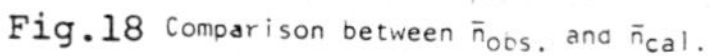

Fig.18 Comparison between $\bar{n}_{obs.}$ and $\bar{n}_{cal.}$

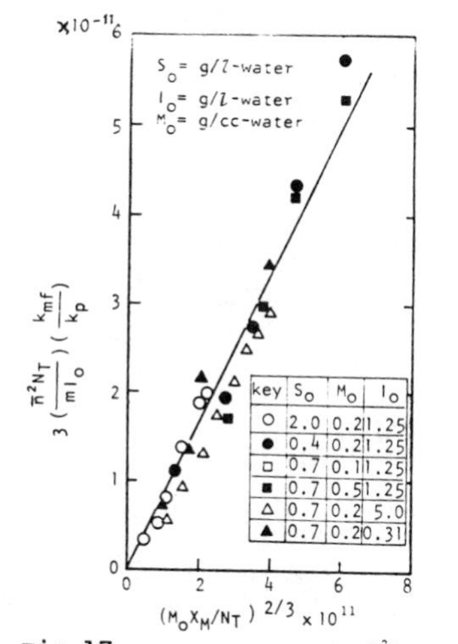

Fig.17 Relationship between $3\left(\frac{\bar{n}^2 N_T}{m I_o}\right)\left(\frac{k_{mf}}{k_p}\right)$ and $\left(\frac{M_o X_M}{N_T}\right)^{2/3}$

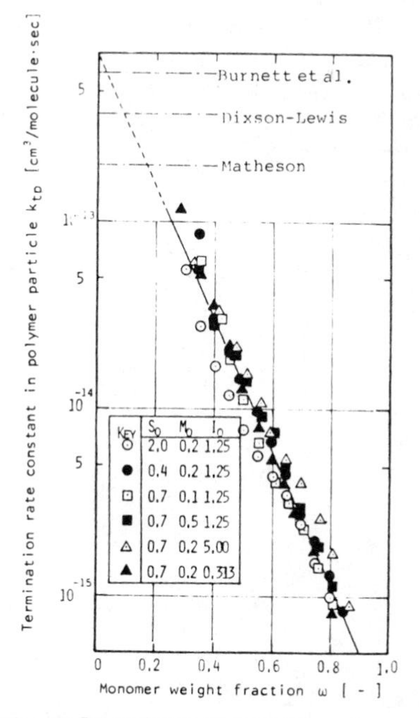

Fig.19 The variation of termination rate constant k_{tp} in polymer particle with polymer weight fraction in polymer particles ω.

values of $\bar{n}$ estimated by using these equations and the observed value of N_T. A good agreement can be seen between the observed and calculated values of $\bar{n}$ over the wide range of experimental conditions and monomer conversion though Eq.(11) was established with the assumption that only particles containing 0,1 and 2 radicals need be considered.

In Fig.20 is shown a comparison between the observed and calculated conversion versus time curves, calculation being done by using the following equation and numerical integration method.

$$\left(\frac{M_o N_A}{k_p M_p}\right) \int_0^{X_M} \frac{d X_M}{M_p \bar{n} N_T} = \int_0^t dt = t \qquad (25)$$

Eq.(25) is derived from Eq.(10). It can therefore be concluded that we are now predictable the variation of the value of $\bar{n}$ and hence monomer conversion versus time relation with satisfactory accuracy as long as the observed number of polymer particles is used.

In conclusion, we may say that one of the most important problems which still remain to be solved is to clarify the mechanism by which polymer particles are formed and to establish a particle nucleation model by which the number of polymer particles produced can be estimated, because the classical theory of Smith and Ewart does not serve for this purpose in the emulsion polymerization of vinyl acetate. In the succeeding section, a new reaction model is presented assuming that the particles are formed from emulsifier micelles, based on our experimental results.

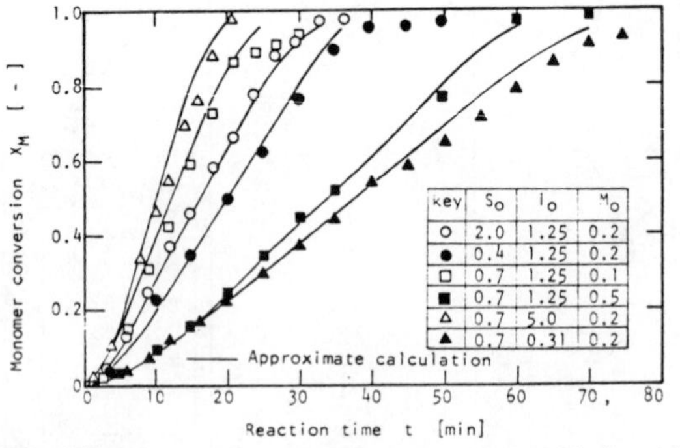

Fig.20 Comparison of observed conversion vs. time relation with that of approximate calculation.

Proposed reaction model

Presentation of basic equation. From the experimental results and arguments given above, a kinetic model can be developed as shown below with the assumptions that polymer particles generate from emulsifier micelles and only particles containing 0, 1 and 2 radicals need be considered for simplicity because the average number of radicals per particle is low compared to unity mostly in the emulsion polymerization of vinyl acetate under ordinarly experimental conditions. In table 1 are shown the elementary reactions participating in the emulsion polymerization of vinyl acetate and their rates considered in our reaction model. We neglect here termination reaction in the water phase. This will be the case, as pointed out by Ugelstad et al. in the emulsion polymerization of vinyl chloride, also in this system. According to the reaction scheme shown in table 1, we have a following set of differential equations by taking material balance for respective species:

Table 1 Elementary reactions of vinyl acetate emulsion polymerization and definition of their rates

Reaction	Reaction Type	Reaction Rate	
Initiation of radicals in the water phase	$I \longrightarrow 2R^*$	$r_i = 2k_d f I_o$	A
Initiation of particles in the water phase	$R^* + m_s \longrightarrow N_1$	$k_1 m_s R^*$	B
Entry of radicals into polymer particles	$R^* + N_n \longrightarrow N_{n+1}$	$k_2 N_n R^*$	C
Propagation reaction in polymer particles	$N_n \longrightarrow N_n$ $(P_j^* + M \rightarrow P_{j+1}^*)$	$k_p M_p \Sigma n N_n$	D
Termination reaction in polymer particles	$N_n \longrightarrow N_{n-2}$ $(P_j^* + P_i^* \rightarrow P_j + P_i)$	$k_{tp} n(n-1) N_n / v_p$	E
Chain transfer to monomer in polymer particles	$N_n \longrightarrow N_n$ $(P_j^* + M \rightarrow P_j + M^*)$	$k_{mf} M_p \Sigma n N_n$	F
Escape of radicals from polymer particles	$N_n \longrightarrow N_{n-1} + R^*$	$k_f n N_n$	G

N_n means the number of polymer particles containing n radicals.

(1) Initiator and monomeric radicals in the water phase R^*:

$$dR^*/dt = r_i + k_f N_1 + 2k_f N_2 - k_1 m_s R^* - k_2 N_T R^* \quad (26)$$

Applying a stationary state method to R^*, Eq.(26) is converted to

$$R^* = (r_i + k_f N_1 + 2k_f N_2)/(k_1 m_s + k_2 N_T) \quad (27)$$

where N_1 = number of particles containing 1 radical in cm^3-water, m_s = number of micelles in cm^3-water, k_1 and k_2 are rate coefficients defined in table 1, respectively.

(2) Total number of polymer particles N_T:

$$dN_T/dt = k_1 m_s R^* = (r_i + k_f N_1 + 2k_f N_2)/(1 + \varepsilon N_T/S_m) \quad (28)$$

where $\varepsilon = (k_2/k_1) M_m$, M_m = aggregation number of micelles, S_m = number of emulsifier molecules forming micelles in cm^3-water.

(3) Number of polymer particles containing 1 radical N_1:

$$dN_1/dt = k_1 m_s R^* + k_2 (N_T - N_1 - N_2) R^* + 2k_f N_2 - k_2 N_1 R^* - k_f N_1 \quad (29)$$

$$= [1 + (\varepsilon N_T/S_m)(1 - 2N_1/N_T - N_2/N_T)](r_i + k_f N_1 + 2k_f N_2)/(1 + \varepsilon N_T/S_m) + 2k_f N_2 - k_f N_1 \quad (30)$$

(4) Number of polymer particles containing 2 radicals N_2:

$$dN_2/dt = k_2N_1R^* - k_2N_2R^* - 2k_fN_2 - 2(k_{tp}/v_p)N_2 \tag{31}$$

$$= [(N_1 - N_2)/N_T]\,(r_i + k_fN_1 + 2k_fN_2)/(1 + S_m/\varepsilon N_T) - 2k_fN_2 - 2(k_{tp}/V_p)N_2 \tag{32}$$

(5) Polymerization rate:

$$dX_M/dt = (k_pM_pM_g/M_oN_A)\,\bar{n}\,N_T = (k_pM_pM_g/M_oN_A)(N_1 + 2N_2) \tag{33}$$

(6) Material balance for emulsifier molecules forming micelles S_m:

In the region of $0 < X_M < X_{Mc2}$,

$$S_m = S_o - [36\pi/\rho_p^{\,2}(1-\psi_c)^2 a_s^{\,3}]^{1/3}(M_oX_M)^{2/3}\,N_T^{1/3} \quad (S_m \geq 0) \tag{34}$$

In the region of $X_{Mc2} < X_M < 1.0$,

$$S_m = S_o - [36\pi/\rho_p^{\,2}(1-\psi_c)^2 a_s^{\,3}]^{1/3}(M_oX_{Mc2})^{2/3}N_T^{1/3} \quad (S_m \geq 0) \tag{35}$$

$$k_v = [36\pi/\rho_p^{\,2}(1-\psi_c)^2 a_s^{\,3}]^{1/3} \tag{36}$$

where ρ_p = density of monomer-swollen polymer particles(g/cm^3), a_s = surface area occupied by an emulsifier molecule(cm^2/molecule), ψ_c = monomer weight fraction in the particles in the range of $X_M \leq X_{Mc2}$ (=0.23).

In this reaction model ε is the most important parameter and means the ratio of radical capture by the particle and the micelle. Therefore, the value of ε affects directly the rate of particle nucleation process. As pointed out in the preceeding section, radicals are very easy to escape from the particles in the emulsion polymerization of vinyl acetate. So, it is quite reasonable to consider that radicals in the emulsifier micelles are also very easy to escape from them and that only part of radicals which entered the micelles are effective for particle nucleation, the others only passing through the micelles without converting them into particles. This phenomenon results in an apparent decrease in the value of k_1, that is, in an increase in the value of ε.

Comparison of experimental and calculated results

(1) consideration on model parameters:

A set of differential equations given above are solved numerically with the Runge-Kutta-Gill method on a digital computer. However, the exact value of ε cannot be evaluated, as shown in the previous paper(8), at the present stage because radical escape from the micelles cannot be estimated quantitatively. Therefore, the exact value of ε must be determined so that the calculated values of N_T agree at best with those observed. The value of k_v can be known from Eq.(36) by measuring a_s by the soap titration method. The values of ε and k_v thus obtained are as follows:

$$\varepsilon = 1.2 \times 10^7 \tag{37}$$ $$k_v = 7.36 \times 10^{14} \tag{38}$$

(2) Comparison of variation in number of particles with progress of polymerization:

Fig.21 shows a comparison between calculation and observation in the variation of the number of polymer particles with progress of polymerization. The theoretical value represents well the tendency of variation in particle number and agrees to the observed value within an experimental error. The value of ε may changes with the chang in emulsifier and initiator concentrations, but it appears that we can use approximately a constant value of ε over the wide range of emulsifier and initiator concentrations. This is very convenient for prediction of particle number .

(3) Effect of emulsifier concentration upon number of polymer particles and progress of polymerization:

The solid lines in Fig.22 represent the calculated conversion versus time curves. A good agreement can be seen between calculation and observation except in the higher range of monomer conversion at lower emulsifier concentration. Fig.23 shows a comparison between the observed and calculated values of the number of polymer particles. In the higher range of initial emulsifier concentration, theoretical predictions are in good agreement with the observed values without regard to the value of CMC, but in the lower range of emulsifier concentration the calculated values are not necessarily in good agreement with those observed if such a high value of CMC observed in the emulsion polymerization of styrene(8) is assumed in this system. Discrepancy between the experimental and calculated values of N_T may have arised from neglecting the homogeneous nucleation of particles in the water phase.

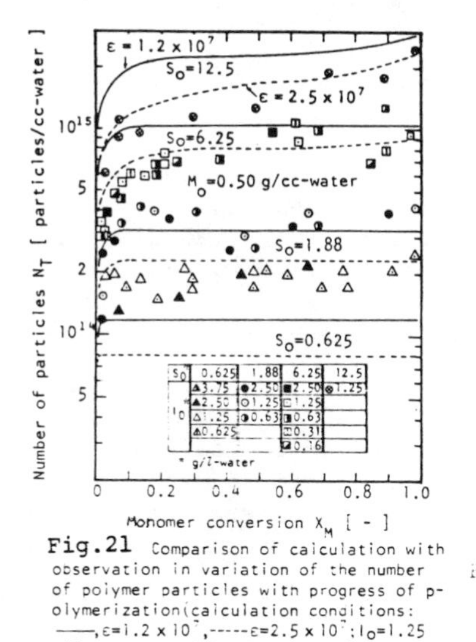

S₀	0.625	1.88	6.25	12.5
I₀	▲3.75	●2.50	■2.50	⊗1.25
	▲2.50	○1.25	□1.25	
	△1.25	◑0.63	◨0.63	
	◬0.625		⊡0.31	
			◩0.16	

Fig.21 Comparison of calculation with observation in variation of the number of polymer particles with progress of polymerization(calculation conditions: ——, ε=1.2 x 10⁷, ----- ε=2.5 x 10⁷: I₀=1.25 g/l-water, M₀ =0.50 g/cc-water,)

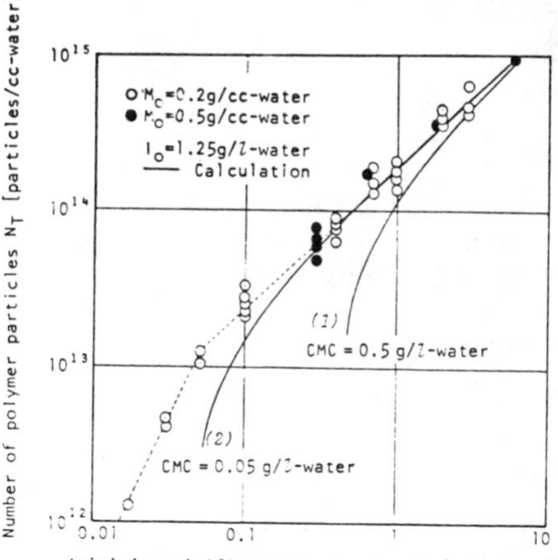

Fig.23 Comparison between the observed and calculated number of polymer particles at various initial emulsifier concentration: —— calculated value at; (1) CMC= 0.5 g/l-water, (2) CMC=0.05 g/l-water with ε=1.2 x 10⁷.

At any rate, further investigation must be done on the mechanism of particle nucleation in the lower lange of initial emulsifier concentration. According to our reaction model, we have:

$$N_T \propto (S_o - S_{CMC})^{0.94} \tag{39}$$

This is in good agreement with Eq.(9) obtained experimentally.

Fig.24 shows a comparison between the observed and calculated values of $\bar{n}$. A good agreement can be seen between them except in the higher range of monomer conversion at lower emulsifier concentration. However, if polymer particles containing more than 2 radicals are taken into consideration, disagreement will vanish even in the higher range of monomer conversion at lower emulsifier concentration.

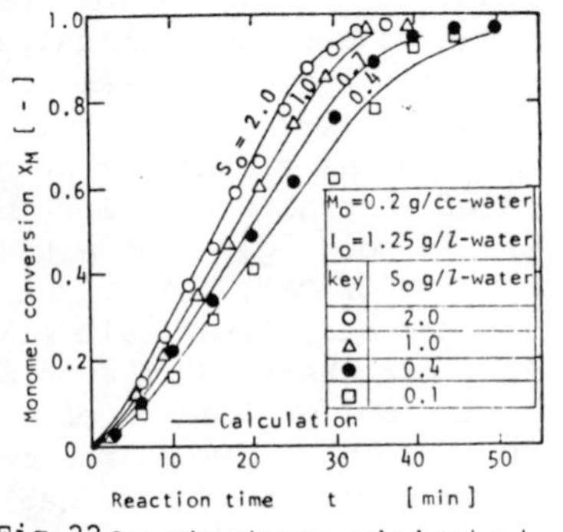

Fig.22 Comparison between calculated and observed conversion vs. time curve at various initial emulsifier concentrations.

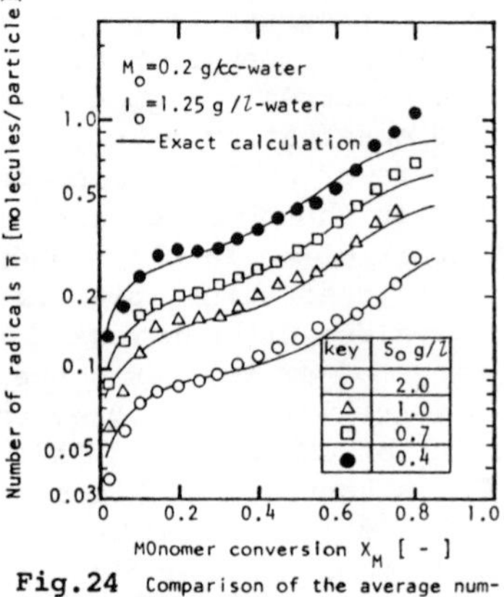

Fig.24 Comparison of the average number of radicals per particle between exact calculation and experiment

(4) Effect of initiator concentration upon number of polymer particles and progress of polymerization:

According to our reaction model, it is found that the number of polymer particles does not increase with an increase in the initial initiator concentration, but may be considered to be rather constant, as shown in Fig.25. Solid line in Fig.25 gives the value calculated with an assumption that CMC is very low, and represents well the tendency of the observed data points. From the straight line the initiator dependence exponent is calculated to be 0.04:

$$N_T \propto I_o^{0.04} \tag{40}$$

Fig.26 shows a comparison between the observed and calculated conversion versus time curves, corresponding to Fig.25. It is seen that the calculated lines agree with the experimental data to over 90 % conversion.

(5) Effect of monomer concentration upon number of polymer particles and progress of polymerization:

In Fig.27 is shown a log-log plot of the number of polymer particles versus initial monomer concentration. The solid line in Fig.27 represents the calculated value. It is concluded that the reaction model proposed here can express the tendency of the experimental data points and the absolute value of N_T. From the straight line in Fig.27, we find:

$$N_T \propto M_o^{0.0} \qquad (41)$$

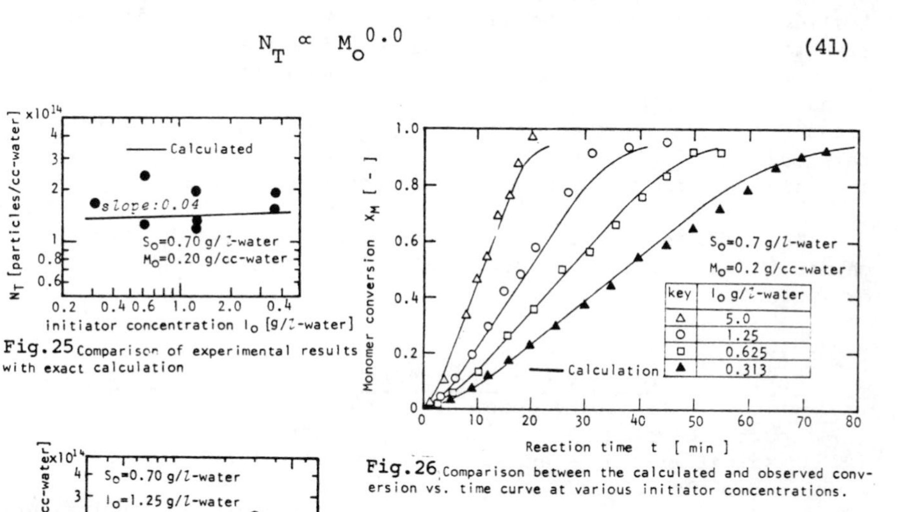

Fig.25 Comparison of experimental results with exact calculation

Fig.26 Comparison between the calculated and observed conversion vs. time curve at various initiator concentrations.

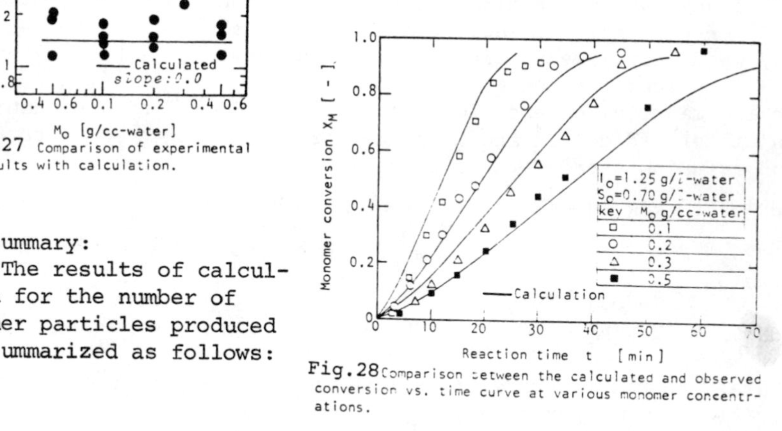

Fig.27 Comparison of experimental results with calculation.

Fig.28 Comparison between the calculated and observed conversion vs. time curve at various monomer concentrations.

(6) Summary:

The results of calculation for the number of polymer particles produced are summarized as follows:

$$N_T \propto S_o^{0.94}\, I_o^{0.04}\, M_o^{0.00} \qquad (42)$$

From the slope of a linear portion of the calculated conversion versus time curves, the following relationship can be obtained for the rate of the polymerization:

$$r_p \propto S_o^{0.15}\, I_o^{0.50}\, M_o^{0.38} \qquad (43)$$

It is concluded that these equations represent well the characteristics of the emulsion polymerization of vinyl acetate with sodium lauryl sulfate as an emulsifier and potassium persulfate as an initiator.

To conclude

In conclusion, we may say that our reaction model developed

by assuming that the polymer particles form from the emulsifier micelles expresses well the characteristics of the emulsion polymerization of vinyl acetate. It must be emphasized here that desorption from the polymer particles and the micelles is the most important and essential mechanism on which our reaction model is based. This reaction model may be applicable to the similar systems such as vinyl chloride and acrylonitrile emulsion polymerization because desorption mechanism is dominant in these systems.

The authors wish to express their gratitude to Mr. K. Nakagawara, Mr. K. Fujita and Mr. S. Sasaki for carring out the experiments.

Appendix I

Derivation of k_f the rate coefficient for escape of monomer radicals: The following assumptions are made;

(1) a particle contains not more than 1 radical.
(2) a radical of chain-length not longer than s can escape and reenter the particle with the same rate.
(3) Instantaneous termination takes place when another radical enter the particle which contain a radical.
(4) we do not make distinction between radicals with or without an initiator fragment.

Taking material balance for the particles N^*_I and N^*_i containing an initiator radical I* and a radical of chain-length i, respectively, and assuming steady state, we have:

$$dN^*_I/dt = k_e N_o I^* - (k_e R^* + k_{mf} M_p + k_i M_p + k_{oI}) N^*_I = 0 \quad (A\ 1)$$

$$dN^*_1/dt = k_e N_o M^*_1 + k_i M_p N^*_I + k_{mf} M_p N^* - (k_e R^* + k_{mf} M_p + k_p M_p + k_o) N^*_1 = 0 \quad (A\ 2)$$

$$dN^*_i/dt = k_e N_o M^*_i + k_p M_p N^*_{i-1} - (k_e R^* + k_{mf} M_p + k_p M_p + k_o) N^*_i = 0 \quad (A\ 3)$$

$$dN^*_s/dt = k_e N_o M^*_s + k_p M_p N^*_{s-1} - (k_e R^* + k_{mf} M_p + k_p M_p + k_o) N^*_s = 0 \quad (A\ 4)$$

where N_o=particle number containing no radical, M_p=monomer concentration in the particles, k_{mf}=transfer rate constant to monomer, k_i=initiation rate constant, k_e=rate constant of radical absorption, k_o= rate constant of radical desorption from the particles, k_{oI}=rate constant of initiator radical desorption from the particle, N^*_i=number of particles containing a radical M^*_i, N*=total number of active particles and R* is

$$R^* = I^* + \sum_{i=0}^{s} M^*_i \quad (A\ 5)$$

Taking material balance for radicals in the water phase and assuming steady atate, we get:

$$dI^*/dt = r_i + k_{oI} N^*_I - k_e N_T I^* = 0 \quad (A\ 6)$$

$$dM^*_i/dt = k_o N^*_i - k_e N_T M^*_i = 0 \quad (A\ 7)$$

$$dM^*_s/dt = k_o N^*_s - k_e N_T M^*_s = 0 \quad (A\ 8)$$

Adding (A 6) to (A 8) leads to:

$$k_e N_T (I^* + M_1^* + \cdots + M_i^* + \cdots + M_s^*) = r_i + k_{oI} N_I^* + k_o N_1^* + \cdots + k_o N_i^* + \cdots + k_o N_s^*) \quad \text{(A 9)}$$

Inserting Eq.(A5) into Eq.(A9) gets:

$$k_e N_T R^* = r_i + k_f N^* \quad \text{(A10)}$$

$$k_f N^* = k_{oI} N_I^* + k_o N_1^* + \cdots + k_o N_s^* \quad \text{(A11)}$$

Eq.(A11) defines k_f.

Considering that polymers with high degree of polymerization can be obtained in the emulsion polymerization, we could assume that:

$$k_e R^*,\ k_{mf} M_p \ll k_p M_p,\ k_i M_p .$$

From Eq.(A 1), (A 2) and $N^* = \bar{n} N_T$, we gets approximately for N_I^*,

$$N_I^* = r_i (1-\bar{n}) / (k_{oI} \bar{n} + k_i M_p) \quad \text{(A12)}$$

By similar treatment we get for N_1^* and N_i^* ;

$$N_1^* = [k_{mf} M_p / (k_p M_p + k_o \bar{n})] N^* + [k_i M_p / (k_p M_p + k_o \bar{n})] N_I^* \quad \text{(A13)}$$

$$N_i^* = \left(\frac{k_p M_p}{k_p M_p + k_o \bar{n}}\right)^{i-1} N_1^* = \left(\frac{k_p M_p}{k_p M_p + k_o \bar{n}}\right)^{i} \cdot \left[\left(\frac{k_{mf}}{k_p}\right) N^* + \left(\frac{k_i}{k_p}\right) N_I^*\right] \quad \text{(A14)}$$

From Eq.(A11), (A14) and $N^* = \bar{n} N_T$, we get:

$$k_f = k_{oI} (N_I^*/N_T) + k_o [(k_{mf}/k_p) + (k_i/k_p)(N_I^*/N_T)] \sum_{i=1}^{s} \left(\frac{k_p M_p}{k_p M_p + k_o \bar{n}}\right)^{i} \quad \text{(A15)}$$

Literature Cited

(1) J. Ugelstad, P.C. Mork, P. Dahl and P. Rangnes, J. Polymer Sci. part c, 27, 49 (1969)

(2) J. Ugelstad and P.C. Mork, Brit. polymer J., 2, 31 (1971)

(3) M. Harada, M. Nomura, W. Eguchi and S. Nagata, J.Chem. Eng. Japan, 4, 54 (1971)

(4) M. Nomura, M. Harada, K. Nakagawara, W. Eguchi and S. Nagata, J. Chem. Eng. Japan, 4, 160 (1971)

(5) N. Friis and L. Nyhagen, J. Appl. Polymer Sci., 17, 2311 (1973)

(6) N. Friis and A. E. Hamielec, J. Polymer Sci., 19, 97 (1975)

(7) Nakaj ma, A., Kobunshi Kagaku, 11, 142 (1954)

(8) M. Harada, M.Nomura, H. Kojima, W. Eguchi and S. Nagata, J. Polymer SCi, 16, 811 (1972)

(9) Sakurada, I., ed.,"vinyl acetate resin", p77 Kobunshi Kagaku Pub., Kyoto (1964)

(10) C.R. Wilke and P. Chang, A. I. Ch. E. Journal, 1, 264 (1955)

(11) D.H. Napper and A.G. Parts, J. Polymer Sci., 61, 113 (1962)

(12) I.M. Kolthoff and I.K. Miller, J.Am. Chem. Soc., 73, 3055 (1951)

(13) Morris, C., and A.G. Parts, Makromol. Chem., 119, 212 (1968)

(14) Matheson, M.S., E.A. Auer, E.B. Bevilaqua and E.J. Hart, J. Am. Chem. Soc., 71, 173 (1949)

(15) Dixson-Lewis, G., Proc. Royal Soc., (London) A198, 510 (1949)

(16) Burnett, G.M., L. Valentine and H.W. Melville, Trans. Faraday Soc., 45, 960 (1949)

8

Determination of Latex Particle Size from Polymerization Rates

CARLTON G. FORCE

Charleston Research Center, Westvaco Corp., North Charleston, S. C. 29406

Several techniques are available for measuring the particle size of aqueous dispersions of materials having diameters considerably smaller than the wavelength of visible light. Of these, electron microscopy is one of the more desirable because the particle images can actually be visualized. However, the equipment required is expensive, measurement of the micrographs is time consuming and tedious and sample preparation is highly technical and fraught with unresolvable sources of error. The minimum deviation which can be expected is 5% and accuracy is usually not nearly this good, particularly with soft particles like polybutadiene or SBR. Such particles must be brominated or otherwise treated to harden them sufficiently to maintain their size and shape when dried on the electron microscope grid. Bromination increases particle size an amount which varies with the type of polymer.

A method that requires inexpensive equipment and produces a satisfactory average particle size for many applications is based on the fact that latex particles will adsorb soap molecules until each particle is fully coated with a monolayer before sufficient soap becomes free in the aqueous serum to produce soap micelles. Determination of the onset of micellization provides a convenient end point to determine the quantity of soap adsorbed. From knowledge of the area each soap molecule occupies on a latex particle, a calculation of the volume to surface area (V/S) average particle diameter of the latex can be made (1). The greatest problem in this technique is accurately establishing the area occupied by a soap molecule on a latex particle. This has been accomplished by determination of soap saturation requirements for a latex whose average particle size is known from another technique. Because of the inherent inaccuracies in latex particle size determination, differences in soap purity and probably unrecognized variabilities in latexes such as the number of charged groups generated in the chains during polymerization, there are values in the literature for the area occupied by an oleic acid molecule ranging from 20.2 Å^2 to 47 Å^2 (1,2,3). Electron microscopy was the most popular method utilized in establishing average latex particle size, but other techniques such as ultra centrifugation were also employed (4) for measuring particle size for the calculation of these soap molecular areas.

In this paper, a means is demonstrated for experimentally determining the percent of monomer conversion at which soap micelles disappear from the system during emulsion polymerization. By applying the mathematics and latex particle surface area which a soap molecule will occupy, developed for the soap titration particle size measurement procedure, the average particle size of the finished latex can be calculated. In fact, under some circumstances it might even be feasible to adjust monomer content, if necessary, during the reaction and short stop at the specific average particle size desired for a given latex.

Experimental

The following recipe was used in polymerizing the polybutadiene latexes for this study.

Butadiene	100
Water	200
Oleic type potassium soap	4.35
Potassium chloride	0.4
Tamol N	0.15
Ferrous hepto sulfate	0.012
EDTA tetra sodium salt	0.0326
Sodium sulfoxylate	0.0335
p-menthane hydroperoxide	0.065
t-dodecyl mercaptan	0.25

The polymerizations were carried out at 5°C in 28 fluid ounce soft drink bottles rotated end over end in a constant temperature water bath.

With this recipe, the initial reaction rate is different than the rate after polymerization has proceeded for a period of time. In order to study this phenomenon, it was necessary to frequently determine percent conversion during the coarse of the reaction.

Samples for the total solids required were obtained every 10 minutes for the first 80 minutes of the polymerization. The sampling period was then extended to 20 minutes for the next 40 minutes and 30 minutes thereafter up to a total polymerization time of 300 minutes. Hourly samples were taken beyond 300 minutes.

Conductivity measurements were also made during some of the polymerizations. For this work, circular electrodes were cut from platinum foil. Each of these was silver soldered to the end of a platinum wire. The platinum wire was inserted in a glass tube shorter than the wire and previously bent in a hot flame to conform relatively closely to the shape of the 28 oz. soft drink bottle used as the polymerization vessel. The platinum wire extending beyond the glass tube was then inserted through the self sealing rubber gasket and a hole punched in the metal snap-on soft drink pressure cap used to seal the bottle. Two electrodes were thus placed in each bottle in positions such that the glass tubing would maintain them a maximum distance apart near opposite sides of the bottle. This distance was necessary for suitable conductivity readings.

In making these measurements, lead wires from a General Radio Co. Impedance Bridge type 1650A were attached to the platinum wires protruding from the bottle with aligator clips. After quickly measuring the conductivity of the solution with the bridge, a sample was removed for solids determination and the bottle was replaced in the polymerization bath for continuation of the reaction.

Results

Plotting the percent conversion against time, as shown in Figure 1, gave straight lines characterizing each of the two rates.

In Figure 2, conductivity is plotted against percent conversion. Conductivity is high during the early stages of polymerization because soap is present in micellar form producing better electrical transport than after it has become primarily attached to more bulky latex particles. Beyond the CMC, conductivity decreases over a few percent conversion to a considerably smaller value. Conductivity remains relatively high in these systems because of the salts other than soap which are present. However, the soap is by far the highest concentration of ionizable salt present and its bonding in a less mobile form to the growing rubber particles causes a considerable decrease in conductivity.

The intersection of the rate curves for the polymerization on which the data for Figure 2 was obtained was at 8.0% conversion. This compares closely with 8.3% conversion at the critical micelle concentration (CMC) as determined by conductivity in Figure 2. Comparative data for other systems are shown in Table I. These results strongly suggest that the intersection of the polymerization rate curves occurs at the CMC of the oleate soap emulsifier.

The most widely accepted theories of emulsion polymerization (5,6) point out that new latex particles are generated only as long as soap in excess of the particle surface adsorption requirements is present in the system. After sufficient particle surface area has developed to adsorb the soap in the aqueous

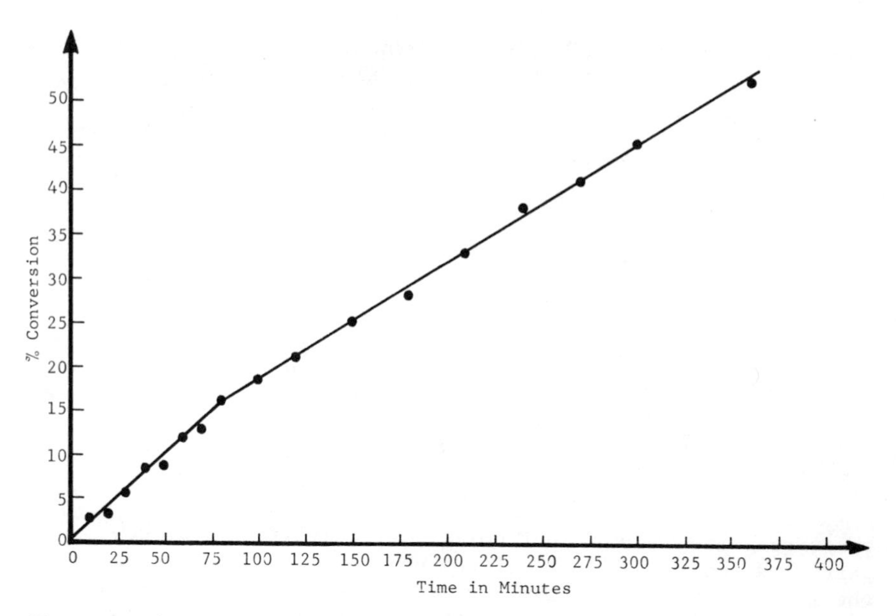

Figure 1. % Conversion vs. time for polymerization using 4.35 parts Westvaco 1480 soap/100 parts butadiene

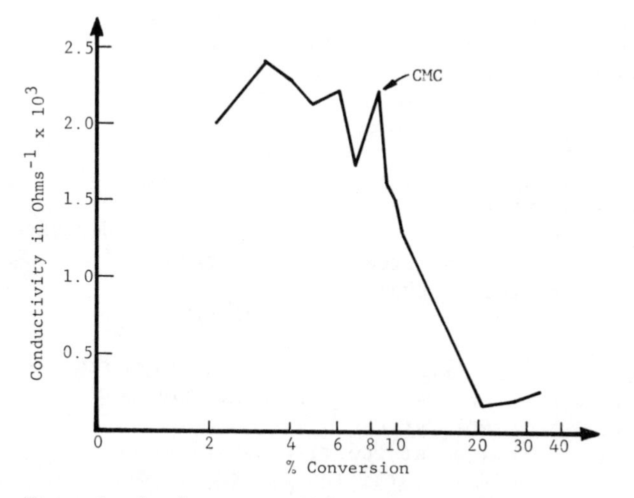

Figure 2. Conductivity vs. % conversion for latex polymerized with oleate soap

TABLE I. COMPARISON OF PERCENT CONVERSION AT CMC BY CONDUCTIVITY MEASUREMENTS AND POLYMERIZATION RATES

% Conversion at CMC	
Conductivity	Polymerization Rate
12.1	12.0
12.2	12.2
7.5	7.4
11.1	11.5
9.5	9.3
8.2	7.6
12.5	12.6
9.2	9.3
13.4	12.7

phase to below its critical point for stabilization of new particles, essentially all of the polymerization continues in the latex particles already formed in the system. For soaps having low CMCs, such as the fatty acid soaps used in this study, it is generally agreed this critical point closely coincides with the CMC. From the following calculations, it can be seen that the surface area occupied by each soap molecule and the percent conversion at the CMC allows determination of the number of particles present in the latex. From this number, the necessary amount of monomer can be calculated to produce any desired final average particle size.

Symbols

A = particle surface area occupied by each soap molecule. $A = 47\ \text{Å}^2$/molecule for potassium oleate (2)
C_a = soap concentration adsorbed to polymer particles
P_c = percent conversion at CMC. $P_c = 15.7\%$ for Figure 1
P_f = percent conversion at end of polymerization. $P_f = 92.2\%$ for Figure 1 latex
S = weight of soap present in polymerization bottle. $S = 5.575$g for Figure 1 latex
Q = acid number of fatty acid. $Q = 191.0$ for Westvaco 1480
C_m = soap concentration free in latex serum at CMC. $C_m = 1.6 \times 10^{-4}$ moles for Westvaco 1480
Avogadros number = 6.02×10^{23} molecules/mole
W_c = weight of polymer present at CMC
W_f = weight of polymer present at end of polymerization
T = surface area at CMC in square angstrums per gm of polymer
D_c = particle diameter at CMC
D_f = particle diameter at end of polymerization
N = number of polymer particles per gm of polymer

G = polymer density. $G = 0.92\ gm/cm^3 \times 10^{-24} cm^3/Å^3 = 0.92 \times 10^{-24} gm/Å^3$ for polybutadiene

Weight of monomer in bottle = 148 gms for Figure 1 latex

Calculations

$$C_a = (S \times \frac{Q}{1000} \div 56.1) - C_m$$

$$C_a = (5.575g \times \frac{191.0}{1000} \div 56.1) - 1.6 \times 10^{-4} \text{ moles} = 0.0195 \text{ moles}$$

$$W_c = 148 \text{ gms} \times \frac{P_c}{100}. \quad W_c = 148 \times \frac{15.7}{100} = 23.2 \text{ gms}$$

$$T = \frac{C_a}{W_c} \times A \times 6.02 \times 10^{23} \text{ molecules/mole}$$

$$T = \frac{0.0195 \text{ moles}}{23.2 \text{ gms}} \times 47\ Å^2/\text{molecule} \times 6.02\ 10^{23} \text{ molecules/mole}$$

$$= 23.7 \times 10^{21}\ Å^2/gm$$

$$N = \frac{1}{V_c G}$$

where V_c = volume of a spherical particle at CMC. $V_c = 1/6\ \pi\ D_c^3$

A_c = surface area of a spherical particle at CMC.

$$A_c = \pi\ D_c^2$$

$$T = A_c N = \frac{\pi\ D_c^2}{1/6\ \pi\ D_c^3 G}$$

$$\text{or } D_c = \frac{6}{T\ G}\ ;\ D_c = \frac{6}{23.7 \times 10^{21}\ Å^2/gm \times 0.92 \times 10^{-24} gm/Å^3} = 275\ Å$$

$$N = \frac{1}{1/6\ \pi\ D_c^3 G} \quad \text{particles/gm}$$

$$N = \frac{6}{3.14 (275\ Å)^3 (.92 \times 10^{-24} g/Å^3)} = 9.99 \times 10^{16} \text{ particles/gm}$$

Total number of particles present in the polymerization at the CMC = $N_t = NW_c$

$$N_t = 9.99 \times 10^{16} \text{ particles/gm} \times 23.2 \text{ gms} = 232 \times 10^{16} \text{ particles}$$

$$W_f = 148 \text{ gms} \times \frac{P_f}{100}$$

$$W_f = 148 \times \frac{92.2}{100} = 136 \text{ gms}$$

Particles/gm at end of polymerization $N_f = \frac{N_t}{W_f}$

$$N_f = \frac{232 \times 10^{16} \text{ particles}}{136 \text{ gms}} = 1.70 \times 10^{16} \text{ particles/gm}$$

$$N_f = \frac{1}{1/6\ \pi\ D_f{}^3 G}$$

$$\text{or } D_f = \sqrt[3]{\frac{6}{N_f\ \pi\ G}}$$

$$D_f = \sqrt[3]{\frac{6}{1.70 \times 10^{16}\ (3.14)(0.92 \times 10^{-24})}} = 496\ \text{Å}$$

In developing an analysis of this type, it is desirable to establish correlation with a widely accepted technique. This was done by determining particle size during polymerization by the polymerization rate technique and submitting the final latex for particle size analysis by electron microscopy.

The particle size calculated from polymerization rates is a volume to surface area particle diameter. The number average particle size is normally determined by electron microscopy. However, any of the possible averages may be calculated from electron microscope data. The pertinent particle diameter formulas for this study are:

$$D_{\text{no. avg.}} = \frac{\sum_i N_i D_i}{\sum_i N_i}$$

$$D_{\text{V/S avg.}} = \frac{\sum_i N_i D_i{}^3}{\sum_i N_i D_i{}^2}$$

N = number of particles having diameter D.

Particle size comparisons for three latexes analyzed by both techniques are shown in Table II.

TABLE II. COMPARISON OF PARTICLE DIAMETERS FROM POLYMERIZATION RATE ANALYSIS WITH MEASUREMENTS BY ELECTRON MICROSCOPY

Measurement	Particle Diameter (Å)		
	Latex 1	Latex 2	Latex 3
Polymerization Rate (47 Å^2/Soap Particle)[2]	376	495	380
E. M. No. Avg.	288	444	302
E. M. V/S Avg.	347	595	391

In view of the problems involved with the particle size measurement of soft particles by electron microscopy and the uncertainty in surface coverage by oleate soap molecules which were discussed in the introduction, agreement between the V/S average electron microscope measurements and polymerization rate measurements for these latexes is quite acceptable.

The effect changes in concentration and composition of a number of the constituents in the polymerization recipe have on latex particle size was determined from the polymerization rate data. In this recipe, the redox initiator system is added as two separate components. First, the p-menthane hydroperoxide is injected into the bottle containing monomer, water, soap, salt and regulator. Then, the ferrous hepto sulfate, versene and sodium sulfoxylate dissolved together in aqueous solution is injected. The effect doubling each of these initiator solutions has on particle size is shown in Table III.

TABLE III. EFFECT OF INITIATOR CONCENTRATION ON LATEX PARTICLE DIAMETER

Initiator Concentration	% Conversion at CMC	Average Particle Diameter (Å)	
		at CMC	at 100% Conversion
Normal	14.2	248	476
Double $FeSO_4$, versene nad sulfoxylate	20.0	349	598
Double peroxide	10.3	180	384

Additional redox solution in the system increases the latex particle diameter. Increasing the peroxide reduces the latex particle diameter.

Materials other than peroxide which are capable of forming free radicals under the conditions of polymerization also reduce latex particle size. These results are shown in Table IV.

TABLE IV. EFFECT OF ADDITIVES CAPABLE OF FORMING FREE RADICALS ON LATEX PARTICLE DIAMETER

Additive	% Conversion at CMC		Average Particle Diameter (Å)			
			at CMC		at 100% Conversion	
	Series 1	Series 2	Series 1	Series 2	Series 1	Series 2
None	20.1	14.2	352	248	600	476
8% Linoleic Acid in Emulsifier	15.5	9.1	271	159	505	354
Hydroquinone	10.7	8.5	187	149	394	338
8% Linoleic Acid + Hydroquinone	10.2	7.0	178	122	382	297
8% Linoleic Acid + Double Hydroperoxide	7.7	3.7	135	65	317	194

This data demonstrates a degree of additivity in effect upon particle diameter between substances which easily form free radicals and linoleic acid containing an active methylene group which can give up hydrogen readily to produce a radical. All of these substances are efficient retarders of the polymerization rate.

Finally, an increase in particle size would be expected from a decrease in emulsifier concentration (5,6). This effect was evaluated in several polymerizations and the particle sizes of the latex obtained were always found to be larger for the samples with the least soap present using the polymerization rate analysis. Typical results are shown in Table V.

TABLE V. EFFECT OF EMULSIFIER CONCENTRATION ON LATEX PARTICLE DIAMETER

Parts of Soap/100 Parts Monomer	% Conversion at CMC		Average Particle Diameter (Å)			
			at CMC		at 100% Conversion	
	Series 1	Series 2	Series 1	Series 2	Series 1	Series 2
4.35	15.7	20.4	375	357	510	606
3.00	12.9	17.3	326	456	645	818

Conclusion

Results from duplicate series of bottle polymerizations are shown in Table IV and V for the purpose of demonstrating the variability which exists. In spite of meticulous care being exercised by a highly experienced operator, this degree of variability in absolute particle size between polymerization series made on different days appears to be inevitable.

Within a given bottle polymerization series, particle sizes of latexes made with different conditions generally showed good correspondence with those made in similar series, but the absolute particle sizes between the series were often quite different. This suggests that the variations are primarily due to inhomogeneities in chemicals such as ferrous sulfate and minor differences in makeup and polymerization technique which would tend to disappear in larger reactions. This is probably a primary reason why latex producers put little or no faith in latex polymers prepared in smaller than 10 gallon reactors. Differences in particle size is one manifestation of a fundamental difference between the polymers. Since these particle sizes are measured by polymerization kinetics, it stands to reason that the variations are probably accompanied by differences in degree of polymerization of the polymer molecules making up the particles (7).

Acknowledgement

The author wishes to thank Mr. Ralph Irick for experimental assistance and Westvaco Corporation for permission to publish this work.

Abstract

Polymerization rate is faster early in the reaction than at later stages. The two rates intersect at the critical micelle concentration (CMC) of the oleate soap emulsifier. Final particle size of the latex can be calculated during polymerization from this measure of the percent conversion at the CMC using knowledge of the latex surface area occupied by each soap molecule. Thus, the opportunity is provided for latex particle size control. The effect on latex particle size of several variables in the system was evaluated. Increasing the concentration of the redox component in the initiator increased particle size. Increasing the concentration of the peroxide or adding other agents capable of decomposing to free radicals reduced latex particle size. Reducing the emulsifier concentration increased the particle size.

Literature Cited

1. Maron, S. H., Elder, M. E. and Ulevitch, I. N., J. Colloid Sci. (1954) 9, 89.
2. Willson, E. A., Miller, J. R. and Rowe, E. H., J. Phys. Chem. (1949) 53, 357.
3. Orr, R. J. and Breitman, L., Can. J. Chem. (1960) 38, 668.
4. Brodnyan, J. G. and Brown, G. L., J. Colloid Sci. (1960) 15, 76.
5. Harkins, W. D., J. Chem. Phys. (1954) 13, 381; J. Am. Chem. Soc. (1947) 69, 1428; J. Polymer Sci. (1950) 5, 217.
6. Roe, C. P., Ind. Eng. Chem. (1968) 60(9), 20.
7. Sundberg, D. E., PhD. Dissertation, U. of Delaware (1970), University Microfilms 71-6448.

9

Polymerization of Acrylonitrile in Ultracentrifugal Fields

ADOLF G. PARTS and ENN ELBING

School of Chemistry, Macquarie University, North Ryde, N.S.W., 2113, Australia

It has been shown (1) that the rate of polymerization of acrylonitrile water solution was lowered if the solution was stirred, or if the polymerization was performed in a centrifugal field of ca 2,000 g. Studies showed that the increase of the centrifugal field to ca 23,000 g (2) further lowered the rate of polymerization. In continuing these studies the polymerization was performed in the Spinco Analytical Ultracentrifuge where the field was increased to 180,000 g (3). The polymerizations in the laboratory centrifuge were studied by taking samples, stopping the reaction and determining gravimetrically the amount of polymer formed. Polymerizations in the ultracentrifuge were examined by frequently photographing the level of the sedimented polymer. After the runs were finished, the polymer was taken out of the cell, washed, dried and weighed. The conversion was calculated. The data are presented in the form of graphs in Fig. 1 of (3) and in the Table I of this publication.

Curve 1 (Fig. 1) in (3) extends to 58 percent conversion (time about 52 minutes) and covers the whole investigated polymerization range. The lowest curve in the same figure refers to the polymerization in the ultracentrifugal field of 180,000 g. In 70 minutes about 6 percent has been polymerized. Table I contains the data about polymerization in the

TABLE I

% Conversion	4	12	20	28	36	44	52
Time min (a)	53	104	148	207	293	424	620
Time min (b)	7.8	15.3	21.0	28.2	35.1	42.9	52.5
a:b	6.8	6.8	7.0	7.3	8.3	9.9	11.8

(a) ultracentrifugal field (b) normal gravitational field

ultracentrifugal field till up to 620 mins. The shape of the percent-conversion-time relation curve is similar to that found in the normal gravitational field. It is S-shaped, with an increase in the conversion rate up to 15 percent (120 minutes) and gradually decreasing thereafter.

Discussion of Coagulation

It is assumed that the polymer molecules which are formed in the solution containing acrylonitrile, hydrogen peroxide, ferric ions, pH=1.8 and at room temperature undergo perikinetic coagulation. The primary polymer particles, or "base" particles coagulate and form latex particles which contain an integer number of base particles. Since von Smoluchowski (4) published the theory of perikinetic coagulation, the experimental data (5) have confirmed the conclusions of the theory. The theory was generalized and the experiments extended by many investigators, amongst whom we will mentione only a few: Chandrasekhar (6), Hidy (7), Hidy and Lilly (8), Martynov and Bakanov (9), Rajagopal (10), Zebel (11), Beeckmans (12) and Rice and Whitehead (13).

The starting point in the above mentioned papers is a solution of colloidal particles which undergoes either a perikinetic or ortokinetic coagulation, or both. If the number and size distribution of particles is known at the start of the coagulation, the number and size distribution of them can be calculated and determined at any later stage. The agreement between the experimental data and those calculated was good - sometimes excellent.

In the polymerization study, however, polymer is being created continuously, the polymer particles coagulate and if in the centrifugal field they will be ejected gradually from the polymerizing solution. This occurs as soon as the latex particles reach big enough dimensions to follow the forces in the centrifugal field. The latex particles move to the bottom of the cell.

The mathematics of coagulation in the centrifugal field during polymerization is thus more complex: the creation of new base particles has to be considered

We were interested in answering the questions: (a) is there any change in the rate of polymerization if performed in a centrifugal field compared with that without the field and (b) how does this rate depend upon the strength of the centrifugal field. Both questions were already attacked in (1).

The magnitude of the centrifugal force at a given distance from the axis of rotation is proportional to the mass of the particle and this varies as a cube of the particle radius. The frictional force between the moving particle and the medium is proportional to the radius of the particle. The velocity of sedimentation of latex particle is thus proportional to the square of the radius of the particle. Bigger particles sediment faster and the solution becomes depleted of them also faster. The higher the centrifugal field, the smaller is the relative amount of bigger latex particles in the solution.

To start with, we assume - to simplify the calculations - that the coagulation constant "k" of the differential equationsof von Smoluchowski remains unaffected by the centrifugal field and that it does not depend upon the size of latex particles. A second assumption is that the base polymer particles are all of the same size, i.e. that they contain the same number of monomer units.

The base particles - N(1) in number per unit volume - are created with the rate α, and they disappear by coagulation with any particle which may contain any number i of base units. The latex particles containing i base units are created when two specific latex particles, one with i-j, the other with j base units collide and coagulate. The latex particles with i base units disappear when they collide and coagulate with any other latex particle.

We write thus a set of differential equations describing the rate of change of latex particles N(i) with i base units as:

$$dN(1)/dt=\alpha-kN(1)[N(1)+N(2)+N(3)+\ldots]=\alpha-kN(1)\sum_i N(i)$$
$$=\alpha-kN(1)N$$

$$dN(2)/dt=(k/2)N(1)N(1)-kN(2)N$$

$$dN(3)/dt=kN(1)N(2)-kN(3)N$$

$$dN(4)/dt=kN(1)N(3)+(k/2)N(2)N(2)-kN(4)N$$

....

$$dN(i)/dt=(k/2)\sum_{s=1}^{i-1}N(s)N(i-s)-kN(i)N$$

....

The set of differential equations of von Smoluchowski has been preserved with the exception of the first equation above, which accounts for the creation of base particles with the rate α.

Adding the left-hand and the right-hand sides of these equations we obtain:

$$dN(1)/dt+dN(2)/dt+...=dN/dt=\alpha-(k/2)N^2 \quad \text{(P)}$$

According to equation (P) the total number N of latex particles, each containing an integral number of base polymer units increases from zero at t=0 asymptotically to the limiting value N_{lim} after long enough time. Then

$$dN_{lim}/dt=0=\alpha-(k/2)N_{lim}^2 \text{ and } N_{lim}=(2\alpha k^{-1})^{1/2}$$

The mathematical solution of the differential equation (P) is

$$(2\alpha k)^{-1/2}\ln[(N_{lim}+N)/(N_{lim}-N)]=t \quad \text{(Q)}$$

The number N of latex particles at time t is thus expressed as a function of α, k, N_{lim} and t. We write explicitly

$$N=(2\alpha k^{-1}) [\exp(t\beta)-1]/[\exp(t\beta)+1] \quad \text{(R)}$$

with $\beta=kN_{lim}=(2\alpha k)^{1/2}$

We repeat the physical picture which yields the equations (Q) and (R): base units of polymer are formed as a result of polymerization at the rate α. They coagulate perikinetically. The total number of latex particles and not yet coagulated base units, as determined by equation (R) can never exceed the limiting value $N_{lim}=(2\alpha k^{-1})$. The latex particles will gradually increase their content of base units and their individual mass by coagulation.

The value of $N(1)_{lim}$, that is the limiting value of base units, can be obtained from the first differential equation of the series on page 3 by applying the limiting conditions:

$$dN(1)_{lim}/dt=0=\alpha-kN(1)_{lim}N_{lim}$$

as

$$N(1)_{lim}=\alpha/kN_{lim}=\tfrac{1}{2}N_{lim}$$

Expressed in words: out of the total limiting latex particles one half are base units; the other numerical half are the coagulated latex particles with more than one base unit in each.

It is possible to obtain the instantaneous value of N(1) from the first differential equation of the set on page 3, inserting the value of N from (R) and solving for N(1). The procedure may be repeated to find N(2) etc. The expressions obtained are quite involved, the calculations seem to have no practical purpose and we will not pursue them any further.

If, however, we make the approximations that all $dN(i)/dt=0$ (Approximations which are the better fulfilled the smaller the value of i) i.e.

$$dN(i)/dt=(k/2)\sum_{s=1}^{i-1} N(s)N(i-s)-kN(i)N=0$$

we can write

$$N(i)N=\frac{1}{2}\sum_{s=1}^{i-1} N(s)N(i-s) \ldots\ldots\ldots\ldots\ldots (T)$$

and obtain by successive application of (T) for some limiting numbers -

$N(2)_{lim}=0.125N_{lim}$; $N(5)_{lim}=0.02734N_{lim}$;

$N(10)_{lim}=0.0092735N_{lim}$ $N(20)_{lim}=0.003215N_{lim}$;

$N(30)_{lim}=0.001739N_{lim}$ $N(40)_{lim}=0.001126N_{lim}$;

An analytical expression for $N(i)_{lim}$ is not available.

The coagulation constant $k=(4/3)(RT/N_A\eta)$ - with T=300K and $\eta=0.0100$ poise - yields the value $K=5.54x10^{-12}$. For the rate of polymerization in the ultracentrifugal field of 180,000 g (Graph 1, lowest curve (2)) we found 0.0454% or $2.27x10^{15}$ acrylonitrile molecules per ccm per second in the early stages of polymerization. The average number of monomer units in the polymer molecule (in the base unit) from the determination of the molecular weight of the formed polymer by Dr. E. Elbing (private communication) was found to be of the order 10^3 at 50% conversion, (hydrogen peroxide - iron (III) ions as initiator) while this was of the order of 10^4 at 0.5% conversion (potassium persulphate as initiator). As a third value for the number of monomers in a base unit we select 10^6 - to be able to compare the results of estreme assumptions. Thus we obtain Table II

TABLE II

Number of monomers per base unit	10^3	10^4	10^6
Number of base units (α) formed per unit time in ccm	$2.27x10^{12}$	$2.27x10^{11}$	$2.27x10^{9}$
$\beta=(2k\alpha)^{\frac{1}{2}}$	5.0	1.6	0.16
$N_{lim}=(2\alpha/k)^{\frac{1}{2}}$	$9.1x10^{11}$	$2.86x10^{11}$	$2.86x10^{10}$
t sec	4.7	15	150

In the last line the time is calculated from the equation (Q) at which the total number of latex particles has reached the value such that $N_{lim}-N=10^{-10}N_{lim}$—that is within 10^{-8} percent of the limited.

The rate of coagulation, as expressed in the value of the coagulation constant is actually considerably greater than that given above, and this is due to three factors -

The rate of coagulation of spherical particles of the same size is independent of their size. However, the rate of coagulation of two kinds of particles with radii r_A and r_B, when $r_A \neq r_B$, is increased by the factor (5)

$$\gamma=(r_A+r_B)^2/4r_Ar_B$$

which for $r_A=5r_B$ is 1.8 and for $r_A=100r_B$ is 25.

The particles are moving in the centrifugal field and this increases the rate of collision between them.

The Cunningham correction factor discussed in (14) and calculated in (11) for some specific cases increases the rate of collisions between particles of unlike size considerably. As an illustration, the coagulation constant between two particles A and B of unlike size

$$r_A=0.1\mu \text{ and } r_B=0.001\mu$$

is about 1600 times that between particles A & B of the same size

$$r_A=r_B=0.1\mu$$

It is difficult to give the functional dependence of the coagulation constant on the sizes of colliding particles and thus to take it into account. The simplified assumptions and the complexity of actual polymerization does not permit a quantitative calculation of k, but we conclude that the actual coagulation constant k_{act} has much greater value than that given on page 5. The time necessary to achieve the nearly limiting size distribution of coagulated particles is shorter than that in Table II.

With this in mind we can satisfy ourselves - for the time being - with rough estimates and assume the validity of fundamental assumptions made. We take (as an example) the case from Table II where the number of monomers per base unit is 10^6. A latex particle which contains 20 base units has a radius of 7.1×10^{-6} cm in coiled up form and moves in the ultracentrifugal field of 180,000 g with a velocity of about 2 cm/min. Once formed it will travel the distance from the top of the cell in the ultracentrifuge to the bottom of it in about one minute. Thus we assume for subsequent estimations that in the cell of the ultracentrifuge only latex particles containing 20 or less base units are present, the bigger ones being thrown out of the solution to the bottom of the cell.

We will calculate the monomer units in polymeric material which are contained in the latex particles with one to 20 base units. To be able to do this, let us assume that the latex with 1,2,3...20 base units has reached the size distribution state which is not far different from the limiting distribution calculated for $dN(i)/dt=0$. The addition of one base unit raises the index i of any particular latex particle by one. Those latex particles which have reached i=21 are assumed to be thrown out of the solution. The limiting number of all latex particles is 2.86×10^{10} per ccm. From those one half have only one base particle, 1/8 have two, 1/16 have three etc. The calculations being performed using formula (S) give the necessary numbers. Each group multiplied by the number of base units contained in each member of the group gives the base units in this group, and all added together yields the total number of base units in the groups. We obtain -

$$\sum_{i=1}^{20} N_i x i = 1.43\times10^{10} \times 2.507\times10^{6} = 3.585\times10^{16}$$

If this state was achieved after 150 seconds (actually earlier!) then during this time out of

$$2.27\times10^{15}\times150 = 34.05\times10^{16}$$

monomer molecules which have polymerized the fraction 3.585:34.05=0.105 or 10.5% are still in solution as latex particles, and about 90% are thrown out of it. After 15 minutes, only 1.6% of polymerised monomer is in the solution, 98.4% is thrown out.

The qualitative picture of the influence of the centrifugal field if the base unit contains only 10^3 or 10^4 monomer units is unaltered. The number of base units, to reach the same number of latex particles must be increased to 20×10^3 respectively 20×10^2 (in lieu of 20) and the member by member calculation of polymer in suspension computer should be used, which we intend to do.

There is additional information available concerning the influence of the formed polymer upon the rate of polymerization. Parts (15) presented percent polymerization-time curves for acrylonitrile. The results are presented in the form of graphs in Fig. 1 to Fig. 5 in (15). The five figures refer to the different values of hydrogen peroxide, concentration of ferric ions and pH of the solution in which the polymerizations were performed. The three curves in each figure refer to the three concentrations of acrylonitrile at the start of polymerization. They were: 1.06M, 0.80M and 0.40M of acrylonitrile. The conversion percentage falls with the concentration of acrylonitrile present in the solution. After 50% conversion the concentration of acrylonitrile of the middle curve has reached the concentration of the lower curve at the start of polymerization. Without exception the rate of polymerization according to the middle curve of every figure is higher than that at the start of the lowest curve of the same figure. This we ascribe to the absence of polyacrylonitrile at the start of polymerization - and a slower rate results.

Summary

The remarkable decrease in the rate of polymerization of acrylonitrile in ultracentrifugal fields is qualitatively explained as due to the elimination of formed polyacrylonitrile from the solution as a result of its sedimentation. The surface of polyacrylonitrile is of vital importance in determining the rate of polymerization of acrylonitrile in water solution.

Literature Cited

1. Moore, D.E.,Parts, A.G.,Makromol. Chem. (1960), 37.108
2. Elbing,E., B.Sc.Honours Thesis, Sydney University, Sydney, N.S.W. Australia
3. Parts, A.G.,Elbing, E., Polymer Preprints, (1975), 16.211
4. Smoluchowski,M.v., Physik.Zeit.(1916), 17.557,583
5. Müller,H.,Kolloidchem. Beihefte (1928), 26, 257, 27.223
6. Chandrasekhar,S., Rev.Modern Phys. (1943), 15.1
7. Hidy,G.M., J.Colloid Sci.(1965), 20.123

8. Hidy,G.M.,Lilly,D.K., J. Colloid Sci. (1965), 20.867
9. Martynov, G.A., Bakanov,S.P., in Deryagin,B., Editor, "Research in Surface Forces". 182.Consult.Bureau,N.York, (1963)
10. Rajagopal,E.S.,Kolloid Zeit., (1959), 167.18
11. Zebel,G., Kolloid Zeit., (1957), 156.102
12. Beeckmans,J.M., Can. J. Chem., (1965), 43.2312
13. Rice, C.L.,Whitehead,R.J., J.Colloid and Interface Sci. (1967), 23.174
14. Davies,C.N., Proc. Phys. Soc.(London), (1945), 57.259
15. Parts, A.G., J. Polymer Sci., (1959) 37.131

10

Radiation Induced Emulsion Polymerization of Vinyl Chloride

J. BARRIAC, R. KNORR, E. P. STAHEL, and V. STANNETT

Department of Chemical Engineering, North Carolina State University, Raleigh, N. C. 27607

High energy radiation has a number of advantages as the initiator for emulsion polymerization. These have been discussed in a previous publication (1). Since the radical flux generated by the radiolysis of the system is essentially independent of temperature, it is ideal for low temperature polymerizations. In the case of vinyl acetate and vinyl chloride where the molecular weights are largely controlled by chain transfer, radiation is, in principle, a favored method of initiation. On the negative side, the direct interaction of the radiation with the polymer may cause excessive branching or structural damage to the polymer. In fact, irradiated polyvinyl chloride is less thermally stable (2) but only at doses above about 0.5 Mrads. The present study of gamma radiation initiated vinyl chloride polymerization in emulsion, however, shows that good conversions can be obtained with less than this total dose.

The mechanism and kinetics of vinyl chloride polymerization in emulsion have been studied extensively, mainly involving the use of chemical initiators, and has been summarized by Talamini and Peggion (3) and by Ugelstad et alia (4). The mechanism shows a gradual change as the number of particles (N) is increased.

Giskehaug (5) and Gatta (6) studied the polymerizations at low numbers of particles using seeded emulsion systems. Their results show a one-half order initiator dependence with very little response to increases in particle number. These results are explained by assuming a high degree of aqueous polymerization with bimolecular termination of aqueous radicals. Polymerizations at intermediate particle numbers (10^{16} - 10^{17} particles/liter) were reported by these authors and by Peggion, Testa and Talamini (7). The results show an increasing particle number dependence and an increase in initiator dependence over the one-half order relationship holding at low particle numbers.

These results suggest that latex particles become increasingly important as reaction sites as N is increased. The higher initiator dependence is a reflection of this trend because at low N the particle size is relatively large. As a result both desorption and termination are greatly hindered for radicals occupying these particles. This can result in an unsteady state latex particle population or equivalently a form of emulsion gel effect as is observed in the present study at low particle number.

Ugelstad et al. (8) have also studied the emulsion polymerization of vinyl chloride at high particle numbers. Their results show that as N is increased from 10^{16} to 10^{20} particles/liter, the particle dependence increases from nearly zero to 0.15-0.30 (depending on interpretation) with an initiator dependence of one-half over the entire range. The value of 0.3 found at high particle numbers appears to be an upper limit on this dependency and as such it is analogous to the dependency found in this study. The authors explain these results by assuming polymerization primarily in latex particles and applying the relationship

$$N_0 >> N_1 >> N_2 >> N_3 \text{ etc.}$$

where N_i represents the number of particles containing i radicals. Based on this relationship, the number of particles containing more than two radicals is considered to be negligible. Thus, by considering only particles containing 0, 1, and 2 radicals the Smith-Ewart recursion equation can be reduced to a simple analytical expression.

In the present study we have established that, with radiation initiation at high particle numbers, the Smith-Ewart recursion formula can be simplified even further. By assuming rapid termination in the latex particles where the average number of radicals per particle, $\bar{n} < 1/2$, a rate expression is derived wherein only particles containing 0 or 1 radical are significant.

Experimental

The vinyl chloride monomer was obtained from the Ethyl Corporation and was 99.9% pure. It was washed with silver nitrate and sodium hydroxide solutions and dried over a calcium chloride column and finally trapped in dry ice - methanol traps.

The emulsifiers were purified before use by recrystallization from alcohol. The ampoules were loaded and degassed under high vacuum before sealing.

The emulsion polymerizations were carried out in vacuum sealed ampoules in a rotating wheel device positioned in the gamma field, providing agitation and assuring radiation uniformity by tumbling the ampoules end over end. The equipment used has been described previously (10).

The polymer was isolated by evaporation of the unconverted monomer and the conversion determined gravimetrically.

Molecular weights were determined by measuring the intrinsic viscosities in cyclohexanone using the equation of Dannusso et alia (9)

$$[\eta] = 2.4 \times 10^{-5} M_n^{0.77}$$

Particle sizes were measured by electron microscopy by the carbon film technique.

A small amount of post polymerization was observed when the samples were left at room temperature. This was eliminated by freezing the samples in liquid nitrogen immediately after the irradiation was completed.

Results

Some typical conversion curves are presented in Figure 1 for 33% solids and with between 0.03 and 3.00 percent of sodium lauryl sulfate calculated on the water content. It can be seen that polymerization, essentially to completion, took place rapidly. For example, at a dose rate of 0.175 Mrads per hour and at the highest emulsifier concentration, the polymerization was complete in less than 45 minutes. The conversion curves at normal soap concentrations were reminiscent of those found with vinyl acetate (1,11) in that they were linear up to nearly 80% polymerization. At the very low emulsifier concentrations, there was an acceleration in the rate at about 50% conversion similar to that found with the gel effect. However, even with only 0.1% emulsifier, this effect had almost disappeared.

Two emulsifiers were studied, sodium lauryl sulfate and sodium dioctyl sulfosuccinate. The latter gave very stable emulsions of the monomer lasting several days whereas the sodium lauryl sulfate emulsion broke in a few minutes. The final latices, however, in general were more stable with the sodium lauryl sulfate and were also more viscous. With both emulsifiers there was always a small amount of coagulum in the form of very fine hard particles. The amount of coagulation was much greater in the case of the sodium dioctyl sulfosuccinate emulsifier and with both emulsifiers decreased linearly with increasing emulsifier concentration. With the reasonable amount of 5% of sodium lauryl sulfate calculated on the monomer content and at 33% solids, the coagulum only amounted to 2.5% of the total polymer whereas the corresponding figure for the sodium dioctyl sulfosuccinate was nearly 8%. The rotating wheel device gave only limited possibilities of varying the agitation condition so no special attention could be given to this variable.

The molecular weight of the polyvinyl chloride appears to be mainly governed by chain transfer to monomer. The molecular weights at a number of conversions, dose rates, monomer-water ratios, and soap concentrations are presented in Table I and appear to be little affected by the reaction variables. The mean number average molecular weight is 1.25×10^5 close to that calculated from the reported values for chain transfer to monomer at 25°C (12).

The particle size of a number of latices were measured using electron microscopy. The size distributions were comparatively narrow, the ratio of volume average diameter to number average di-

TABLE I

Molecular Weight of Polyyinyl Chloride Produced by Radiation Induced Emulsion Polymerization at 25°C

Dose rate Mrads/hr	% Emulsifier (on water)	Water/monomer ratio	Percent Conversion	Mn x 10^{-5}
0.175	Sodium lauryl sulfate			
	0.5	2	14.4	1.11
	1.0	2	70.0	1.15
	2.0	2	29.6	1.34
	3.0	1	8.8	1.25
	3.0	1	31.1	1.24
	3.0	1	65.0	1.26
	3.0	1	65.0	1.27
	3.0	1	89.5	1.22
	3.0	2	63.6	1.36
	3.0	3	61.4	1.11
	5.0	2	47.1	1.28
	10.0	2	59.0	1.27
0.072	1.0	2	70.0	1.12
0.0025	1.0	2	95.0	1.15
	Sodium dioctyl sulfosuccinate			
0.0025	1.0	2	95.0	1.18
0.0720	1.0	2	80.0	1.18
0.1750	1.0	2	80.0	1.14
			AVERAGE	1.21

ameter varying from 1.07 to 1.80. Some typical values are given in Table II and some typical distribution curves in Figure 2. The distributions appear to be quite narrow at low conversions but at higher conversions all the values lie between 1.4 and 1.8 and no general trends can be observed.

The data given in Figure 1 shows the dependency of the rate on the emulsifier concentration. This dependency has been determined in more detail and plots of the linear rates against the emulsifier concentration are given in Figure 3. With sodium lauryl sulfate and with sodium dioctyl sulfosuccinate, the rates are proportional to the 0.32 power of the emulsifier concentration within the entire range studied. The soap dependence is less than predicted by Smith-Ewart theory and somewhat higher than found with vinyl acetate with chemical initiation (11a), with radiation initiations orders of about 0.3 were found by O'Neill (11b). The number of particles was also measured and these are shown plotted against the soap concentration in Figure 4. In the case of 0.5% sodium lauryl sulfate, the number of particles were measured at 35%, 75%, and 96% conversions. The values were found to be 5.8, 6.0, and 6.5 x 10^{15} particles per cubic centimeter showing only a small (~12%) increase in the number of particles with conversion. There was found to be a dependence of the number of particles on the first power of the soap in the case of sodium lauryl sulfate and 1.3 power with sodium dioctyl sulfosuccinate and in neither case was there any break in the slopes.

The influence of the monomer to water ratio on the polymerization rate was studied with sodium lauryl sulfate as the emulsifier. The conversion curves for the case of 3% emulsifier are shown in Figure 5. In Figure 6 the linear conversions for ten minutes of irradiation at 0.175 Mrads per hour are plotted against the water-monomer ratio for 1, 3, and 5% emulsifier. All three sets of data show a linear dependence of the rate on the ratio, in other words, the rate per cubic centimeter of water phase is independent of the monomer-water ratio.

The rate and the number of particles were studied at three dose rates, 0.0025, 0.072 and 0.175 Mrads per hour. The results are plotted for the rate studies in Figure 7 and show a 0.4 order of dependence of the rate on the dose rate. The studies involving the number of particles found are summarized in Table III. The maximum change in the number of particles across a seventy-fold change in dose rate was less than three-fold and no trend in numbers with the dose rate was observable. The dependence of the rate on the number of particles is shown in Figure 8. The rate was found to be proportional to the 0.33 power of N but tending to a much lower order at low particle numbers.

Finally, the temperature dependence of both the rates and the numbers of particles were determined by carrying out measurements at 6°C, 25°C, and 40°C. The Arrhenius plots are presented in Figures 9 and 10. The activation energies were found to be 4.0 Kcals per mole for both emulsifiers studied in the case of the rates and

TABLE II

Particle Size Distribution (Ratio of Volume to Number Average) for Polyvinyl Chloride Latices Produced by Radiation Induced Emulsion Polymerization

Emulsifier	Dose Rate	Temperature °C	Percent Conversion	d_{vol}/d_n
1% sodium	0.072	25	70	1.54
Lauryl sulfate	0.0025	25	95	1.76
	0.1750	40	70	1.70
0.5% sodium	0.1750	6	20	1.07
Dioctyl	0.175	40	80	1.53
Sulfosuccinate	0.072	25	80	1.40
	0.0025	25	95	1.81

Water-monomer ratio = 2.0

$\bar{d}_v$ 0.04 - 0.05 μ

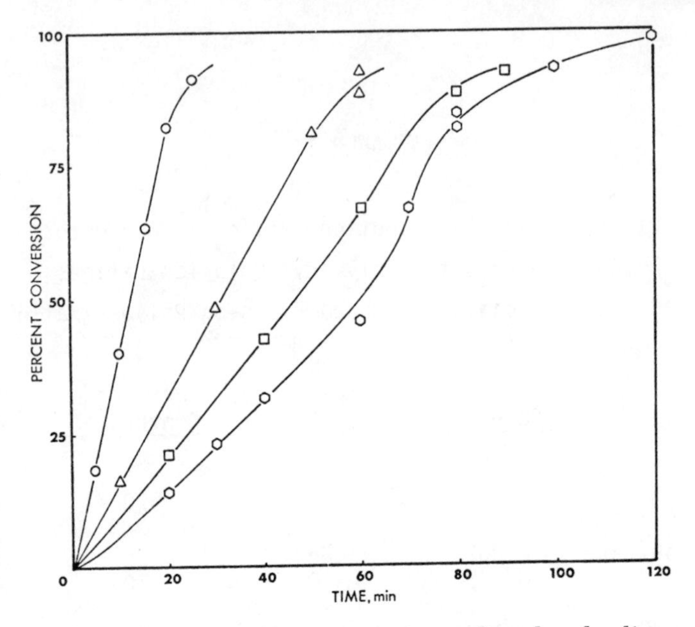

Figure 1. Conversion curves at various sodium lauryl sulfate (SLS) concentrations. Dose rate 0.175 Mrads/hr, temperature 25°C, water to monomer ratio = 2.0. ○, *3.0% SLS;* △, *1.0% SLS;* □, *0.1% SLS;* ⬡, *0.03% SLS. N varies between* 5×10^{17}–3.3×10^{19} *particles/l.*

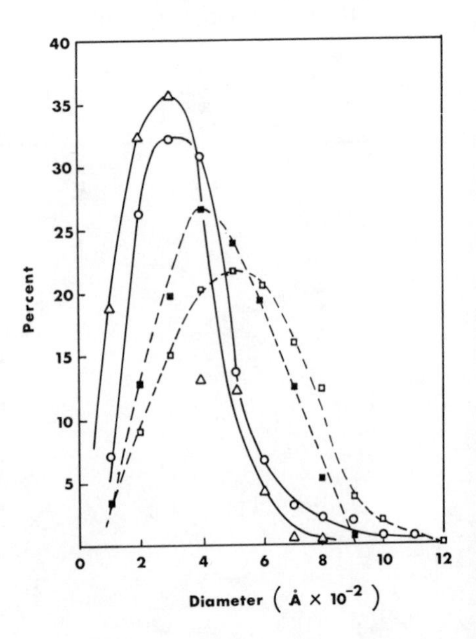

Figure 2. Particle size distributions of polyvinyl chloride latices prepared at 25°C. ○, *SLS 1%, dose rate 2500 rads/hr, 97.9% conversion;* △, *SLS 1%, dose rate 72,000 rads/hr, 80.3% conversion;* □, *NaOSS 0.5%, dose rate 2500 rads/hr, 95.0% conversion;* ■, *NaOSS 0.5%, dose rate 72,000 rads/hr, 85.0% conversion. NaOSS = sodium dioctyl sulfosuccinate.*

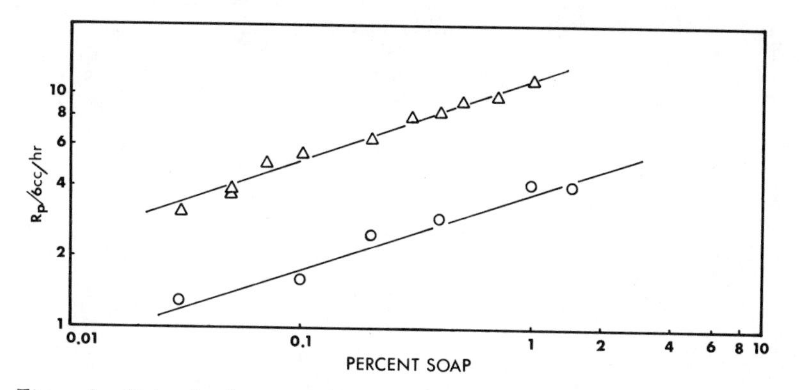

Figure 3. Rate of polymerization vs. the soap concentration. Dose rate 0.175 Mrads/hr, temperature 25°C, water-to-monomer ratio = 2.0. △, SLS; ○, NaOSS; slopes, 0.32.

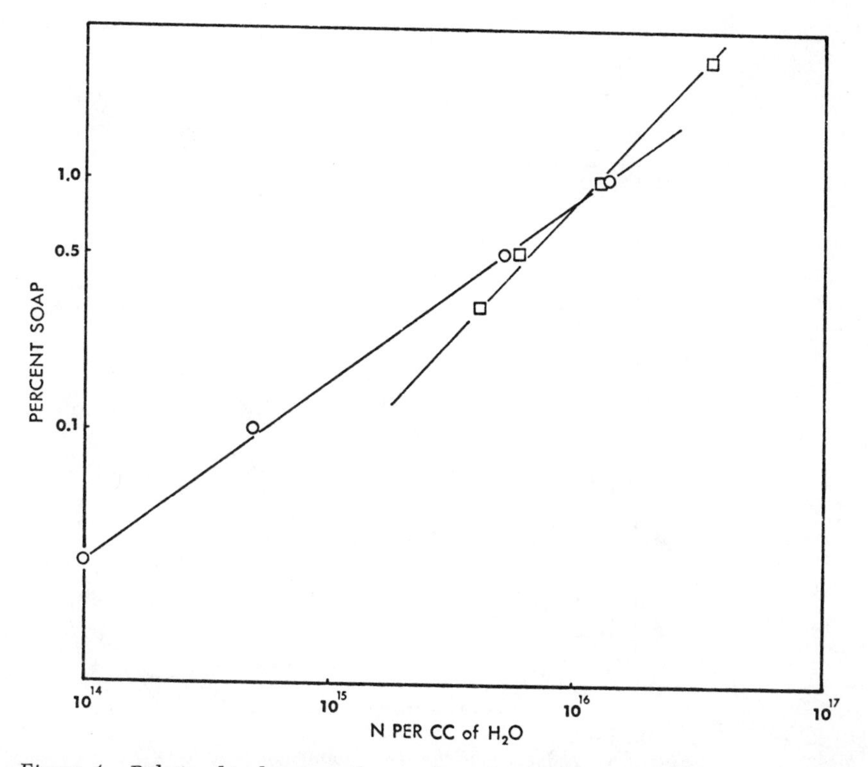

Figure 4. Relationship between the number of particles, N, per cc and the soap concentration. Dose rate 0.175 Mrads/hr, temperature 25°C, water-to-monomer ratio = 2.0. △, SLS slope = 1.0. ○, NaOSS slope = 0.75.

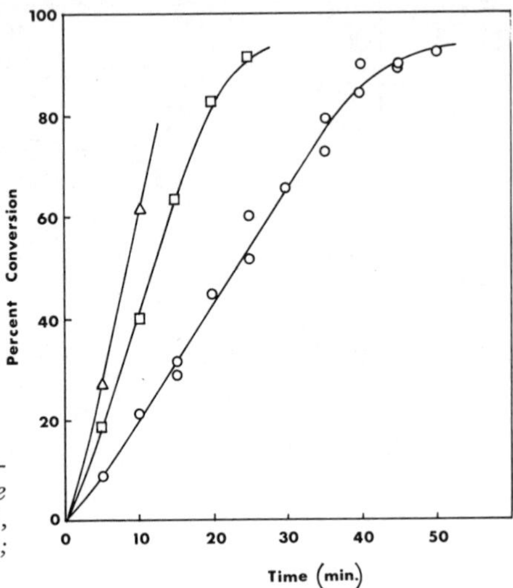

Figure 5. Influence of the water-to-monomer ratio on the percent conversion. Dose rate 0.175 Mrads/hr, temperature 25°C, 3.0% SLS water-to-monomer ratios. △, 3.0; □, 2.0; ○, 1.0.

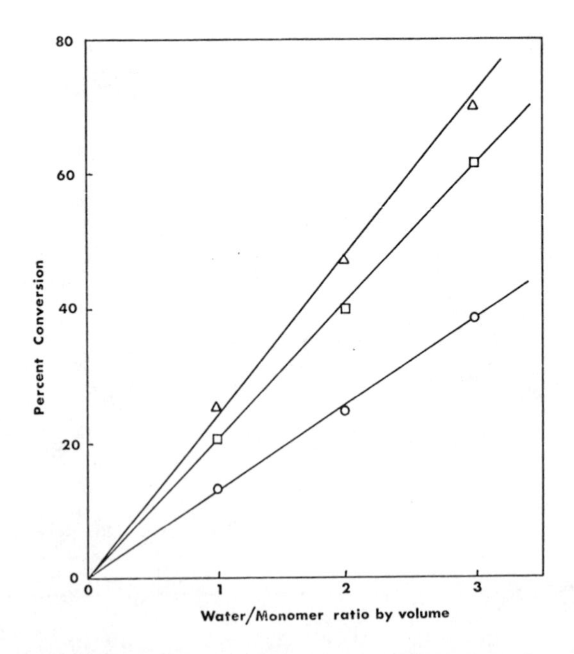

Figure 6. Influence of the water-to-monomer ratio on the percent conversion. Dose = 0.0292 Mrads (~ 10 min at 0.175 Mrads/hr), temperature 25°C. SLS: △, 5%; □, 3%; ○, 1%.

TABLE III

Effect of Dose Rate on the Number of Particles Formed

Emulsifier	Dose rate Mrads/hr	Percent Conversion	Rate gram/cc/hr	No. of particles x 10^{-19}/liter of water
1% sodium	0.0025	98.0	0.18	1.19
Lauryl sulfate	0.072	80.3	0.56	3.5
	0.175	87.0	0.92	1.2
0.5% sodium	0.0025	95.0	0.14	0.65
Dioctyl	0.072	80.0	0.47	0.93
Sulfosuccinate	0.175	80.0	0.47	0.55

Temperature = 25°C

Water-monomer ratio = 2.0

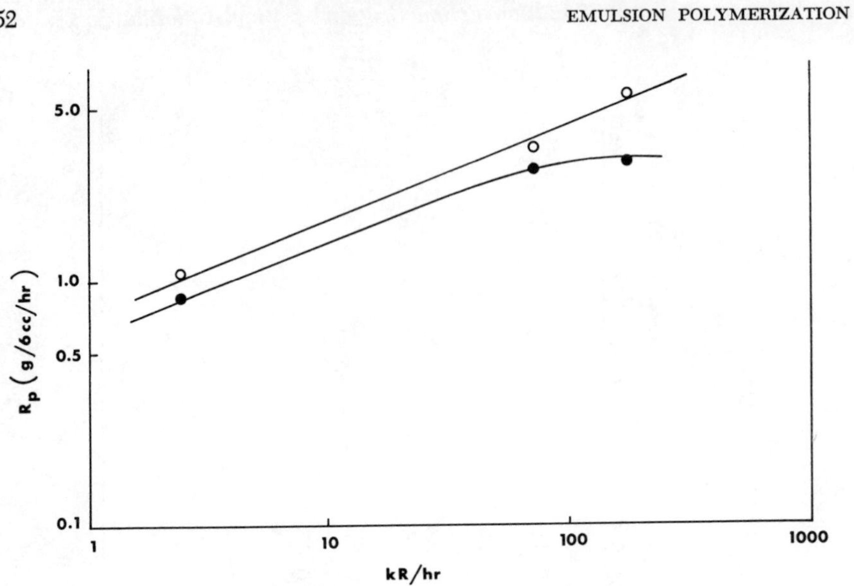

Figure 7. Dependence of the rate on the dose rate. Temperature 25°C, water-to-monomer ratio = 2.0. ○, 1% SLS; ●, 0.5% NaOSS.

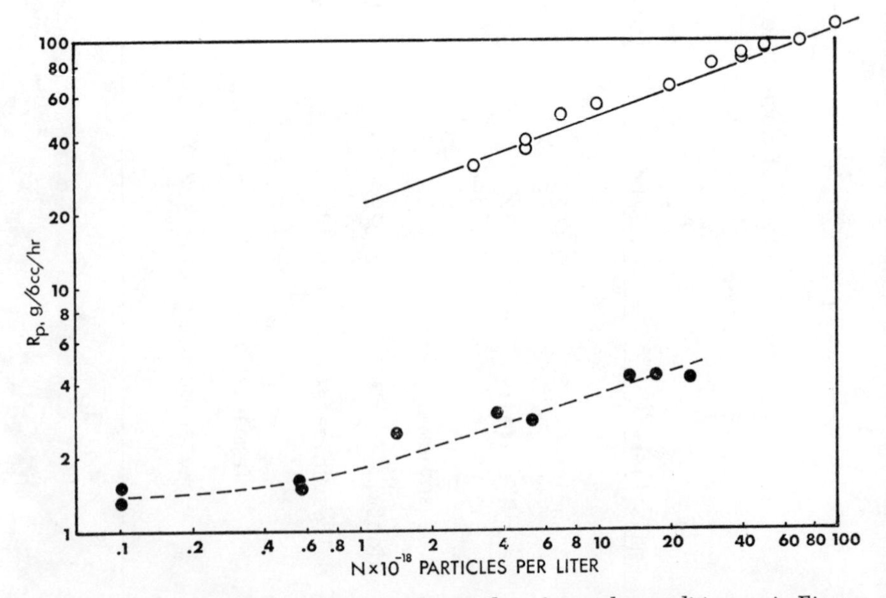

Figure 8. Dependence of the rate on the number of particles conditions as in Figures 3 and 4. ○, SLS slope = 0.33; ●, NaOSS slope = 0.33 tending to zero at low N.

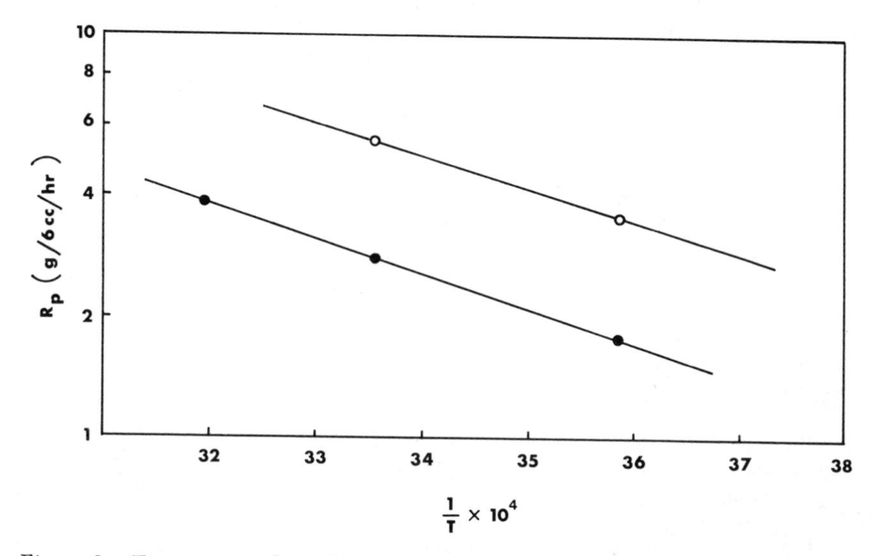

Figure 9. Temperature dependence of the rate. Dose rate 0.175 Mrads/hr, water-to-monomer ratio = 4.0. ○, *1% SLS;* ●, *0.5% NaOSS. Ea = 4.0 Kcals/mol.*

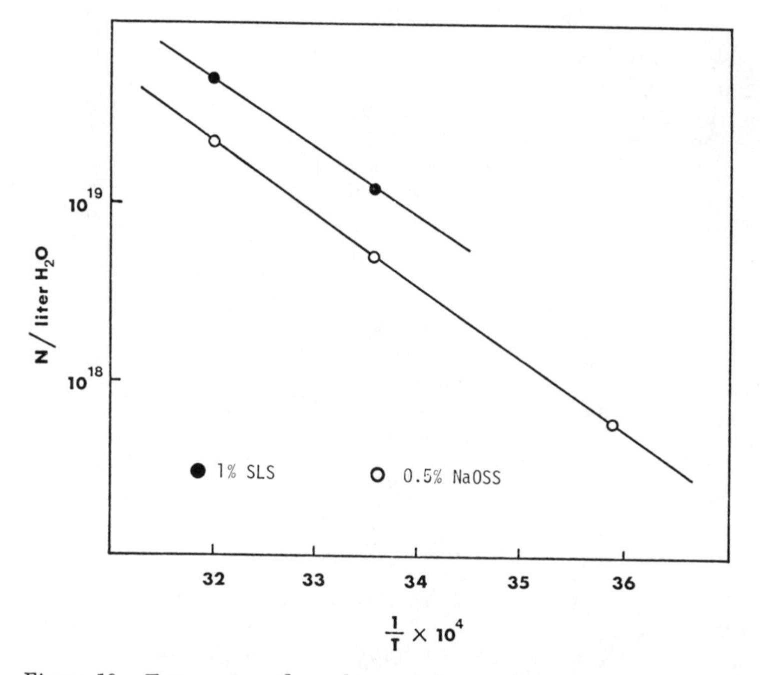

Figure 10. Temperature dependence of the number of particles. Dose rate = 0.175 Mrads/hr, water-to-monomer ratio = 2.0

17.0 Kcal per mole for both emulsifiers for the number of particles. The emulsifier concentrations used for these experiments were in the range where the rates were dependent on about the one-third power of the number of particles and the high activation energy for the number of particles was not, therefore, reflected in a similar value for the rate itself. G (monomer) values for a number of the polymerizations conducted have been calculated. Values of from 50,000 to 120,000 were found calculated on the total energy absorbed. These were with the highest dose rate used. At the lowest dose rates much higher values were obtainable. Thus, at 2500 rads per hour G (monomer) values exceed 600,000 at 5% emulsifier content.

Discussion

The results obtained in this study can be interpreted in terms of existing emulsion theory. Specifically, we find that in the range of reaction conditions employed in this study Smith-Ewart's Case I is closely approximated.

In the present work we maintain that there is a steady state population of aqueous radicals in equilibrium with a corresponding particle phase population. This equilibrium can be expressed as

$$\text{absorption} = \text{rate desorption} + \text{rate initiation}$$

$$\text{rate initiation} = \text{rate termination}$$

This leads to the well known Smith-Ewart recursion equation which has been solved by Stockmayer (14) extended by O'Toole (15) and further by Ugelstad *et alia* (16) to the case where radical reabsorption occurs. This expression has been solved in detail by computer for this paper in the manner presented by Ugelstad *et alia* (16). In particular the average number of radicals per particle, $\bar{n}$, has been related to the quantities m and α', defined below, for the case where no aqueous termination or growth takes place

$$m = k_d \cdot \frac{v}{k_t}$$

where k_d is the rate constant for desorption of radicals from a particle, v is the volume of the polymer particle and k_t the rate constant for mutual termination in the particle. m, therefore, is a measure of the rate of desorption relative to the rate of termination and

$$\alpha' = \frac{\rho_w \cdot v}{N \cdot k_t}$$

where ρ_w is the rate of radical production in the aqueous phase.

The relationship between $\bar{n}$, α', and m is shown in Figure 11 for the case of no aqueous polymerization or termination as shown by Ugelstad et alia (16). The emulsion polymerization of styrene is well represented in this figure. This monomer can be characterized as one where desorption is very slow relative to termination. It, therefore, would be best described by an $\bar{n}$ vs. α' relationship with a low m value, such as curve 1 in Figure 11 ($m = 10^{-5}$). Styrene, of course, is the one monomer where Smith-Ewart Case II applies over a wide range of reaction conditions. Gardon (17) has shown however that $\bar{n}$ becomes greater than 1/2 at low particle numbers (i.e. high α') where conditions are such that radicals are absorbed faster than they can terminate or desorb.

Similarly, literature data suggest that at high particle numbers (low α') $\bar{n}$ falls below 1/2 giving an approximation of Smith-Ewart's Case I. A possible example of this phenomenon is given by Smith (18) in a study of seeded polystyrene emulsion polymerization initiated by potassium persulfate. The rate of polymerization per particle remained constant for particle concentrations of 10^{15}-10^{17} particles/liter but decreased at concentrations of 10^{17} to 5×10^{18} particles/liter. The Case I situation develops in this instance because for any given latex particle the time interval between two consecutive radical absorption increases as the total number of latex particles increases. As this time interval increases the probability of radical desorption also increases and a point is reached where the overall rate of desorption becomes significant. This leads to a reduction in $\bar{n}$ as desorption increases. It should be pointed out that Smith interpreted his results in terms of inadequate diffusion from the styrene layer to the aqueous phase. The present explanation however appears more probable to the present authors.

Figure 11 gives, in addition, the initiator dependence at constant particle number as the slope of log $\bar{n}$ vs. log α'. Thus, it is suggested at the extremes of high and low particle numbers (or α') that initiator dependence increases from zero in the Case II range to 1/2 where $\bar{n} < 1/2$ and where $\bar{n} > 1/2$. The zero order initiator dependency is well known for Case II polymerizations and the 1/2 order initiator dependency at $\bar{n} > 1/2$ has been reported by numerous authors. An initiator dependence of 1/2 has been observed by Ugelstad (8) for vinyl chloride emulsions where $\bar{n} \ll 1/2$.

It is suggested that the chemically initiated emulsion polymerization of vinyl chloride can be represented by curve 3 in Figure 11 ($m > 1$). Desorption plays such a dominant role with this monomer that $\bar{n}$ rarely exceeds 1/2. This is true not only because the desorption constant, k_d, is high but also because with vinyl chloride and other comparatively water soluble monomers the number of latex particles formed per unit of surfactant is very high. Thus, while 10^{16} - 10^{17} particles/liter is typical for styrene emulsions, particle numbers as high as 10^{19} particles/liter are common with vinyl chloride. As a result, the rate of radical absorption per particle tends to be low relative to the rate of initiation while desorption per particle tends to be enhanced. This effect is reflected in a low value for α' and, more significantly, a high ratio of desorption to absorption.

We have suggested that in styrene emulsion polymerizations $\bar{n}$ can fall below 1/2 if particle numbers are high. Although this situation shows a similarity to vinyl chloride polymerization there is a very significant change in the distribution of radicals within latex particles as the ratio of desorption to termination (i.e. m) is increased. Calculations of radical distribution versus m have been carried out based on the Smith-Ewart recursion formula. These calculations show that for small values of m (less than ~10^{-3}) the probability of finding a particle containing two or more radicals is negligible. This is clearly the case for styrene depicted by curve 1 in Figure 11. The result is quite reasonable when one considers that for styrene following Smith-Ewart Case II kinetics the latex particles contain either one or zero radicals and $N_0 = N_1$. Thus, our calculations demonstrate that this situation is extended to $\bar{n} < 1/2$ where $N_0 > N_1$.

At higher values of m, specifically, and where m is increased to 0.10 and $\bar{n} = 0.02$ radical distribution calculations show that 2% of occupied particles contain 2 radicals while .02% contain 3 radicals or more. This is outside the range of validity of Smith and Ewart's Case I model but is well described by Ugelstad's model for vinyl chloride polymerization wherein the Smith-Ewart recursion formula is solved by considering only latex particles N_0, N_1, and N_2.

When m is increased beyond 0.1 the radical distribution is not significantly changed as long as $\bar{n}$ is in the range 0.02 or less. At higher values of $\bar{n}$, however, the distribution of radicals broadens considerably. At $\bar{n} = 0.1$, for example, 18% of occupied particles contain two or more radicals while 1.3% contain 3 or more radicals. At this distribution about 90% of termination occurs in particles containing two radicals and 9% of termination occurs in particles containing three radicals. Under these conditions the polymerization is outside the range of validity for Ugelstad's N_2 model as well as Smith and Ewart's Case I model.

This mathematical analysis combined with our experimental findings and survey of vinyl chloride literature suggests that at very high particle numbers (10^{19} - 10^{20} particles/liter) a limiting case is reached wherein $\bar{n} < 1/2$ and radicals are distributed so that no particle contains more than 1 radical. This limiting case is evidenced in styrene polymerizations at high particle numbers. It is also evidenced in Ugelstad's data (8) for persulfate initiated polymerization at the highest particle numbers studied. Calculations of radical distribution confirm this, showing N_2 population to be negligible for all m when $\bar{n} < 10^{-3}$.

The radiation polymerization of vinyl chloride is also described by this limiting case. The radiation polymerization of vinyl chloride, however, is somewhat different in character from chemically initiated systems. These differences are specifically

1. The initiating species are predominantly .OH and H. radicals which impart much less surface activity than

the negatively charged sulfate or sulfonate radicals from persulfate initiators. These are faster in diffusing and presumably more efficiently absorbed by particles.

2. Reaction temperature is typically 20-25°C compared to 50°C or more for chemically initiated systems. This means lower chain transfer activity and lower aqueous monomer solubility.
3. The ionic strength of the aqueous phase is lower because no inorganic salts are required. This should increase latex stability over persulfate catalyzed polymerizations. More and smaller particles will be stabilized yielding faster rates of absorption and particle termination.

The combined effect of these special characteristics is to (1) enhance the rate of particle phase termination relative to the rate of radical absorption and desorption (i.e. reduce m and α'); (2) shift the equilibrium between aqueous and particle phase radicals by reducing the level of aqueous radical activity. The result, we believe, can be depicted by an $\bar{n}$ vs. α' relationship such as curve 2 in Figure 11. The situation is analogous to the styrene example where $\bar{n} < 1/2$ except that conditions for Case II are never met.

As stated above, this situation for $\bar{n} < 1/2$ is a limiting case of the Smith-Ewart recursion formula where only N_0 and N_1 need be considered. The rate expression for this can be derived from differential mass balances on N_1 and total radical population, P_T. For N_1 a steady state is assumed where

$$(dN_1)/(dt) = (k_a P_w N) - (k_d/r)N_1$$

for $N \sim N_0 >> N_1$ and where r is the particle radius and P_w the concentration of radicals in the aqueous phase

Assuming rapid termination a steady state radical population can be described by

$$(dP_T)/(dt) = R_i = 2k_a P_w N_1 \sim 0$$

It is assumed that aqueous phase termination is negligible. The expression $k_a P_w N_1$ assumes termination for each radical entry to an occupied particle. Combining these equations to solve for N_1 gives

$$N_1 = \left(\frac{N\ R_i}{2k_d/r}\right)^{1/2}$$

If aqueous polymerization is taken to be negligible then the rate of polymerization can be expressed:

$$R_p = k_p M_p \left(\frac{N\ R_i}{2k_d/r} \right)^{1/2}$$

At a fixed monomer to water ratio $N \alpha 1/r^3$, this leads to $R_p \propto N^{1/3}$.

Although the rate equation predicts a very slight autoacceleration proportional to $r^{1/2}$, this would not be detectable experimentally except possibly at low conversions. This is in good accord with the apparently linear conversion curves found in this study (Figure 1). Some autoacceleration is found in our data but this is only at low particle numbers and high conversion. This behavior is believed to be a form of gel-effect within larger latex particles.

The dose rate dependence found in this study is 0.40, identical to that predicted from Smith-Ewart Case II theory. The Smith-Ewart prediction, however, arises from a nucleation step wherein

$$N \propto R_i^{0.40}$$

A 0.4 initiator dependence arises when this is combined with a rate expression of the form

$$R_p = k_p M_p \bar{n} N$$

The rate of initiation has no effect on $\bar{n}$ which is considered a constant equal to 1/2. Thus, if the particle number is held constant and the rate of initiation is varied, one finds that rate is zero order with respect to initiator.

The case for vinyl chloride emulsion in this study is precisely the opposite of this because in the nucleation stage the particle number appears to be independent of dose rate while $\bar{n}$ varies according to the expression $\bar{n} \propto R_i^{0.4}$.

The small discrepency in initiation rate dependency between radiation (0.4) and chemically initiated systems (0.5) is interpreted as an artifact of the radiation initiation process. We have interpreted this in terms of a loss of initiation efficiency due to a low concentration of monomer in the aqueous phase. A similar effect has been observed in a study by Acres and Dalton (19) where the intensity exponent was shown to be directly related to monomer concentration.

The particle number dependency expected from the rate equation is 0.33. This is in good accord with the results of this study (Figure 8) and , significantly, corresponds well with the upper limit in the data of Ugelstad (8) for persulfate initiated polymerization. The particle number was shown to be approximately proportional to the soap concentration and independent of the initiation rate. This differs considerably from the .6 power soap dependency and .4 power initiator dependency found by Smith and Ewart. We suggest this difference is the result of radical desorption during the nucleation stage.

According to Smith-Ewart theory, once a primary radical is captured by a micelle, that micelle grows continuously throughout the nucleation stage and absorbs surfactant from inactive micelles. Growth continues until all micellar surfactant is adsorbed by newly formed polymer particles. Particle number is dependent on initiation rate because the continuing growth of the active micelles places a limit on the nucleation time period. Thus, the higher the initiation rate the greater will be the number of micelles activated in the allotted time.

In polymerizations where desorption is very high, all micelles grow simultaneously. This is because radical absorption is followed almost immediately by desorption with only a brief micellar growth period. As a result, all micelles experience an equal but intermittant growth. The number of latex particles is determined solely by the number of micelles initially present. The number of micelles, m, is given by the expression

$$m \propto (s-s_c)$$

where s is the surfactant concentration and s_c is the critical micelle concentration. Thus, we find the observed first order relationship between surfactant concentration and particle number. In addition, since all micelles grow simultaneously and at approximately the same rate, the initiation rate has little effect on the number of growing micelles but serves only to determine the duration of the nucleation stage.

Finally, the experimental values of $\bar{n}$ (from the polymerization rates) are plotted against the particle volumes in Figure 12. With the largest size particles the assumptions in the Case I model begin to fail and $\bar{n}$ begins to increase rapidly with particle size. Specifically, the effect of increasing particle size is to reduce desorption and radical termination and, thereby, increase the probability of a particle being occupied by more than one radical. Further evidence of this trend appears in Figure 8. The slope of the log R_p vs. log N curve shown in this figure deviates from the expected 0.33 dependency at the lowest particle numbers studied.

Acknowledgments

We would like to thank Mr. E. Oda for his contributions to the initial phases of this work and the U. S. Atomic Energy Commission, Division of Isotopes Development for their financial support.

Abstract

The emulsion polymerization of vinyl chloride, initiated by gamma radiation from a Cobalt-60 source, has been studied in detail. Good conversions to high yields were obtained.

The dependence of the rate of polymerization was to the 0.4

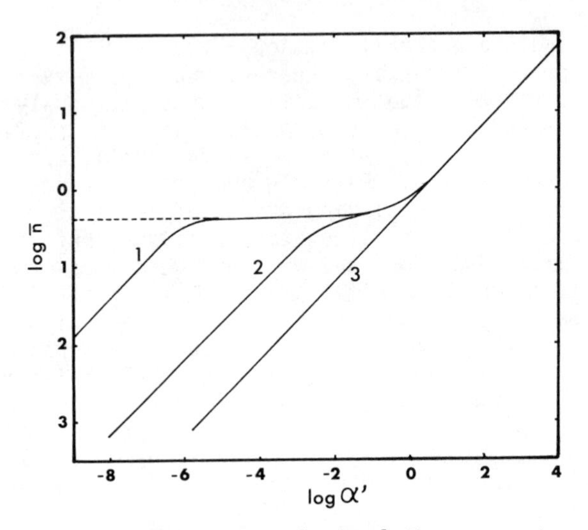

Figure 11. Solutions of the Smith–Ewart recursion equation for the case of no aqueous propagation or termination. Dotted line m = 0 (Smith–Ewart Case II). Curve 1 (m = 10^{-6}) depicts typical styrene-like polymerization. Curve 2 (m = 0.01) depicts radiation initiated emulsion polymerization of vinyl chloride. Curve 3 (m > 1.0) depicts chemically initiated emulsion polymerization of vinyl chloride.

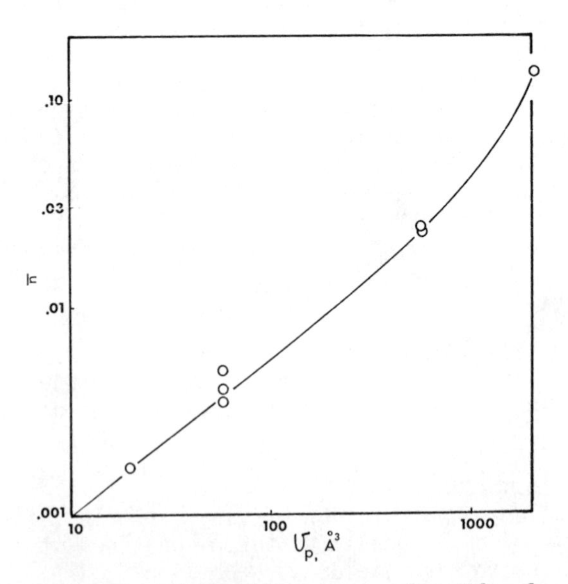

Figure 12. Calculated $\bar{n}$ vs. measured particle volume, showing deviation from the simplified Smith–Ewart Case I model at large particle volumes

power on the dose rate and to the 0.3 power on the emulsifier concentration. The number of particles was constant from 35% conversion to completion and was dependent to about the 1.0 order on the emulsifier concentration but was essentially independent of the dose rate.

The molecular weights were roughly constant at a given temperature and about equal to those calculated from the chain transfer to monomer values. Some post effects were observed.

The results have been interpreted along the lines of a modified Smith-Ewart Case I kinetic model.

Literature Cited

1. Stannett, V., Gervasi, J. A., Kearney, J. J., and Araki, K., J. Appl. Poly. Sci. (1969), 13, 1175.
2. Henderson, N. K., Yamakawa, S., Stannett, V., and Bengough, W. I., J. Macromol. Sci. - Chem. (1975), A9(3), 415.
3. Talamini, G. and Peggion, E., Ch. 5, "Vinyl Polymerization," ed. Ham, G., Vol. I, Part I, Marcel Dekker, Inc., New York, N. Y., 1967.
4. Ugelstad, J., Lervik, H., Gardinovacki, B., and Sund, E., Pure and Appl. Chem. (1971), 26(2), 121.
5. Giskehaug, K., Soc. Chem. Ind. (London)(1966), Monograph Number 20, 235-248.
6. Gatta, G., Benetta, G., Talamini, G., and Vianello, G., A.C.S. Adv. in Chem.(1969), Series 91, 158.
7. Peggion, E., Testa, F., and Talamini, G., Macromol. Chem. (1964), 71, 173.
8. Ugelstad, J., Mork, P. C., Dahl, P., and Rangnes, P., J. Poly. Sci. (1969), C27, 49.
9. Danusso, F., Moraglio, G., and Gazzera, S., Chem. Ind. (Milan) (1954), 36, 883.
10. Ishigure, K., O'Neill, T., Stahel, E. P., and Stannett, V., J. Macromol. Sci. - Chem. (1974), A8(2), 353.
11.a Litt, M., Patsiga, R. A., and Stannett, V., J. Poly. Sci. (1970), A1(8), 3607.
11.b O'Neill, T., Pinkava, J., and Hoigne, J., Proc. Third Tihany Symposium on Radiation Chem. (1971), 1, 713, Budapest (see also Reference 1).
12. "Polymer Handbook," 2nd Edition, ed. J. Brandup and E. H. Immergut, John Wiley, New York, N. Y., p. II-60, 1974.
13. Smith, W. V. and Ewart, R. H., J. Chem. Phys. (1948), 16, 592.
14. Stockmayer, W. H., J. Poly. Sci. (1957), 24, 314.
15. O'Toole, J., J. Appl. Poly. Sci. (1965), 9, 1291.
16. Ugelstad, J., Mork, P. C., and Aasen, J. D., J. Poly. Sci. (1967), A1(5), 2281.
17. Gardon, J. L., J. Poly. Sci. (1968), A1(6), 665, 687, and 2853.
18. Smith, W. V., J. Am. Chem. Soc. (1948), 70, 3695.
19. Acres, G. J. K., and Dalton, F. L., J. Poly. Sci. (1963), A1(1), 3009.

11

Aqueous-Phase Polymerization of Butadiene Initated by Cobalt (III) Acetylacetonate

D. C. BLACKLEY and W. F. H. BURGAR

National College of Rubber Technology, The Polytechnic of North London, Holloway, London N7 8DB, England

The purpose of this paper is to give a brief summary of an investigation which has been carried out into the aqueous-phase polymerisation of butadiene initiated by cobalt(III) acetylacetonate in the presence of surfactants such as sodium dodecylbenzene-sulphonate and sodium decyl sulphate. It is intended that fuller details of the investigation will be published elsewhere in due course.

Several papers have appeared in the literature in recent years showing that certain metal acetylacetonates can function as initiators for the polymerisation of vinyl and diene monomers in bulk and solution (1 - 12). Results for the kinetics of bulk and solution polymerisation are consistent with the view that the reaction occurs by a free-radical mechanism. The usual free-radical kinetics are operative, but an unusual feature is that, in some cases, certain additives such as chlorinated hydrocarbons have an activating effect upon the reaction by inducing more rapid decomposition of the initiator (2,11,12,13). Other additives which have been reported as promotors for the polymerisation include pyridine(14) and aldehydes and ketones(15). The complexity of the reaction in the presence of such additives is evident from the fact that chloroform has been reported to be an inhibitor for the polymerisation(3).

Kastnig and his co-workers(2) appear to have been the first to report that metal acetylacetonates activated by chlorinated hydrocarbons are capable of acting as initiators for the emulsion polymerisation of butadiene. Unfortunately the polymerisation recipes and rates of polymerisation were not published, but the data given for the structure of the polybutadienes obtained indicate that some degree of stereoregulation can be effected by this type of initiating system. This conclusion has not been confirmed by subsequent investigations.

Blackley and Matthan(16) have reported that iron(III) acetylacetonate is a very effective initiator for the aqueous-phase polymerisation of butadiene using sodium dodecylbenzenesulphonate as surfactant, but that little polymerisation occurs when isoprene is used as the monomer. No promoters were deliberately added to these reaction systems. Blackley(17) has reported that cobalt(III) acetylacetonate is also an effective initiator for the aqueous-phase polymerisation of butadiene in the absence of added promoters. In each case, the polymer obtained had essentially the microstructure which would be expected if the reaction occurred by non-stereoregulated free-radical polymerisation. In the case of the polybutadienes obtained using cobalt(III) acetylacetonate, the material appeared to be extensively crosslinked.

At the commencement of the investigation to which this present paper relates, the intention was to study in much greater detail than hitherto the catalysis of the aqueous-phase polymerisation of butadiene by iron(III) acetylacetonate. It was soon found, however, that this is in fact a most unsuitable system for precise study. The catalytic ability of iron(III) acetylacetonate seems to depend very much upon the method of preparation, and, indeed, considerable batch-to-batch variations have been observed between materials made by nominally identical procedures and from the same starting materials. Attempts to purify this compound by techniques such as recrystallisation and vacuum sublimation have led to even greater variations in catalytic activity. It should be added that the many samples of the chelate which have been prepared have frequently differed in appearance from one another, and that a broad correlation has been observed between appearance and catalytic activity. It appears that the darker, reddish-mauve variants of the compound appear to be more active than the less highly-coloured variants. It seems that Blackley and Matthan(16) had fortuitously used one of the more active variants in the earlier investigation.

In view of the difficulties encountered using iron(III) acetylacetonate, attention was turned to the cobalt analogue. This compound was known to show some catalytic activity. Furthermore, a simple preparative procedure is available which seems to give a well-defined crystalline product which is reproducible from batch to batch, both as regards appearance and catalytic ability. For this reason, a detailed investigation of systems containing this initiator has been undertaken as, in effect, a "model" for aqueous-phase diene polymerisations using this class of initiator.

In the initial stage of the investigation which this paper summarises, a range of experiments was carried out using beverage bottles as reaction vessels. It was during the course of these experiments that the variability of iron(III) acetylacetonate was discovered, and also the effectiveness of the cobalt analogue. Other indications from this early series of experiments were as follows:

(1) Polymerisation in the monomer phase occurs to a negligible extent. Most of the polymerisation takes place in the aqueous phase, the presence of which is essential for the occurrence of reaction.

(2) The polymerisation rate depends very much upon the initial pH of the aqueous phase.
(3) The polymerisation rate depends very much upon the nature of the surfactant used. In particular, reaction occurs more readily in the presence of sodium dodecylbenzenesulphonate than in the presence of potassium oleate.
(4) Initiation is not due to the presence of adventitious impurities, such as peroxides, which might form a redox couple with the metal compound.
(5) Addition of mercaptans in an endeavour to suppress gel formation leads to severe retardation of polymerisation.

These preliminary experiments established that systems comprising essentially butadiene, water, sodium dodecylbenzenesulphonate and cobalt(III) acetylacetonate were suitable for detailed study. Later work showed that it was desirable to replace the alkylbenzenesulphonate surfactant by sodium dodecyl sulphate.

Materials

1. Cobalt(III) acetylacetonate. Cobalt(III) acetylacetonate was prepared by reacting cobalt(II) carbonate with acetylacetone in the presence of hydrogen peroxide at 90 - 95°C. The green solid which formed was filtered off, washed with water, and then dried for ten minutes at 110°C. The compound was purified by dissolving in boiling benzene, filtering, adding petroleum ether, and then cooling to -20°C. Deep green crystals formed; these were filtered off and dried under vacuum at 40°C. Finally, the compound was recrystallised from methanol, dried, and then stored in the dark under nitrogen. The product melted with decomposition at 205 - 207°C. Its composition agreed with the theoretical expectation.

2. Butadiene. Butadiene was obtained from two sources, namely I.C.I. and Air Products Ltd, in both cases the purity being approximately 98%. It was subjected to two distillations before being polymerised, once at room temperature and once at approximately -40°C.

3. Sodium dodecylbenzenesulphonate. The material supplied by Marchon Products Ltd under the name "Nansa HS 85/S" was used. This is stated to contain a mixture of sodium alkylbenzenesulphonates, the average chain length of the alkyl group being about twelve carbon atoms. The proportion of active material is said to be about 85%, the remainder comprising approximately 12% of sodium chloride and sulphate, 1% of hydrocarbon oil, and 3% of water.

The crude material was purified by a lengthy procedure in which the surfactant was dissolved in a mixture of isopropanol and water at 50°C, the solution treated with sodium metabisulphite to destroy peroxidic impurities, and then refluxed with saturated sodium carbonate solution. On cooling, the mixture separated into two layers, the lower of which contained the inorganic impurities and was in consequence discarded. Treatment with sodium carbonate was repeated twice. The alkaline top layer was then neutralised with the appropriate alkylbenzenesulphonic acid, supplied by Marchon Products Ltd. After dilution with water, petroleum spirit was added in order to extract hydrocarbon oil impurities. The aqueous solution of surfactant was then evaporated to dryness, and the final traces of water, petroleum spirit and isopropanol were removed by heating in a vacuum oven at 50°C to constant weight.

4. Sodium dodecyl sulphate. The material supplied by Marchon Products Ltd under the name "Empicol LX" was used. This is stated to contain some 85% of active material. The crude material was purified by first dissolving in hot methanol to form a 5% solution, and filtering to remove inorganic salts. The filtrate volume was then reduced to half by evaporation, under nitrogen. The solution was cooled to -30°C in order to separate the surfactant as a solid phase, which was then filtered off, washed thoroughly with methanol, acetone and diethyl ether in order to remove fatty materials. The solid was then powdered and dried in a vacuum oven at 40°C for ten minutes.

Experimental Procedure For Polymerisations

Polymerisations were carried out under vacuum in dilatometers containing a glass-covered agitation element which was actuated by external reciprocating magnets. The reaction system comprised two phases, one of which was principally water and the other principally monomer. The initiator partitioned itself between these two phases. Although a surfactant was present in the aqueous phase, the rate of agitation within the dilatometers was insufficient to maintain a stable monomer emulsion. It was, however, sufficient to ensure that the aqueous phase remained saturated with monomer, as judged by the criterion that the rate of polymerisation was independent of the rate of agitation under the conditions of agitation employed. The polymerisation temperature was generally 50°C.

The following procedure was adopted for filling the dilatometers: The aqueous phase, which had previously been purged with nitrogen, was introduced into the dilatometers and then out-gassed to 10^{-5}mm mercury pressure. Out-gassed butadiene was then distilled into the dilatometers. Finally, the dilatometers were purged with nitrogen, evacuated to a pressure of 10^{-5}mm mercury, and then sealed in the usual way. It will appear from results summarised below that considerable care is necessary in preparing the reaction systems if reproducible results are to be obtained.

The compositions of the various reaction systems used are given with the results to which they refer.

Results of Polymerisation Experiments

1. Effect of storage time and presence of oxygen. Prior to carrying out the polymerisation reaction, the reaction vessels were stored for varying lengths of time at -30°C. Early experiments using dilatometers which had been filled out and out-gassed in the presence of residual air rather than nitrogen, revealed unexpected variations in the rate of polymerisation. It was found that the rate of polymerisation appeared to correlate with the time of storage at -30°C before the reaction was initiated. A series of experiments was therefore undertaken to investigate this effect systematically. The results are shown in Figure 1, from which it is seen that the rate of polymerisation increases sharply with storage time over a period of days, eventually reaching a plateau after about ten days storage. The reaction rate in dilatometers filled under residual nitrogen rather than residual air is somewhat higher than the rate eventually attained in dilatometers filled under air; it also appears that in this case the rate is very much less dependent upon storage time. The indications are that oxygen acts as an inhibitor or retarder for this type of polymerisation, and that, in order for polymerisation to proceed at an appreciable rate, it is first necessary for residual oxygen to be removed by reaction with some component of the reaction system during the storage time.

That oxygen has a severe retarding effect upon the polymerisation is confirmed by the results summarised in Figure 2. For this experiment, a number of dilatometers containing a given reaction system were out-gassed in the conventional manner, and the dilatometers and vacuum line to which they were attached then filled with oxygen and slowly re-evacuated. The dilatometer taps were closed in turn, so that each reaction system contained oxygen at a different, but known pressure. The dilatometers were then sealed. Two dilatometers containing the same reaction system were thoroughly purged with nitrogen before out-gassing and sealing. The results obtained show clearly the sensitivity of the reaction rate to residual oxygen, and also give an indication of the reproducibility in reaction rate which is attainable when the reaction system is purged with nitrogen before being out-gassed.

The results reported here provide the reason for adopting the rather elaborate procedure described above for the preparation of most of the reaction systems used in this investigation. It was considered essential to eliminate the complex effects associated with the presence of adventitious oxygen, and the evidence presented here suggests that this objective can be attained by using the procedure described.

2. Effect of aqueous phase: monomer ratio. A series of experiments was carried out to investigate the effect upon polymerisation rate of varying the ratio of aqueous phase to monomer phase, the composition of both phases being kept nominally constant.

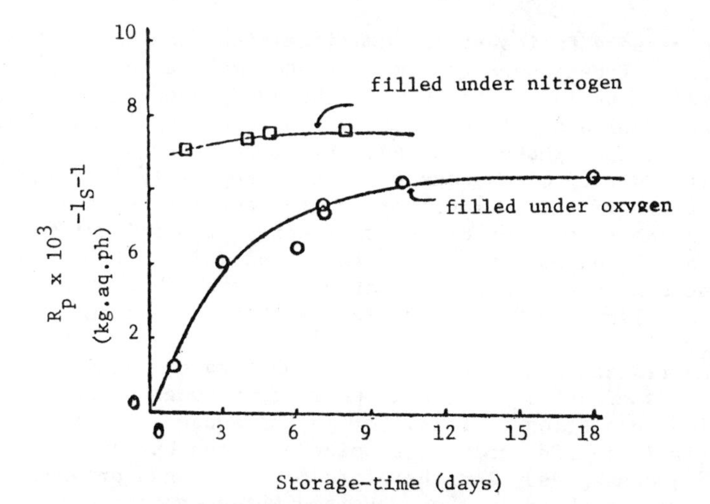

Figure 1. Dependence of the polymerization rate, R_p, on "storage time" (time elapsing between sealing the dilatometer and placing it in the water bath at 50°C to start polymerization). Recipe: 500 g of butadiene and 10 g of C_o (III) (acac)$_3$ per 1000 g aqueous solution, the latter containing 25 g/l. sodium alkyl benzenesulfonate and with the pH adjusted to 10.1 with NaOH.

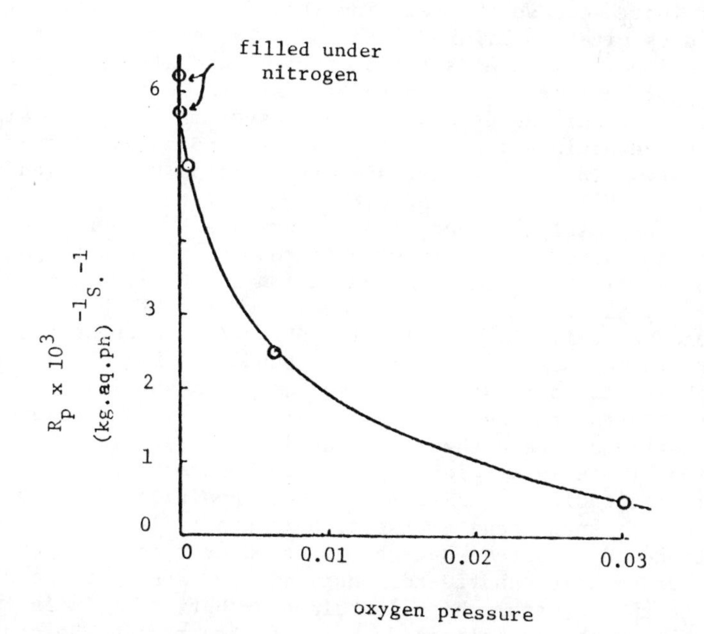

Figure 2. Dependence of the polymerization rate, R_p, on the pressure of oxygen in the dilatometer. Recipe same as Figure 1.

It is necessary to insert the qualification "nominally", because the actual situation is to some extent complicated by the partitioning of the initiator between the two phases.

The results are given in Figure 3. They show that, except at very low aqueous phase: monomer ratios, the rate of polymerisation per unit quantity of aqueous phase is essentially independent of the phase ratio. This implies that the total rate of conversion of butadiene to polybutadiene is directly proportional to the volume of the aqueous phase, if the compositions of the two phases are kept constant. This observation is consistent with the view that the polymerisation takes place within the aqueous phase of the system.

The reduction in rate per unit quantity of aqueous phase which occurs at low ratios of aqueous phase to monomer phase may be due to serious depletion of initiator in the aqueous phase. The initiator is considerably more soluble in the butadiene phase than in the aqueous phase, and therefore may have been present largely in the monomer phase in those systems which contained large volumes of butadiene.

3. Effect of pH. The profound effect which the initial pH of the aqueous phase has upon the rate of polymerisation is illustrated by the results summarised in Figure 4. Little polymerisation is observed when the pH is below about nine. As pH is increased above ten, so the reaction rate increases rapidly, passing through a maximum in the region pH 11 - 12. As the pH is further increased, so the reaction rate falls somewhat. Although the results given in Figure 4 are for a system which contained sodium dodecyl sulphate as the surfactant and also a certain quantity of an inorganic electrolyte (sodium sulphate), very similar behaviour has been observed in systems which contain sodium dodecylbenzenesulphonate as surfactant and no added inorganic electrolyte. In these latter systems, some polymerisation appeared to occur at pH's below eight, although the rate was low. As when sodium dodecyl sulphate was used as surfactant, the rate was observed to increase sharply with increasing pH, and to pass through a definite maximum in the region of pH 11.

As expected, the pH of the latex produced by the reaction correlates closely with the initial pH of the reaction system. However, rather unexpectedly it was found (see Figure 4) that the final pH is always about two units lower than the initial pH. Similar effects are observed in persulphate - initiated emulsion polymerisations, where the reduction in pH is attributed to the formation of bisulphate ions by reaction between water molecules and sulphate radical ions formed by decomposition of the persulphate. No such ready explanation is available in this instance.

It should be noted that the reaction systems to which these results refer were unbuffered. Many of the later rate measurements were made upon systems which contained pH buffers. It is therefore presumed that no significant drop in pH occurred in the course of

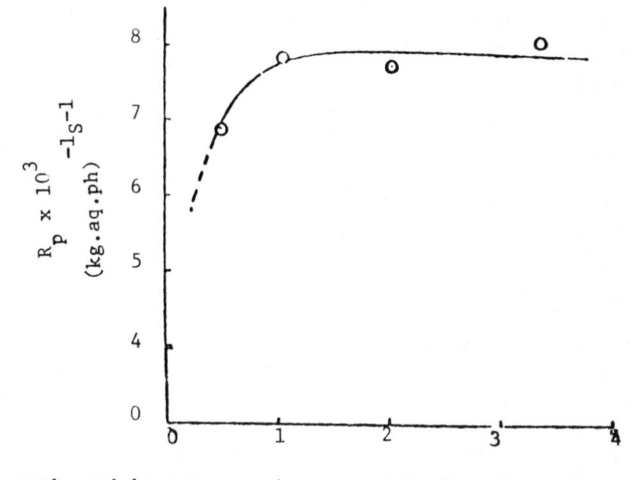

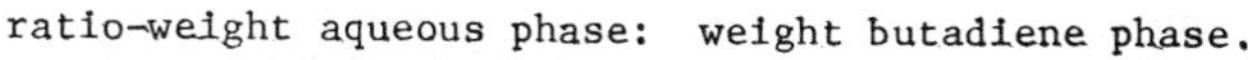

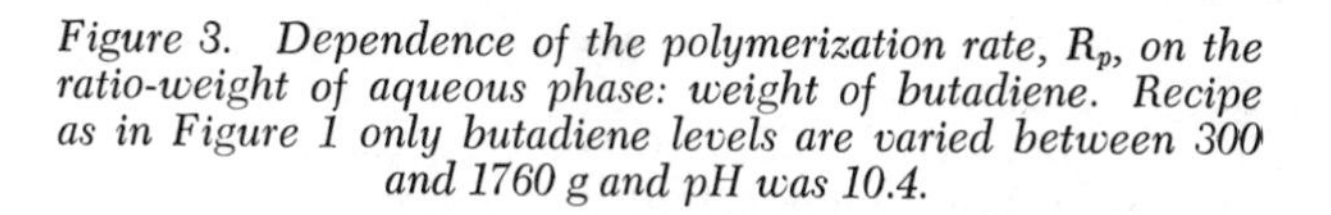

Figure 3. Dependence of the polymerization rate, R_p, on the ratio-weight of aqueous phase: weight of butadiene. Recipe as in Figure 1 only butadiene levels are varied between 300 and 1760 g and pH was 10.4.

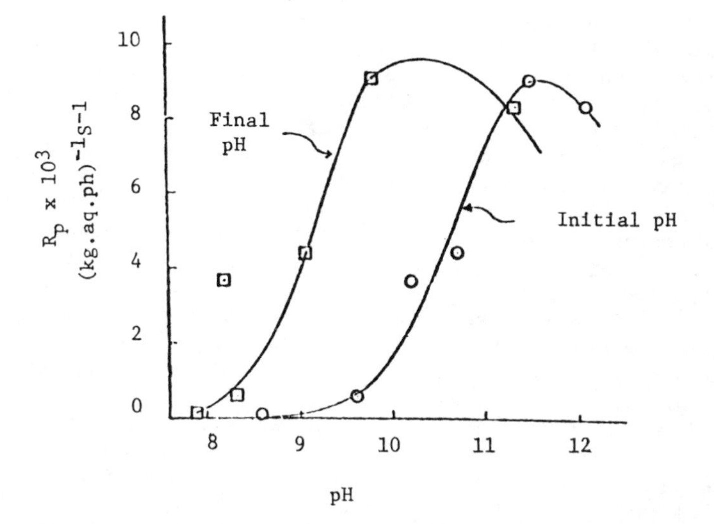

Figure 4. Dependence of the polymerization rate, R_p, on the pH of the aqueous phase at constant surfactant concentration. Recipe analogous to that of Figure 1 except the aqueous solution contained 10 g/l. Na_2SO_4, and the pH varied.

these latter reactions, although no check has been made to establish if this is in fact the case.

4. Effect of surfactant level. Much of the earlier work on this reaction system was carried out using sodium dodecylbenzenesulphonate as the surfactant. In the course of experiments which were intended to investigate the effect upon polymerisation rate of varying the surfactant level, it was discovered that purified sodium dodecylbenzenesulphonates are apparently able to act as initiators of free-radical emulsion polymerisation in the absence of other added initiating substances. Furthermore, as the results summarised in Figure 5 show, the rate of polymerisation in the absence of added initiator is directly proportional to the concentration of sodium dodecylbenzenesulphonate in the aqueous phase. The reason for this behaviour is obscure; it is not apparently due to the presence of peroxides in the surfactant. However, it is perhaps significant that an emulsion polymerisation reaction in which the rate of polymerisation is first-order with respect to surfactant level is consistent with Smith-Ewart "Case 2" kinetics for a system in which the surfactant functions as an initiator as well as a micelle generator.

Because of possible complications arising from the surfactant acting as an initiator, subsequent experiments were carried out using sodium dodecyl sulphate in place of sodium dodecylbenzenesulphonate. This surfactant does not appear to act as an initiator. Figure 6 summarises the variation of rate of polymerisation with level of sodium dodecyl sulphate at various pH levels. Two effects are immediately apparent from these results. In the first place, it is clear that the rate at any given surfactant level rises sharply with increasing pH over the range pH 9 - 11. But secondly, it appears that, as the surfactant level is raised, so the rate of polymerisation tends to pass through a shallow maximum. The effect is especially noticeable at high pH. This result was unexpected, but has been observed in several series of experiments. It is believed that the maximum arises from the operation of two conflicting tendencies. Both tendencies are believed to arise from the increased number of micelles which are present as the surfactant concentration is increased. On the one hand, an increased number of micelles implies an increased number of potential reaction loci and therefore an enhanced rate of polymerisation. On the other hand, as will appear from solubility and decomposition studies to be discussed below, the amount of initiator solubilised within the micelles, and thereby rendered inactive, will increase as the number of micelles increases. The rate of initiation is therefore reduced as the surfactant concentration increases. It is believed that the maximum in the variation of polymerisation rate with surfactant concentration arises from the balance of these two tendencies.

It is perhaps significant that, at a given pH and low surfactant concentration, the variation rate of polymerisation with

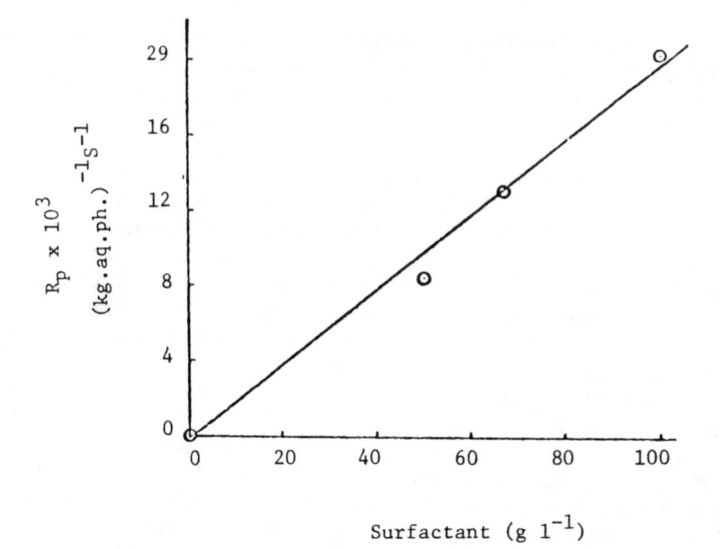

Figure 5. Dependence of the polymerization rate, R_p, on the surfactant concentrations, S, in the absence of cobalt (III) acetylacetonate. Recipe as for Figure 1 except the surfactant concentration is varied at pH 11.3.

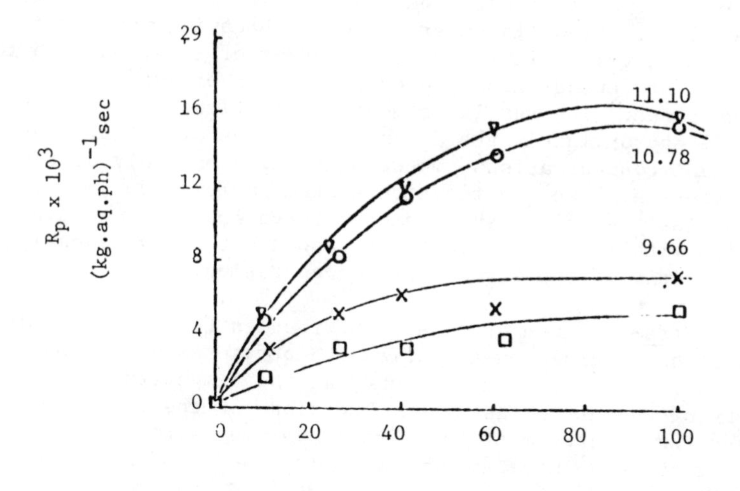

Figure 6. Dependence of the polymerization rate, R_p, on surfactant concentration, S, at constant pH (values shown alongside curves). Recipe as in Figure 1 except the aqueous solution contained 10 g/l. Na_2SO_4 and 2.1 g/l. $NaHCO_3$. The NaOH and surfactant content varied.

surfactant concentration is consistent with an order of reaction of 0.6 (see Figure 7). This order of reaction is in turn consistent with a micellar initiation mechanism of the type envisaged by the Smith-Ewart theory.

5. Effect of adding inorganic electrolyte. One of the principal impurities likely to be associated with surfactants such as sodium dodecyl sulphate and sodium dodecylbenzenesulphonate is inorganic electrolyte such as sodium sulphate. It was therefore considered desirable to investigate the effect of such salts upon the rate of polymerisation. The results of this series of experiments are summarised in Figure 8. Sodium dodecyl sulphate was the surfactant and sodium sulphate the additive. The results show that the reaction rate rises sharply with increasing electrolyte addition until a plateau is reached. Thereafter, further additions of electrolyte have little effect upon the rate of polymerisation. It therefore appears that, if it is desired to obtain reproducible rates of polymerisation with this type of reaction system, it is preferable to control both the pH and the ionic strength of the aqueous phase. These two objectives are most readily achieved by adding buffers to the system. A buffer comprising sodium carbonate and sodium bicarbonate in an appropriate ratio has been found to be suitable.

6. Effect of initiator level. The effect of increasing amounts of cobalt(III) acetylacetonate upon the rate of polymerisation is complex. Whilst the rate always appears to increase as the initiator level increases (in contrast to the behaviour observed when the surfactant level is increased), the order of reaction with respect to initiator depends upon the concentration of surfactant in the aqueous phase. The results summarised in Figure 9 show that the order is approximately the Smith-Ewart value of 0.4 at high surfactant concentrations, whereas it falls markedly as the surfactant level in the reaction system is lowered. Figure 10 illustrates the rather surprising observation that the order of reaction with respect to initiator appears to vary linearly with the logarithm of the surfactant concentration.

7. Effect of temperature. The results for the determination of rates of polymerisation at various temperatures are summarised in Figure 11, from which it is seen that the temperature variation of rate is of the Arrhenius type. The slope of the relationship between the logarithm of the rate of polymerisation and the reciprocal of the absolute temperature corresponds to an overall activation energy of polymerisation of 15.8 kcal $mole^{-1}$. This value is very close to that which has been reported for the aqueous -phase polymerisation of butadiene initiated by rhodium chloride(16). However, in view of the marked dissimilarity in reaction product, and therefore presumably in reaction mechanism, it seems likely that the correspondence is fortuitous.

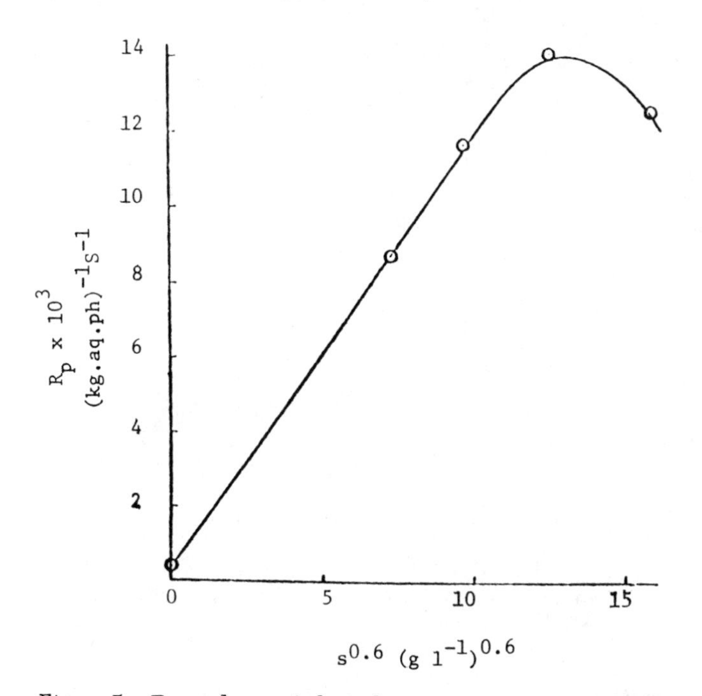

Figure 7. Dependence of the polymerization rate, R_p, on $S^{0.6}$. Recipe as in Figure 1 but with 10 g/l. Na_2SO_4, 2.1 g/l. $NaHCO_3$ at pH 10.9, with surfactant content varied between 0–100 g/l.

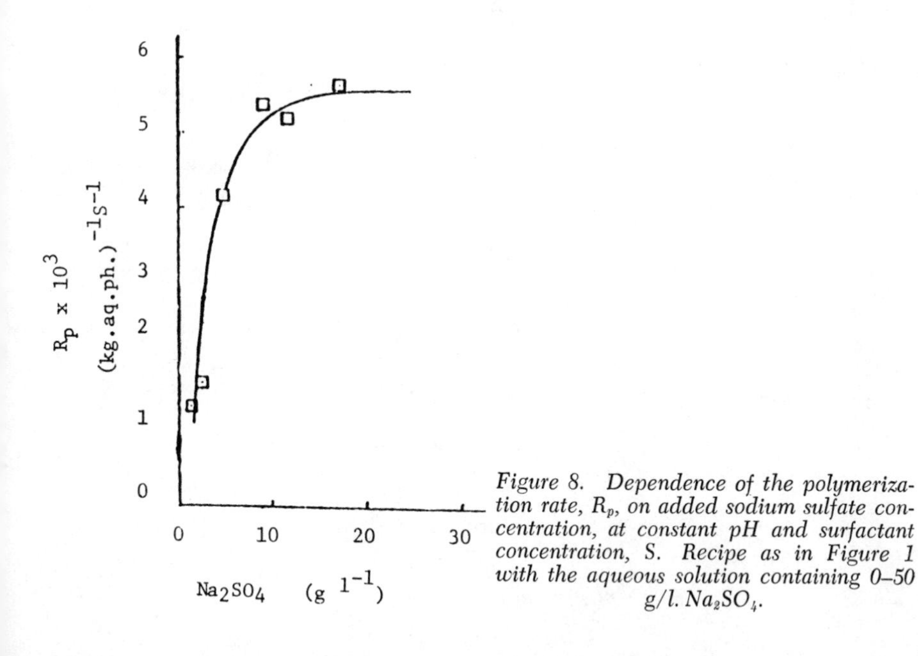

Figure 8. Dependence of the polymerization rate, R_p, on added sodium sulfate concentration, at constant pH and surfactant concentration, S. Recipe as in Figure 1 with the aqueous solution containing 0–50 g/l. Na_2SO_4.

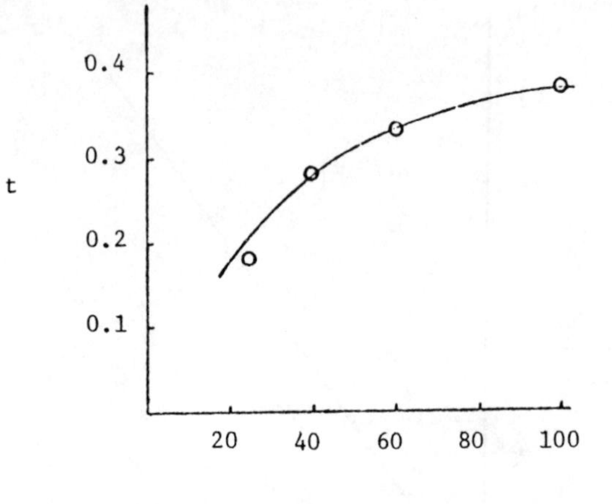

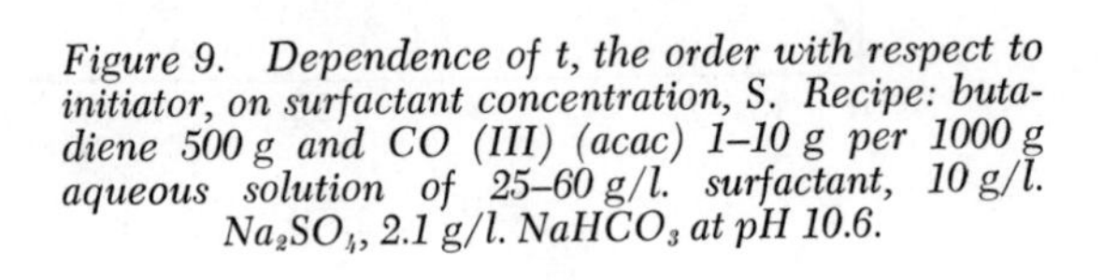

Figure 9. Dependence of t, the order with respect to initiator, on surfactant concentration, S. Recipe: butadiene 500 g and CO (III) (acac) 1–10 g per 1000 g aqueous solution of 25–60 g/l. surfactant, 10 g/l. Na_2SO_4, 2.1 g/l. $NaHCO_3$ at pH 10.6.

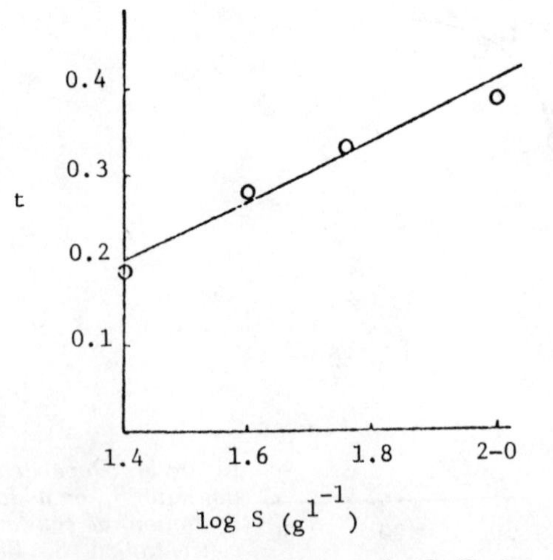

Figure 10. Dependence of t on log S. Recipe as in Figure 9.

Other Experiments

In an endeavour to seek a deeper understanding of some of the features of what is undoubtedly a rather complex reaction system, a number of ancillary experiments have been carried out. These include the following:

(1) studies of the kinetics of the decomposition of cobalt(III) acetylacetonate in various media;
(2) studies of the solubility of cobalt(III) acetylacetonate in aqueous media, and of its partitioning between aqueous and hydrocarbon media;
(3) experiments using C^{14-} labelled cobalt(III) acetylacetonate.

1. Decomposition kinetics of cobalt(III) acetylacetonate. The decomposition of cobalt(III) acetylacetonate in various media has been studied by following the change in concentration by means of ultraviolet spectrophotometry. The following facts have been established concerning the decomposition of this compound at 50°C:

(a) Little decomposition of cobalt(III) acetylacetonate occurs it it is dissolved in a non-polar solvent such as hexane. It appears that the presence of polar molecules such as water is essential if significant decomposition is to occur. It therefore seems likely that the aqueous-phase polymerisation of butadiene is initiated through the decomposition of cobalt(III) acetylacetonate which is dissolved in the true aqueous phase. That fraction of the initiator which is dissolved in the monomer phase is presumably ineffective. So also, presumably, is the fraction which is solubilised within the surfactant micelles, since the interiors of the latter are essentially non-polar in nature.

(b) In aqueous media, decomposition occurs at a rate which is first order with respect to the concentration of cobalt(III) acetylacetonate. Furthermore, the rate constant for the decomposition depends upon both the pH of the aqueous medium and the surfactant concentration. The effect of pH is shown in Figure 12, and that of surfactant concentration in Figure 13. In both cases, log c_t/c_o is plotted as a function of t, where c_o denotes the initial concentration of cobalt(III) acetylacetonate, and c_t denotes its concentration after a time t. The linearity of these plots shows that the decomposition of the initiator is first order over a wide range of reaction conditions. Figure 12 shows that the first-order rate coefficient for the decomposition rises sharply as the pH is increased from about 9.5 to 11.6. This observation implies that the initial step of the initiator-decomposition reaction is probably interaction with hydroxyl ions. It seems that here is the most likely explanation for the pronounced effect which the initial pH of the

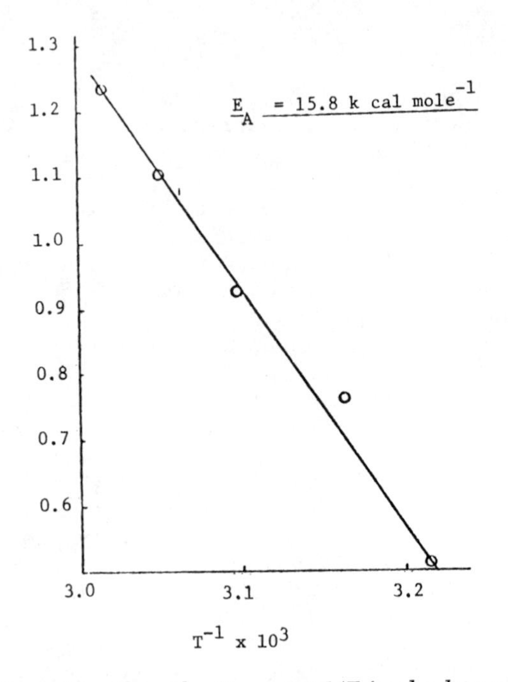

Figure 11. Plot of log R_p against $1/T$ for the determination of the overall activation energy of polymerization, E_A. Recipe differs from Figure 1 in that the aqueous solution contains 2.1 g/l. $NaHCO_3$ and pH 10.6.

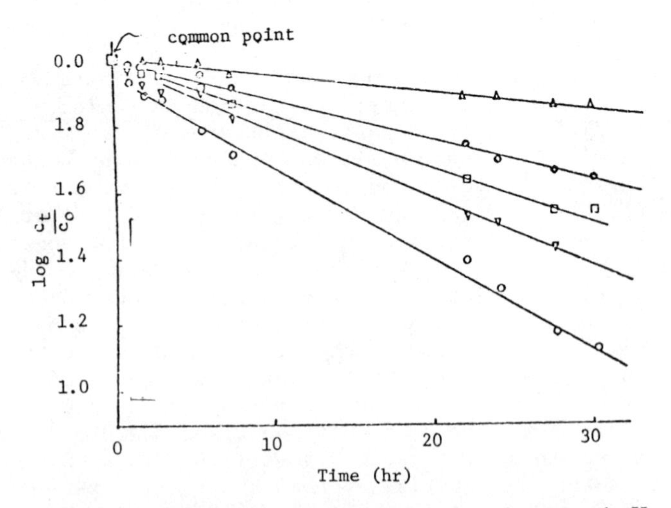

Figure 12. Plot of log C_t/C_o against time as a function of pH varied with buffer solutions; surfactant = 0 g/l. ; temperature = 60.5°C. The lines are associated, from top to bottom, with the following pH and $k_c \times 10^6$ (sec^{-1}) values: 11.62 and 18.6, 10.84 and 14.0, 10.5 and 10.9, 10.28 and 8.0, 9.56 and 4.1.

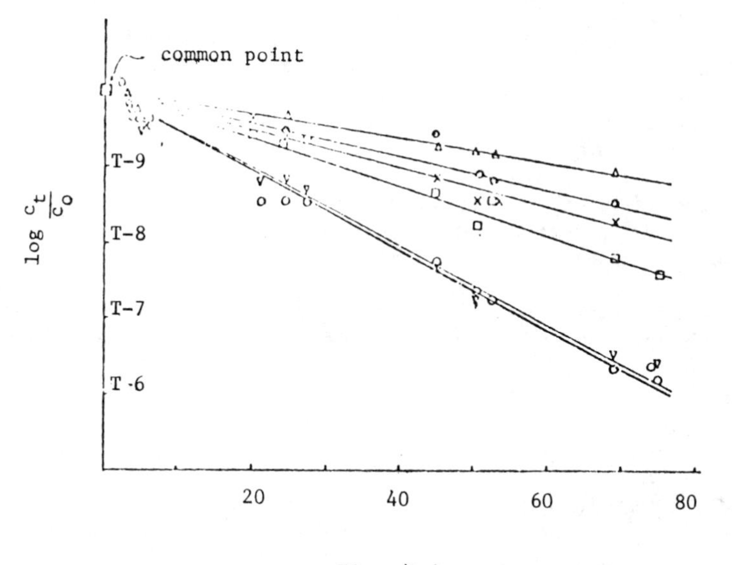

Symbol	0	▽	□	x	●	△
S (gl^{-1})	0	10	25	40	60	100
$k_d \times 10^6 (sec^{-1})$	3.31	3.20	2.01	1.60	1.37	0.94

Figure 13. Plot of log C_t/C_o against time as a function of surfactant concentration, S. Decomposition in buffered surfactant solutions; pH = 11.62; temperature =49.7°C.

aqueous phase has upon the rate at which the aqueous-phase polymerisation of butadiene takes place. Figure 13 shows that the effect of surfactant is to retard the decomposition of the initiator. The most obvious explanation for this effect is that an increased concentration of surfactant causes an increased amount of surfactant to be solubilised within the surfactant micelles, and therefore to be rendered unsusceptible to decomposition. Certainly, the results of solubility measurements which are reported below support the view that the surfactant does solubilise appreciable quantities of initiator. Whether or not this is the true explanation for the retardation of initiator decomposition by surfactant, the fact that the surfactant behaves in this way explains why the rate of the aqueous-phase polymerisation of butadiene passes through a maximum as the level of surfactant in the system is increased.

(c) The precise manner in which the first-order rate constant for the decomposition of cobalt(III) acetylacetonate in aqueous media at 50°C varies with the pH and the surfactant concentration is complex. Empirically, it has been found that the joint variation can be well represented by the equation -

$$k_d = 1.28 \times 10^{-5} \, [OH^-]^{0.24} \quad 1 - 0.15S^{0.5} \quad [OH^-]^{0.1}$$

where k_d denotes the first-order rate coefficient in sec^{-1}, and S denotes the surfactant concentration in grams per litre. It has not so far been possible to deduce any mechanistic implications from this equation, other than the qualitative implications which have been discussed in the previous paragraph.

Results for the variation of k_d with temperature are shown in Figure 14. Data are given for decomposition in aqueous media at several pH values in the absence of surfactant, and also for an aqueous medium at pH 10.84 to which sodium dodecyl sulphate had been added. It is clear that the variation of k_d with temperature follows the Arrhenius equation in all the cases studied, and that, in contrast to k_d itself, the energy of activation is relatively insensitive to both pH and the presence of surfactant. There is some indication that the energy of activation increases slightly with increasing pH, and that the addition of surfactant at a given pH reduces the energy of activation. However, taking the results as a whole, they are consistent with an energy of activation of 32.9 ± 1.6 kcal $mole^{-1}$. This appears to be approximately the energy required for the scission of the Co-O bond.

Spectroscopic evidence suggests that the initial product of the thermal decomposition of cobalt(III) acetylacetonate in aqueous media is cobalt(II) acetylacetonate, and therefore (presumably) an acetylacetone ligand also. The presence of the latter has not been detected, and so it is believed that it undergoes rapid decomposition. The presence of cobalt(II) acetylacetonate has been

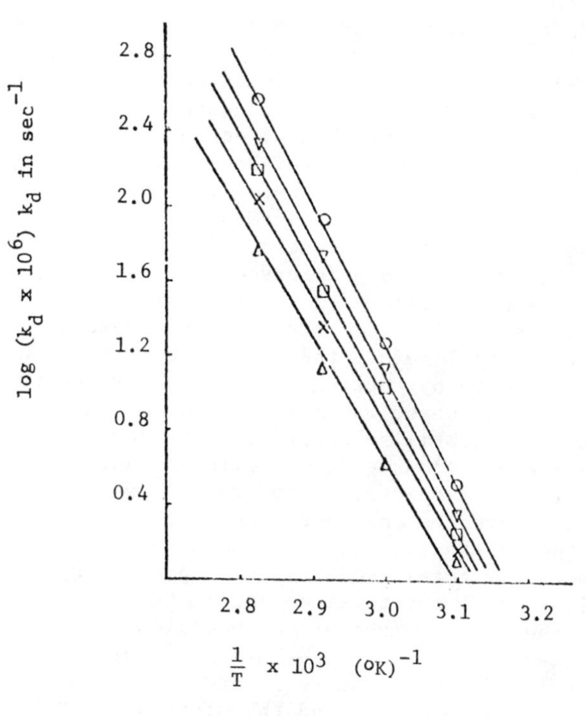

key	pH	s (gl^{-1})	E_D k cal $mole^{-1}$
0	11.62	0	34.5
	10.84	0	33.6
	10.50	0	32.7
X	10.84	25	31.3
	9.56	0	31.3

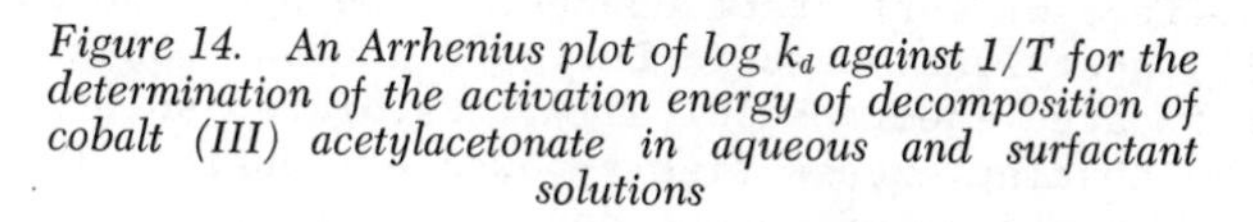

Figure 14. An Arrhenius plot of log k_d against 1/T for the determination of the activation energy of decomposition of cobalt (III) acetylacetonate in aqueous and surfactant solutions

detected, but this compound seems to decompose more rapidly than does the cobalt(III) analogue.

2. Partition and solubility measurements on cobalt(III) acetylacetonate. The following procedure was adopted for investigating the approximate manner in which cobalt(III) acetylacetonate partitions between butadiene and an aqueous phase containing sodium dodecyl sulphate: Known weights of aqueous phase and cobalt(III) acetylacetonate were placed in weighed 1-cm diameter glass tubes to the open end of which were attached glass cones. The tubes were attached to a vacuum line, the aqueous phase out-gassed, and then the required volumes of butadiene distilled in. The tubes were then sealed, re-weighed, and immersed in a water-bath at 50°C. The tubes were shaken and allowed to equilibrate for ten minutes. They were then plunged into liquid nitrogen to freeze the contents, and the tubes broken to liberate the frozen contents. The butadiene fraction was removed by cutting through the frozen cylinder with a knife. The aqueous phase was melted into a beaker, diluted to a known extent with distilled water, and the concentration of cobalt(III) acetylacetonate in the diluted aqueous solution determined by ultraviolet spectrophotometry. It was then a simple matter to calculate the concentration of the compound in the undiluted aqueous phase, and hence, knowing the weight of the aqueous phase, the total amount in that phase. By difference, it was possible to know the total amount in the butadiene phase, and hence, from the known weight of the butadiene phase, the concentration in that phase.

The results given in Figure 15 are for two aqueous phases, one of which had a pH of 8.5 and the other a pH of 11.1, and both of which contained the same concentration of sodium dodecyl sulphate. Two points are immediately apparent from these results: (a) the normal partition law is obeyed, at least over the range of concentrations investigated, and (b) the partition coefficient is independent of the pH of the aqueous phase. The first of these observations implies that cobalt (III) acetylacetonate has the same molecular complexity in both the aqueous and the butadiene phases. These results also show that the initiator partitions in such a way that its concentration in the butadiene phase is considerably greater than that in the aqueous phase; in fact, the partition coefficient for this particular aqueous phase at 50°C is approximately 6.54.

The effect of surfactant concentration upon the solubility of cobalt(III) acetylacetonate in buffered aqueous phases at 50°C was also investigated spectroscopically. Saturated solutions were prepared by shaking excess finely-powdered cobalt(III) acetylacetonate with solutions of sodium dodecyl sulphate for approximately one hour at 50°C. It appeared that one hour was sufficient time for the attainment of equilibrium.

The results of these experiments are shown in Figure 16, from which it is apparent that the saturation solubility of the initiator

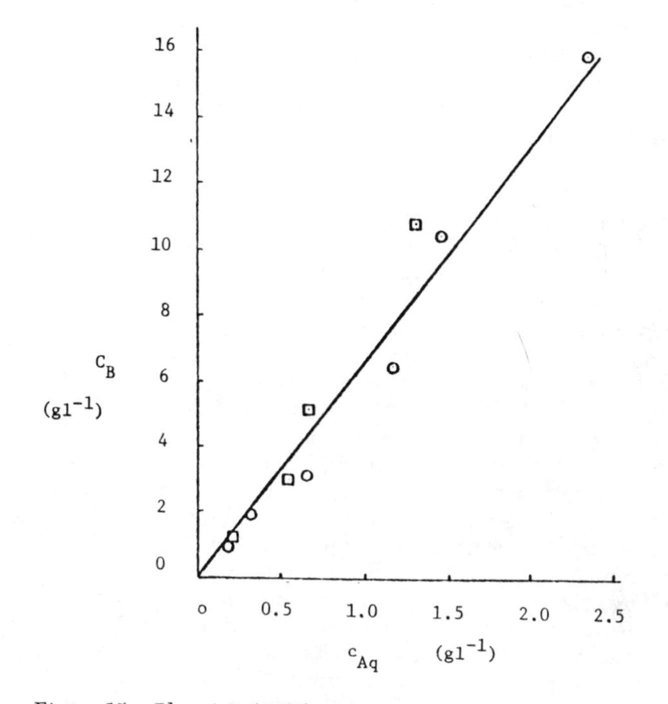

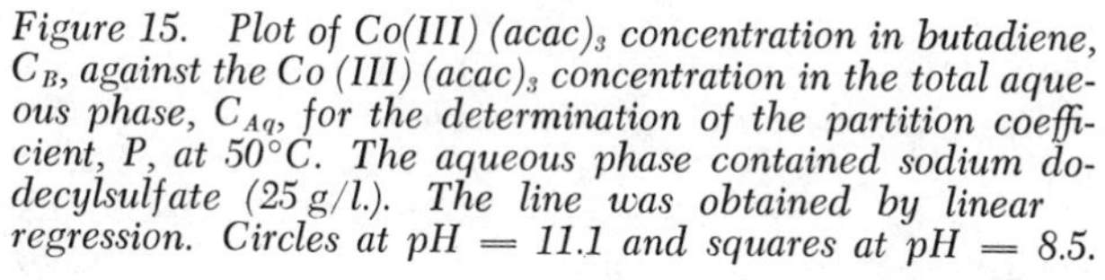

Figure 15. Plot of Co(III) $(acac)_3$ concentration in butadiene, C_B, against the Co (III) $(acac)_3$ concentration in the total aqueous phase, C_{Aq}, for the determination of the partition coefficient, P, at 50°C. The aqueous phase contained sodium dodecylsulfate (25 g/l.). The line was obtained by linear regression. Circles at pH = 11.1 and squares at pH = 8.5.

increases as the surfactant concentration is increased. This is most naturally interpreted as solubilisation of the initiator within the surfactant micelles. In order further to interpret these results, it is necessary to attempt to distinguish the weight of cobalt(III) acetylacetonate which is present in the surfactant micelles (denoted by W_M in Figure 16) from that which is present in solution in the true aqueous phase (denoted by W_A in Figure 16). The approach adopted was to assume that W_A is proportional to the volume of the true aqueous phase (which was calculated by subtracting the estimated micellar volume from the total volume of the aqueous phase), and then to obtain W_M by difference between W_A and the total weight of cobalt(III) acetylacetonate known to be present in unit volume of the aqueous phase. The interesting observation has been made that W_M is approximately proportional to the square root of the surfactant concentration. The results are given in Figure 17. Two sets of data are shown in this graph. The first set (the points) are for W_M as a function of the total surfactant concentration in the aqueous solution. The second set (the points) are for W_M as a function of what we have termed the "effective" surfactant concentration, that is, the actual surfactant concentration minus the minimum concentration of surfactant which is apparently necessary for solubilisation to occur at all. (For reasons which are not understood, this concentration is considerably in excess of the critical micelle concentration.) Clearly, the proportionality between W_M and the square root of the surfactant concentration is obeyed more exactly if the latter has been corrected by first subtracting that concentration of surfactant which apparent contributes nothing to the solubilising power of the solution. As far as is known, this is the first occasion upon which direct proportionality between the concentration of substance solubilised and root concentration of solubilising surfactant has been reported. Various models have been investigated in an attempt to explain this relationship, but without success. The solubilisation of cobalt(III) acetylacetonate by aqueous solutions of sodium dodecyl sulphate certainly appears to contrast with that of, say, hydrocarbon liquids by solutions of fatty-acid soaps, where the relationship between soap concentration and amount of liquid solubilised is more nearly one of direct proportionality.(18)

3. Experiments using C^{14}-labelled cobalt(III) acetylacetonate. Polymerisation experiments using C^{14}- labelled cobalt(III) acetylacetonate as initiator have failed to produce polymers which contain significant amounts of C^{14} activity. Whilst the observations which have been made are not entirely conclusive, they suggest that the initiation mechanism does not result in the acetylacetone ligand becoming chemically bound to the polymer.

Possible Mechanisms for Initiation of Polymerisation

As has been shown above, certain features of the aqueous-phase

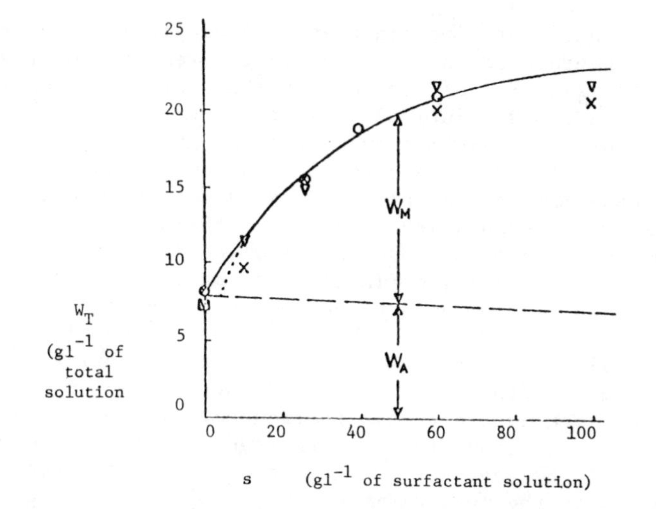

Figure 16. Dependence of the saturated solubility of Co (III) $(acac)_3$, W_T, in buffered surfactant solutions on the surfactant concentration. The concentration of Co (III) $(acac)_3$ soluble in the true aqueous phase, W_A, is illustrated by the dotted line. W_H is the concentration of Co(III) $(acac)_3$ soluble in the surfactant micelles. The pH values are as follows: △, 11.62; +, 10.84; □, 10.28; and ○, 9.56.

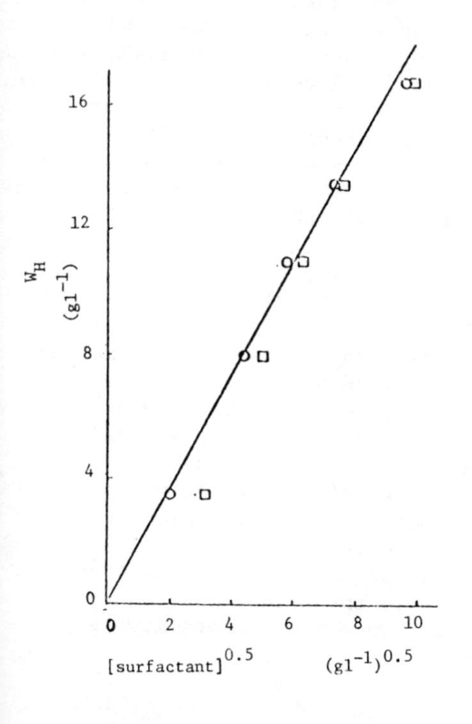

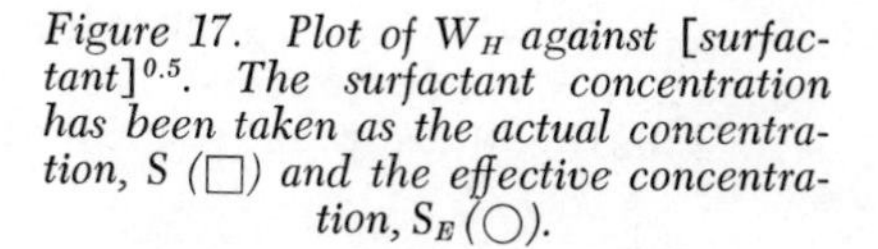
Figure 17. Plot of W_H against $[surfactant]^{0.5}$. The surfactant concentration has been taken as the actual concentration, S (□) and the effective concentration, S_E (○).

polymerisation of butadiene initiated by cobalt(III) acetylacetonate can be interpreted in the light of the results for the decomposition kinetics of the initiator and of its solubility behaviour in aqueous solutions of surfactants. There still remains the problem of the basic mechanism whereby the polymerisation reaction is initiated. There seems to be no reason for dissenting from a mechanism which was proposed by one of the present authors some years ago(17) and which agrees in certain respects with the mechanisms which have been proposed for the initiation of bulk and solution free-radical polymerisations by metal acetylacetonates. The free-radical activity is conceived as being generated by an "internal redox" reaction, in which the central metal ion is reduced to a lower valency state (in this case cobalt(II)) by an electron transfer from the ligand, the latter then acquiring a free-radical site by electronic rearrangement. Polymerisation then proceeds by propagation from the free-radical site so generated. In the case of the acetylacetonates, a drift of electrons from the ligand moiety to the central metal ion is presumably facilitated by the presence of a "quasi-aromatic" system of conjugated unsaturation within the ligand.

Even assuming that the "internal redox" reaction is the primary source of the radical activity through the agency of which the polymerisation reaction occurs, there remain many possible pathways by which initiation is effected in an emulsion system. These possibilities arise because of the presence of species, such as water and surfactant, with which the primary radical can interact. Three possibilities are illustrated in Figure 18. In each case, the radical activity is associated in the first instance with a species in which cobalt(III) has been reduced to cobalt(II) and the acetylacetonate ligand has rearranged to give a free radical on the methylenic carbon atom. In the first possibility, the monomer reacts directly with this species, and propagation then proceeds in the normal way. The consequence of such a mechanism would be that the polymer produced would contain both cobalt (albeit perhaps more loosely bound than in an acetylacetonate) and a moiety derived from acetylacetone. In the second possibility, the species which results from the internal redox reaction interacts with another molecule in the reaction system (such as water) in such a way that the radical-bearing entity is displaced from the metal complex. The displaced radical-bearing entity then reacts with monomer to initiate the polymerisation. The consequence of such a mechanism would be that an acetylacetone moiety, but not cobalt, would become chemically incorporated in the polymer produced. In the third possibility, the radical-bearing entity is displaced as in the second possibility, but undergoes a transfer reaction with, say, water molecules to produce, say, hydroxyl radicals which then become the effective initiating species. Such a mechanism is consistent with the production of a polymer which contains neither cobalt nor acetylacetone residues. To the extent that our experiments using C^{14}- labelled cobalt(III) acetylacetonate give any

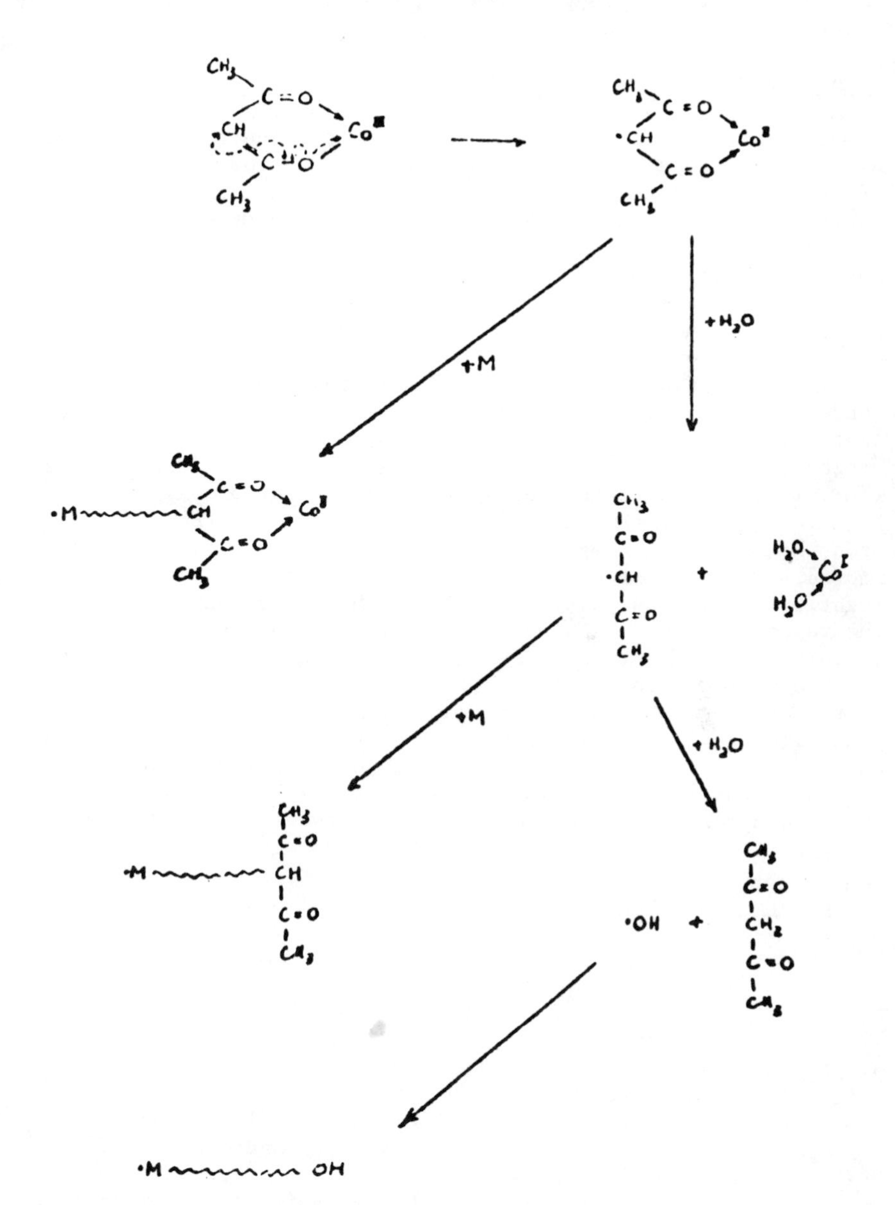

Figure 18. Possible mechanisms which cobalt (III) acetylacetonate initiates aqueuos-phase polymerization

indications at all, they suggest that a mechanism of the third type is probably operative.

Acknowledgements

One of us (W F H B) wishes to thank the Inner London Education Authority and The Polytechnic of North London for granting a period of study leave during which some of the work described in this paper was carried out. Thanks are also due to Dr E W Duck (The International Synthetic Rubber Company) for helpful discussions during its progress.

Abstract

A description is given of the main features of the aqueous-phase polymerisation of butadiene initiated by cobalt(III) acetyl-acetonate in the presence of surfactants such as sodium dodecyl-benzenesulphonate and sodium dodecyl sulphate. It is shown that the polymerisation rate increases as the amount of oxygen in the system decreases, and that, if oxygen is present, the rate depends upon the length of time for which the reaction system has been stored. The rate of polymerisation is directly proportional to the quantity of aqueous phase. Initial pH has a profound effect upon the reaction rate. The dependences of rate upon the concentrations of initiator and surfactant are complex. Results are presented for the decomposition of cobalt(III) acetylacetonate in aqueous solution in the presence and absence of surfactant, and also for the partitioning of the initiator between hydrocarbon and aqueous phases and its solubility in various aqueous phases. The information obtained from these ancillary investigations has consdierably enhanced understanding of some of the features of the polymerisation reaction.

Literature Cited

1. Arnett, E M, Morris, A, and Mendelsohn, M A, J. Am. chem. Soc., (1962) 84, 3821.
2. Kastnig, E G, Naarman, H, Reis, H, and Berding, C, Agnew. Chem. Internat. Edn., (1965) 4, 322.
3. Izawa, S, Shimizu, A, and Kubo, M, Sci. Rep. Toyo Soda Co., (1966) 10, 123.
4. Bamford, C H, and Lind, D J, Chem. Comm., (1966) 792.
5. Otsu, T, Minamii, N, and Nishikawa, Y, J Macromol. Sci. - Chem., (1968) 42, 905.
6. Isawa, Z, and Shibamiya, T, J. Polym. Sci. Part B, (1968) 6, 721.
7. Allen, P E M, and Goh, T H, Eur. Polym. J., (1969) 5, 419.
8. Allen, P E M, Ayling, D, and Goh, T H, Eur. Polym. J., (1970) 6, 1587.

9. Ghanem, N A, Yehia, A A, Moustafa, A B, and Rizk, N A, Eur. Polym. J., (1971) 7, 717.
10. Zafar, M M, Yousufzai, A H K, Mahmood, R, and Hussain, S A, J. Polym. Sci. Part A-1, (1972) 10, 1539.
11. Urehara, K, Nishi, T, Matsumura, T, Tamura, F, and Murata, N, Kogyo Kagaku Zasshi, (1967) 70, 191.
12. Urehara, K, Nishi, T, Matsumura, T, Tamura, F, and Murata, N, Kogyo Kagaka Zasshi, (1967) 70, 755.
13. Kaeriyama, K, and Yamazaki, Y, Kogyo Kagaku Zasshi, (1971) 74, 1718.
14. Uehara, K, Kataoka, Y, Tanaka, M, and Murata, N, Kogyo Kagaku Zasshi, (1969) 72, 754.
15. Kaeriyama, K, Bull. Chem. Soc. Japan, (1970) 43, 1511.
16. Blackley, D C, and Matthan, R K, Br. Polym. J., (1970) 2, 25.
17. Blackley, D C, Proceedings of 4th International Rubber Symposium, (1969) 15.
18. Stearns, R S, Oppenheimer, H, Simon, E, and Harkins, W D, J. Chem. Phys., (1947) 15, 496.

12

Emulsion Redox Copolymerization of Vinyl Ferrocene

F. LOUIS FLOYD

Glidden-Durkee Div., SCM Corp., Strongsville, Ohio 44136

Only recently has the field of metallocenes become of interest to polymer chemists. Ferrocene (I), being the oldest, most readily available, and best characterized of the class, was the logical choice for initial studies. Ferrocene and vinyl ferrocene (II) can be thought of as analogs of benzene and styrene, respectively, undergoing many of the same kinds of reactions, including polymerization of vinyl ferrocene. Two excellent reviews of the

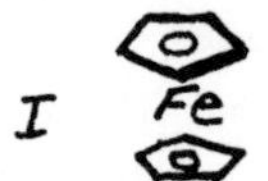

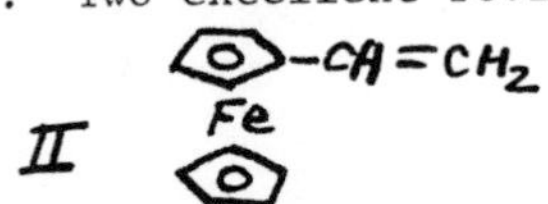

use of metallocenes in general, and ferrocene in particular, in polymers are those of Pittman(1) and Neuse(2).

Poly(vinyl ferrocene) was first prepared by Arimoto and Haven in 1955, (3) a process for which was patented in 1958 (4). Vinyl ferrocene was found to be much less reactive than styrene, but still readily polymerizable with azo initiators. Peroxide initiators were, in general, not effective, causing the ferrocene nucleus to oxidize instead (1).

Although there are numerous references to the emulsion polymerization of vinyl ferrocene, they all appear to emanate from a single source (4). These workers polymerized vinyl ferrocene alone, and with styrene, methyl methacrylate, and chloroprene. No characterization was reported other than elemental analysis. The molding temperatures reported (150 - 200°C) correspond to the Tg range indicated by Pittman (1) for similar copolymers. The initiation system was preferably azobisisobutyronitrile, although potassium persulfate was also used. Organic peroxides were contraindicated, due to oxidation problems with the ferrocene moiety.

Because of the reported property advantages of the ferrocene moiety in polymers, this work was undertaken to ascertain if vinyl ferrocene could be redox copolymerized with other common monomers without serious problems.

Experimental

The current work was concerned with the emulsion copolymerization of vinyl ferrocene with butyl acrylate, styrene, and methacrylic acid to yield a film-forming composition of Tg~0-5°C.

Using the literature recommendations, initial work was centered on azo initiators due to the claim that the peroxides in redox systems oxidized the vinyl ferrocene rather than polymerized it. Initial experiments resulted in slow reactions and poor conversions, even after 8 - 12 hours' reaction time. Reasoning that redox systems might work if the reducing agent were kept in excess in the reaction flask, a tBHP/Formopon catalyst system was examined. This initiator system resulted in complete conversion within 2 - 3 hours and served as a prototype recipe in all subsequent work.

Example: Synthesis of Latex Containing 5% Vinyl Ferrocene

The following ingredients were employed in the synethesis of a conventional emulsion polymer:

Part A: 1312. gm deionized water
49.6gm Aerosol AY-65[5] (65%)
16. gm Formopon[6]

Part B: 856. gm styrene
832. gm butyl acrylate
32. gm methacrylic acid
80. gm vinyl ferrocene

Part C: 720. gm deionized water
22.4gm Triton X-405[8] (70%)

Part D: 22.4gm tBHP(70%)[9]
400. gm deionized water

Part A is charged to a 3-neck, morton-style, 5-liter flask and purged with nitrogen while heating to 60± 1°C. Part B is emulsified into Part C by slow addition while stirring on a high-speed bench stirrer. This monomer emulsion (ME) is likewise purged with nitrogen as is Part D.

10% of ME is added to the reaction flask after 30 minutes' purging for all parts and equilibrated by stirring for 5 minutes. 10% Part D is added to initiate polymerization, and ME started on gradual addition such that total addition time is 2 hours. Part D is added in 10% aliquots during said time to maintain the exotherm. The reaction is maintained at a minimum of 60°C, with the exotherm allowed to run its course (<10°C rise). At completion, the latex is cooled to room temperature, adjusted to pH 9.0 - 9.5 with conc. aq. NH_3 and filtered through cheesecloth.

Results for this experiment were as follows:

pH: 9.35
viscosity: 11 cps (LVT Brookfield, spindle #1, 60 RPM)
density: 8.55 lbs./gal.
solids: 41.8 wt. pct.
conversion: 100%
coagulum: nil
particle diameter: 1050Å (light scattering)
t_g: 3°C (TMA)

Other Examples

In addition to samples containing 1% and 5% vinyl ferrocene and a BA/S/MAA control, two additional samples were prepared containing 1% and 5% ferrocene additive (added as solution in monomer mix) since this represented a substantial cost savings in the use of the ferrocene moiety ($85/lb. for vinyl ferrocene; ~$5/lb. for ferrocene). All systems processed without difficulty and were carried to 100% conversion within 2-3 hours. No oxidation of the ferrocene moiety was observed, nor were there any conversion rate differences among the five systems. Table I summarizes the five emulsions prepared for this study.

Oxidation of the ferrocene moiety to the ferricinium (III) moiety did take place in the presence of excess oxidizing agent. Such a reaction is easily detected during emulsion polymerization by the change in color from pale yellow to pale muddy green. If reversed promptly with reducing agent, no irreversible reactions appear to occur.

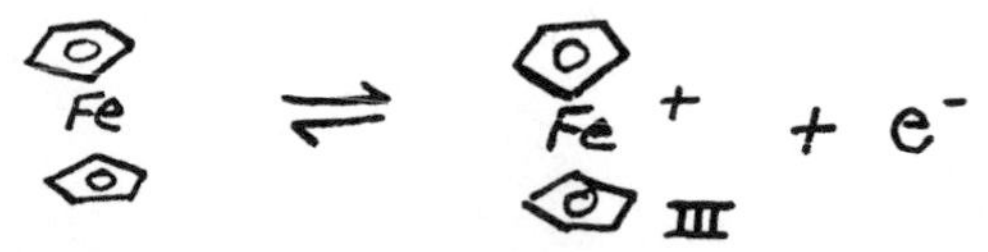

Thus, redox initiation with organic peroxides was demonstrated to be feasible given that the reducing agent is held in excess in the reaction mixture.

Characterization of Latices

Spectroscopy. Infrared and UV spectra were run on the 5 latex samples described in Table I to provide information for interpretation of subsequent application test results, and as a guide for GPC detector settings. In both cases, the absorption of the ferrocene moiety is largely obscured by the absorption of the parent polymer, thus limiting these techniques for definite analysis. There were some characteristic (though weak) peaks at 810 cm^{-1}, 1000 cm^{-1}, 1100 cm^{-1}, and 4000 cm^{-1} in the IR which do correlate to the spectrum of ferrocene itself, but these are too weak to allow reliable quantitative assignments (see Figure 1). The UV curves show the expected slight shift in absorption to longer wavelengths for the VFe polymers, but were otherwise not particularly illuminating (see Figure 2). Note that two of the curves are shifted up the ordinate using the scale on the right to facilitate viewing.

Chromatography

The main question to be answered in this section was whether vinyl ferrocene copolymerized into the latex.

Thin Layer Chromatography. Thin layer chromatography was the first choice due to the simplicity of operation. Initial experiments indicated that ferrocene and vinyl ferrocene would elute on a silica gel substrate with a variety of organic solvents, while the parent polymer (A) was largely unaffected. The surprising part was that physical mixtures of ferrocene or vinyl ferrocene with polymer sample (A), solvated in CH_2Cl_2 could not be separated by any elution solvent. Tables II & III summarize the initial work, while Table IV is a listing of the elution solvents tried without success in separating ferrocene from polymer sample A.

This curious result suggested the formation of a complex between the ferrocene and the polymer ligand. While unexpected, it was easily rationalized via structural examination (ferrocene considered electrophilic; BA, MAA considered nucleophilic). As a result, it was speculated that ferrocene could be incorporated into such a polymer as A as an additive at a considerable cost savings over copolymerization of vinyl ferrocene. More work is required to verify this speculation, of course.

Gas Chromatography. GC analysis on a high temperature column showed that the ferrocene could be easily separated from polymers D and E. Table V summarizes the results, showing that the vinyl ferrocene is apparently copolymerized into the polymer backbone.

Gel Permeation Chromatography. The five samples described in Table I were subjected to GPC analysis in an effort to further clarify the nature of the copolymerization of vinyl ferrocene in samples B and C. Figure 3 depicts the results in terms of molecular size (hydrodynamic volume) distribution. The curves are shifted arbitrarily up the ordinate for ease of viewing. From these curves, it appears that a basically normal molecular weight distribution exists, although a slight low molecular weight tail is observable. It also appears that incorporating vinyl ferrocene had the effect of reducing both molecular size (weight) and polydispersity as is seen in Table VI. Curiously, the ferrocene additive did not exert the same effect.

TABLE I. LATEX PROPERTIES

Sample No.	Composition BA	S	MAA	VFe	Pct. N.V.	pH	Viscosity (cps)	Particle Diameter	Latex Color
A	52	46	2	0	39.7	9.26	11	1400Å	milky white
B	52	45	2	1	38.9	9.35	10	1300Å	pale yellow
C	52	41	2	5	41.8	9.35	11	1850Å	med. yellow
D	52	46	2	a	39.2	9.22	11	1600Å	pale yellow
E	52	46	2	b	32.7	9.20	11	1400Å	med. yellow

a. 1% ferrocene incorporated in monomer mix; styrene level not adjusted.
b. 5% ferrocene incorporated in monomer mix; styrene level not adjusted.

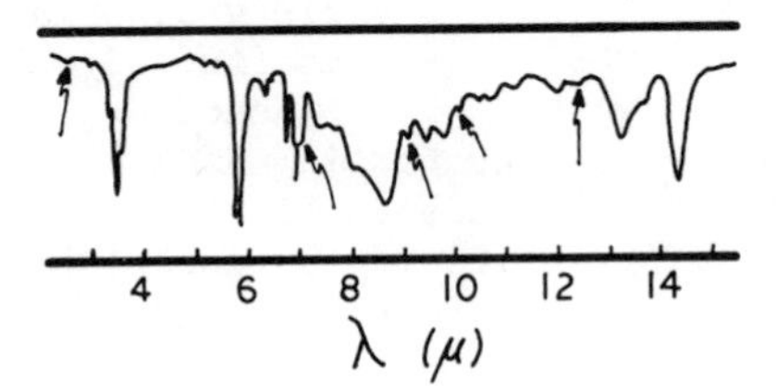

Figure 1.

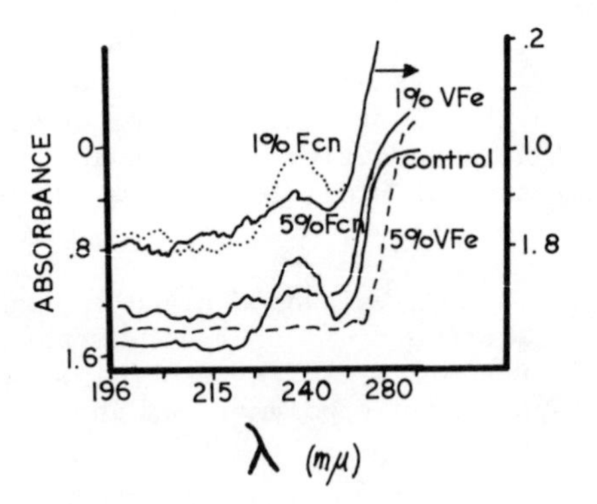

Figure 2.

TABLE II. TLC OF FERROCENE AND VINYL FERROCENE

Elution Solvent	R_f Values Ferrocene	VFe	Comments
tolune	.9	.9	no separation
n-butanol	1.0	1.0	no separation; solvent front stalled at 5 cm
n-propanol	.9	.9	no separation
abs. ethanol	.9	.9	no separation
abs. methanol	.8	.9	separation, but streaming of ferrocene

TABLE III. TLC OF FERROCENE-SPIKED POLYMER

Elution Solvent	Results
abs. methanol	no movement of sample spot
trichloro ethylene	streaming of polymer; no separation
methylene chloride	streaming, but no separation
chloroform	streaming, but no separation
benzene	sl. movement of sample spot; no separation

TABLE IV. SOLVENTS STUDIED IN TLC EXPERIMENTS

n-propanol	pyridine	toluene
acetone	ethyl acetate	trichloro ethylene
ethanol (abs.)	p-dioxane	chloroform
methanol (abs.)	carbon tetra-chloride	n-butanol

TABLE V. GC RESULTS

Sample No.	Ferrocene Present	Ferrocene Found	Vinyl Ferrocene Present	Vinyl Ferrocene Found
A	0	0	0	0
B	0	0	1	0
C	0	0	5	0
D	1%	1%	0	0
E	5%	5%	0	0

Table V does not necessarily disprove the complex theory of the TLC experiments, but does, at least, demonstrate the limited thermal stability of such a complex if, in fact, it exists.

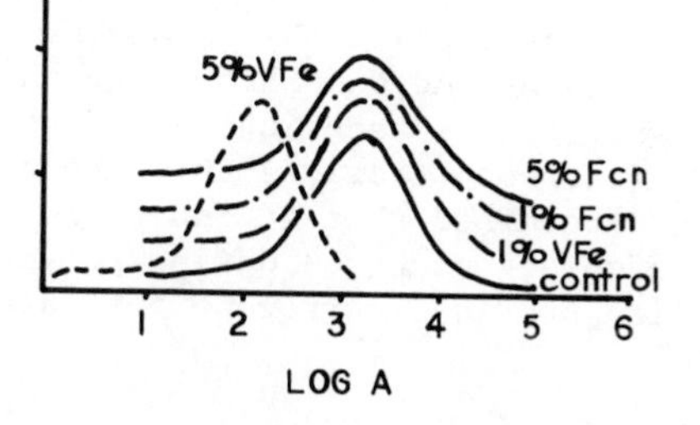

Figure 3.

Thermal Analysis

Thermal stability of the vinyl ferrocene and ferrocene containing polymers plus unmodified polymer and neat ferrocene controls was determined by thermal gravimetric analysis at a heating rate of 10°/min. The results are depicted in Figure 4. Apparently vinyl ferrocene imparts greater thermal stability to the BA/S/MAA polymer than the ferrocene additive. In fact, 1% vinyl ferrocene allows about a 50°C additional rise in temperature before decomposition relative to the unmodified polymer. Curiously, a substantial exotherm is present at 400 - 450°C (bump on curve) for the vinyl ferrocene modified polymers, but does not appear for the ferrocene modified polymers.

For further clarification, DSC runs were made for the same systems at 15°C/min. heating rate. These results are found in Figure 5. Note the confirmation of the very strong exotherm for the vinyl ferrocene systems, as in the TGA curves. The curves in Figure 5 are shifted on the ordinate by arbitrary amounts to facilitate viewing. The presence of multiple decomposition peaks from about 300°C up suggest a stepwise decomposition of the parent polymer. The greater thermal stability of the vinyl ferrocene and ferrocene modified polymers would appear to be largely due to a suppression of the first of these steps at ~300°C, rather than a simple shift in general to higher temperatures. In this regard, again vinyl ferrocene is seen to be more effective than ferrocene. The absence of the melting and vaporization peaks of ferrocene in the ferrocene modified polymers strongly suggests that the ferrocene is in fact dissolved in the polymer. The additional complexities in the curves of the modified polymers above 350°C are real, but without adequate explanation at this time.

Summary

In contrast to the claims of the literature, vinyl ferrocene (available commercially) was found to be a very reactive monomer in the terpolymer system butyl acrylate/styrene/methacrylic acid. It was further found, again in contrast to the claims in the literature, that vinyl ferrocene could be emulsion polymerized via organic peroxide Redox catalysis.

Emulsion polymers containing 1% and 5% copolymerized vinyl ferrocene were prepared as were controls containing 1% and 5% of ferrocene additive and a straight BA/S/MAA reference emulsion. The resulting polymers were characterized by IR and UV spectroscopy; gel permeation, thin layer and gas chromatography; and TGA and DSC thermal analysis. These techniques confirmed that vinyl ferrocene was, in fact, copolymerized into the experimental polymers. It was further determined that no inhibition by, nor oxidation of the ferrocene moiety occurred during polymerization. Interestingly, incorporation of vinyl ferrocene into a BA/S/MMA

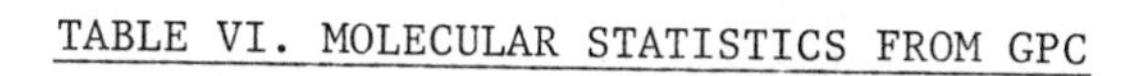

TABLE VI. MOLECULAR STATISTICS FROM GPC

Sample	Wt. Average Size (Å)	Polydispersity Index (Wt. Average/No Average)
A	3933	20
B	3183	16
C	2074	8
D	4605	20
E	4090	27

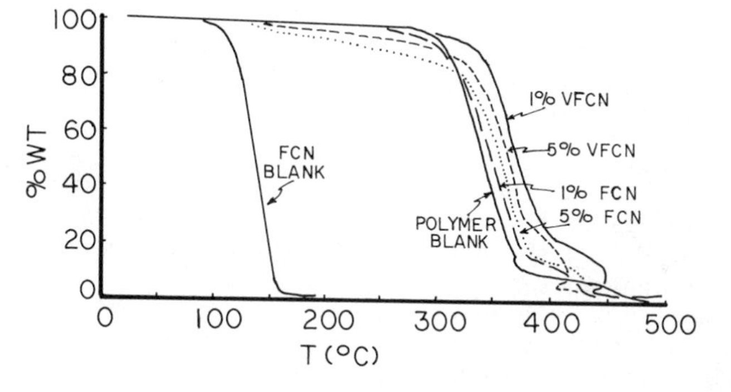

Figure 4. TGA

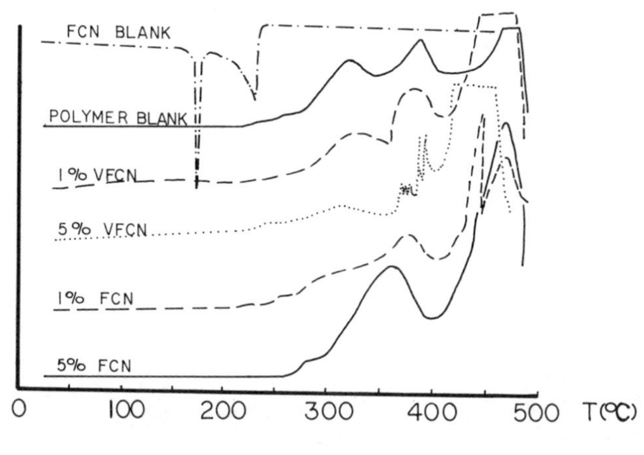

Figure 5. DSC

polymer causes a reduction in molecular weight and polydispersity. Ferrocene appears to form a weak complex with the polymer BA/S/MAA which prevents separation by thin layer chromatography, but not by high temperature gas chromatography. DSC and TGA indicated an enhancement in thermal stability over the unmodified polymer with vinyl ferrocene being more effective than ferrocene. The enhanced thermal stability appears to be more due to a suppression of the first step in the step-wise degredation of the parent polymer rather than a simple shift to higher temperatures.

Acknowledgments

The author gratefully acknowledges the assistance of Mr. H. D. Swafford for IR, UV, and GC characterization, Dr. T. Provder for GPC studies, and Mr. R. M. Holsworth for assistance with the TGA and DSC work. These inputs contributed significantly to the successful conclusion of this project.

Literature Cited

1. Pittman, C. U., Jr., Chem. Tech, July (1971), p. 416 - 423.
2. Neuse, E. W., Advances in Macromolecular Chemistry, Vol 1, Academic Press, New York: (1968), pp. 1 - 138.
3. Arimoto, F. S., and Haven, A. C. Jacs, 77, 6295 (1955).
4. U. S. Patent 2, 821,512; 28 January (1958).
5. Sodium diamyl sulfosuccinate, from American Cyanamid.
6. Sodium sulfoxylate formaldehyde, from Rohm and Haas.
7. Obtained from Arapahoe Chemical Co.
8. Octylphenol polyethoxy ethanol of EO=40, from Rohm and Haas.
9. Tert-butyl hydroperoxide from Lucidol Div., Pennwalt Corp.

13

Molecular Weight Development in Styrene and Methyl Methacrylate Emulsion Polymerization

HOWARD L. JAMES, JR.
Standard Oil of Ohio, Cleveland, Ohio 44128

IRJA PIIRMA
Institute of Polymer Science, The University of Akron, Akron, Ohio 44325

The "ideal" concept of emulsion polymerization was built on the assumption that the monomer was water insoluble and that in the absence of chain transfer, the number average degree of polymerization, x_n can be related to the rate processes of initiation and propagation by the steady-state relationship $x_n = 2\ R_p/R_i$. Since R_i and R_p are both constant and termination is assumed to be instantaneous during the constant rate period described by Smith-Ewart kinetics, the above equation predicts the generation of constant molecular weight polymer. Data has been obtained which agrees with Smith-Ewart (1,2), but there is also a considerable amount of data which shows an increase in molecular weight with conversion during the constant rate period (3,4). Also, the molecular weight distribution has often been found to be very broad (3) contrary to the theoretically predicted most probable distribution.

Molecular weight characterization of polymers requires various molecular weight averages and, preferably, a complete molecular weight distribution, such as that obtained by fractionation. Since these techniques are quite time consuming, molecular weight distributions were seldom obtained on a routine basis. Therefore dilute solution viscometry, coupled with Mark-Houwink equations for the calculation of $\overline{M}_v$, has been the main molecular weight identification method. The recent advent of gel permeation chromatography (GPC) has allowed a rapid, convenient and reliable determination of molecular weight average and molecular weight distribution. Krackeler et al. (5), in comparing the results of GPC with those obtained using a fractional precipitation technique, found a good comparison between the two techniques.

Taken in part from the M.S. Thesis of H. L. James, granted at The University of Akron, June 1974.

Since both the various molecular weight averages and the molecular weight distribution of a polymer are extremely sensitive to any chain transfer reactions, in this investigation an attempt was made to keep the polymerization ingredients as pure as possible. The emulsifier system, consisting of a nonionic as well as an ionic surface active agent, had been tested for its chain transfer activity and found "safe." This surface active agent combination also has been found to yield a latex with quite a narrow particle size distribution (2). The desirability of using a monodisperse particle size latex in kinetic studies has been discussed by several researchers (2,3,6).

Experimental

Materials and Polymerization. Styrene and methyl methacrylate were obtained from commercial sources and were distilled to remove inhibitor. After distillation, the monomers were stored, under nitrogen, in a refrigerator. For the mixed emulsifier system, Emulphogene BC840(GAF), tridecyloxy-polyethylene-oxyethanol, was used as the nonionic constituent, and sodium lauryl sulfate (K and K Labs) was used as the ionic constituent. The sodium lauryl sulfate was at a concentration below its cms whereas the BD-840 was at a concentration above its cmc. This emulsifier system has been shown to yield mixed micelles (7), having a low ionic change (2), which produce latices with rather narrow particle size distributions (2, 8). Potassium persulfate (Fisher Scientific) was used as the initiator and potassium hydroxide was added to insure efficient decomposition of the persulfate.

The following recipe was used for all polymerizations in this work:

Material	Parts by Weight
Monomer	100.0
Water	180.0
Emulphogene BC-840	5.0
Sodium Lauryl Sulfate	0.05
Potassium Persulfate	0.30
Potassium Hydroxide	0.075

The polymerizations were carried out in 4 oz. glass bottles with metal screw caps which were perforated and lined with self-sealing butyl gaskets. The tightly capped bottles, after being purged with nitrogen, were clamped into a water bath thermostatted at 50°C and rotated end-over-end at 45 r.p.m. Percent conversion, at various reaction times, were determined gravimetrically.

Characterization. The polymer samples to be characterized were obtained by pouring various percent conversion samples into a large excess of methanol which contained a sufficient quantity of Pennstop RC 1866 to completely shortstop the latex. The precipitated polymer was then washed several times with both methanol and water and dried in a vacuum oven. Tenth of a percent polymer solutions were prepared for both viscometry and gel permeation chromatography techniques. The solution process required 2 to 4 days depending on the polymer molecular weight and the solvent. Dissolution was aided by gentle hand swirling at intervals of several hours.

i. Gel Permeation Chromatography - The GPC data were obtained in dimethylformamide solution using a Waters Associates Model 100 GPC equipped with a differential refractive index detector and five Styragel columns having the following nominal exclusion limit designations: 10^7 Å, 10^5 Å, 10^6 Å, 3×10^5 Å, and 3×10^4 Å. Measurements were made at 60°C using a flow rate of 1 ml/min. Sample concentrations were 0.1% (w/v) for both the polystyrene calibration standards and the polymers obtained in this work. Prior to injection into the GPC, the solutions were filtered through a 5μ filter. This filtering process was carried out very slowly so as not to shear degrade the polymer. The injection time was 120 sec. in all cases. The calibration curve used was constructed by using polystyrene standards from three different suppliers; Pressure Chemical Co., Waters Associates, and Duke Standards. The molecular weights of these standards ranged from 51,000 to 7.1 million. Figure 1 shows the GPC chromatogram for the standard with a molecular weight of 7.1 million.

Weight and number average molecular weights, uncorrected for instrument spreading, were calculated using the following relationships:

$$\overline{M}_w = \Sigma(H_i M_i)/\Sigma H_i \qquad \overline{M}_n = \Sigma H_i/\Sigma(H_i/M_i)$$

where H_i is the height of the chromatogram, measured from the baseline, at the ith elution count and M_i is the molecular weight at this count, determined by using the polystrene calibration curve. As is common practice, it was assumed that the peak maximum of the calibration standards corresponded to the $\overline{M}_w$ of the standard. The universal calibration technique (9) was not attempted in the GPC study of these high molecular weight polymers due to the difficulties encountered by Slagowski et al. (10). They found that both types of calibration curves, the conventional log $\overline{M}_w$ vs. elution volume and the long $[\eta]$ M vs. elution volume, showed distinct breaks in linearity in the 10^7 g/mole region.

ii. Viscometry - Viscosity measurements of all the polymers were carried out in toluene at 30°C in a Cannon Ubbelohde dilution viscometer. The correction term for kinetic energy was less than 0.2% and was therefore neglected in calculations of intrinsic viscosity.

Although viscometry can be a quite simple and rapid method for obtaining information on polymers, polymers having molecular weights greater than about 5 x 10^6 g/mole are known to shear degrade (11, 12). This non-Newtonian behavior of polymer solutions has been discussed in the literature for both high molecular weight polystyrene (12, 13, 14) and poly (methyl methacrylate) (15, 16).

The intrinsic viscosities obtained in the Cannon Ubbelohde viscometer were corrected for shear effects by using the curve shown in Figure 2. This curve was obtained by using a Zimm-type low shear viscometer (17) to determine the intrinsic viscosities of four emulsion polymerized polystyrene samples (2). The curve was linear over the range of intrinsic viscosity results for polystyrene samples encountered in this study.

Results and Discussion

The emulsion polymerization at 50°C of styrene and methyl methacrylate using concentrations of ingredients as listed in experimental section gave polymerization rates of 11.5%/hr and 33.6%/hr respectively. Figures 3 and 4 show smooth, continuous rate curves indicating the presence of three distinct stages in these polymerization rates. The three stages were designated interval I, II, and III similar to Gardon (18). Interval I represents the particle-forming period, interval II the polymerization period, and interval II the period following the constant rate. Although our investigation is primarily concerned with the molecular weight development during interval II, we have shown data covering all intervals. From Figures 3 and 4 it can be seen that a constant rate period extends from about 25 to 60 percent conversion in the case of styrene and from 13 to 35 percent conversion for methyl methacrylate.

The data for the molecular weight averages, $\bar{M}_w$ and $\bar{M}_n$, obtained by gel permeation chromatography for the samples of polystyrene and poly(methyl methacrylate) are shown in Table I and II, respectively. Although the values are not absolute, they provide the ability to study the molecular weight development in the emulsion polymerization of these monomers in a relative sense. Figures 5 and 6 provide graphical representation for the data.

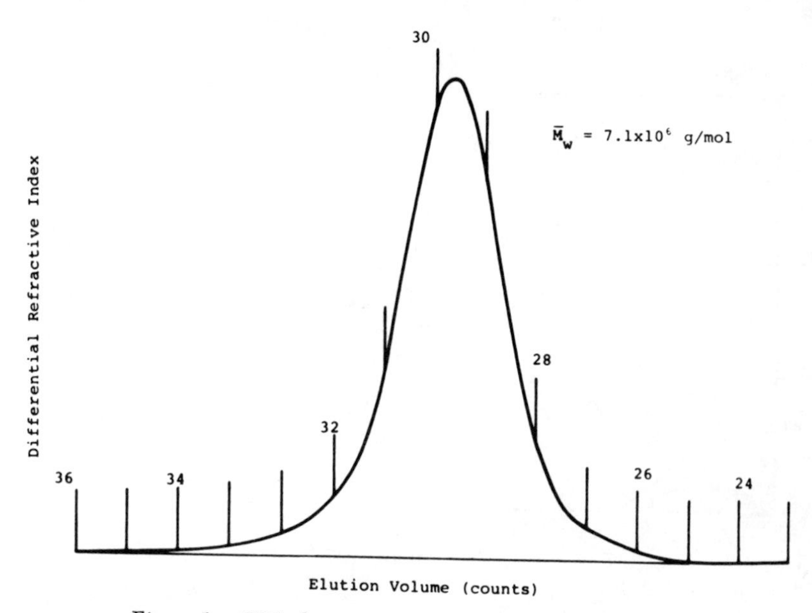

Figure 1. GPC chromatogram of a polystyrene standard

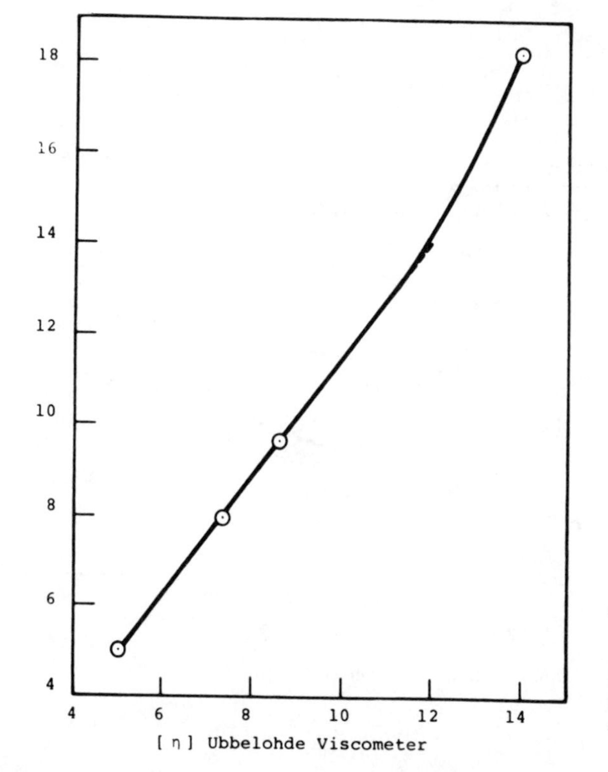

Figure 2. Calibration curve for shear-rate effect (polystyrene in toluene at 30°C)

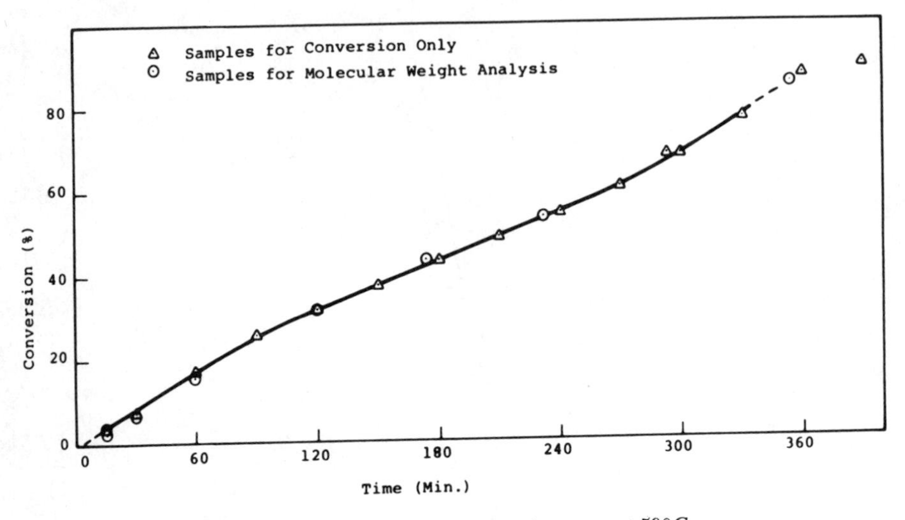

Figure 3. Polymerization rate of styrene at 50°C

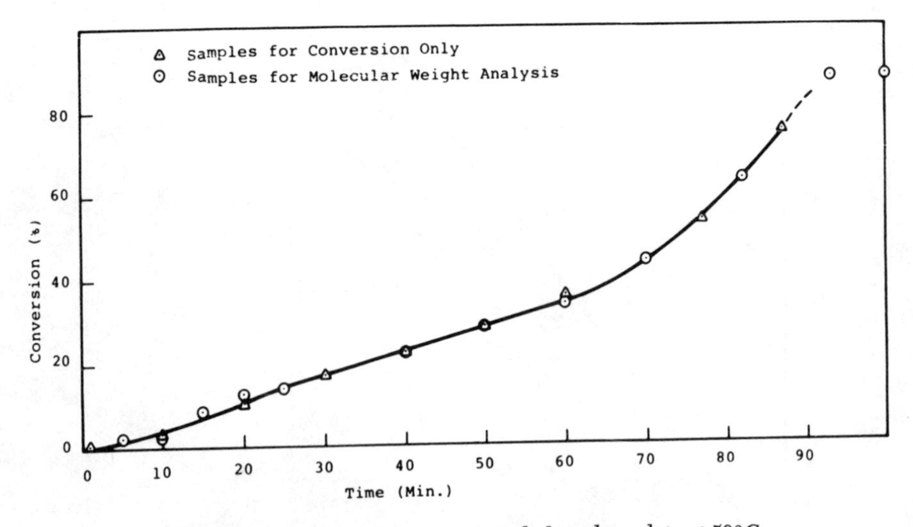

Figure 4. Polymerization rate of methyl methacrylate at 50°C

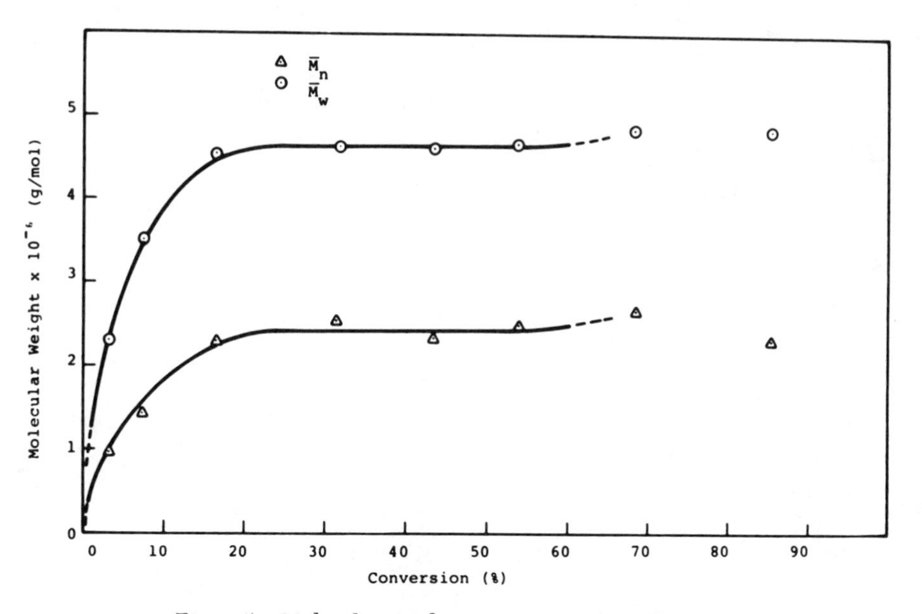

Figure 5. Molecular weight vs. conversion for polystyrene

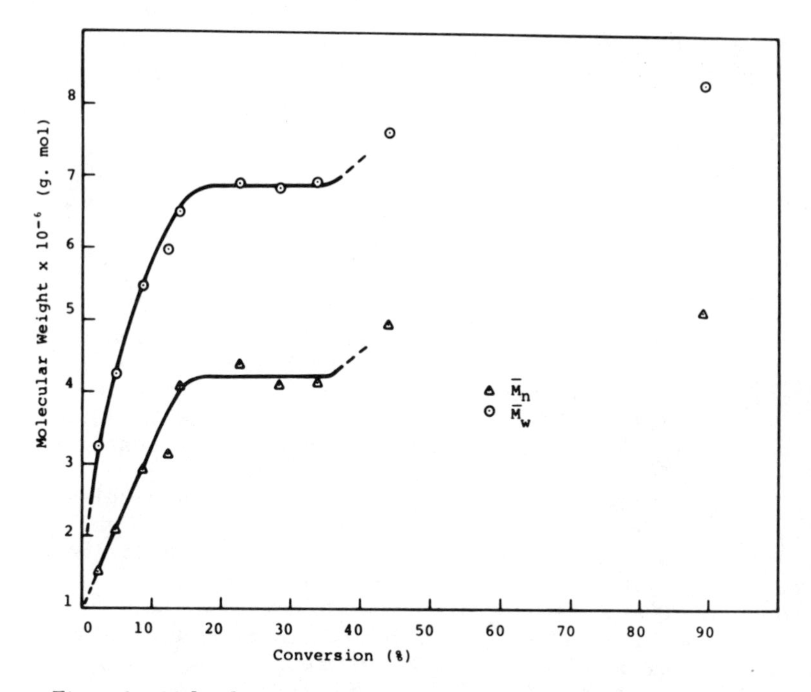

Figure 6. Molecular weight vs. conversion for poly(methyl methacrylate)

Table I

Molecular Weights for Polystyrene Determined by GPC

Conversion %	$\overline{M}_w \times 10^{-6}$ (g/mole)	$\overline{M}_n \times 10^{-6}$ (g/mole)
3.05	2.31	0.98
7.28	3.52	1.44
16.44	4.55	2.31
31.71	4.64	2.56
43.18	4.61, 4.64	2.29, 2.41
53.63	4.74, 4.62	2.51, 2.48
68.19	4.98, 4.72	2.61, 2.75
85.06	4.87, 4.88	2.25, 2.43

Table II

Molecular Weights for Poly(Methyl Methacrylate) Determined by GPC

Conversion %	$M_w \times 10^{-6}$ (g/mole)	$M_n \times 10^{-6}$ (g/mole)
2.42	3.24	1.52
4.71	4.18, 4.30	2.08, 2.09
8.73	5.46	2.94
12.10	5.97	3.14
13.93	6.50	4.09
22.45	6.90	4.40
28.24	6.83	4.10
33.84	6.93	4.14
43.92	7.66, 7.60	4.96, 4.94
87.86	8.31	5.15

The change in intrinsic viscosity with conversion is shown in Figures 7 and 8. Although the data in Figure 8 appears to be quite scattered, all values of intrinsic viscosity between and including 13.9 and 33.8% conversion agree to ± 3.1% of the average value of 5.21 dl/g, well within the experimental error of the method.

In carrying out molecular weight development studies in emulsion polymerization, care must be taken that the emulsifiers do not act as chain transfer agents and thus lower the molecular weight averages of the samples. Kamath (2) and Wang (7) have shown that Emulphogene BC-840 does not act as a chain transfer agent. The high values for intrinsic viscosities obtained in this study also indicate that BC-840 is "safe."

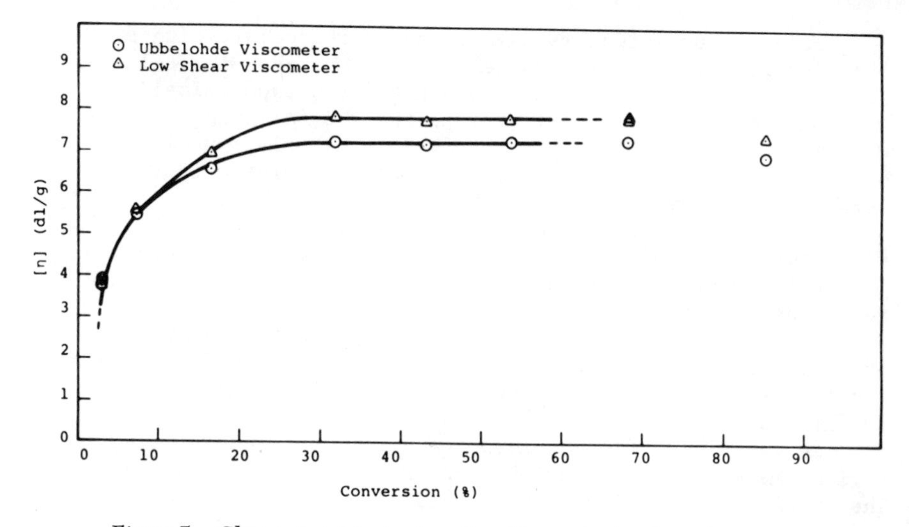

Figure 7. Change in intrinsic viscosity with conversion for polystyrene

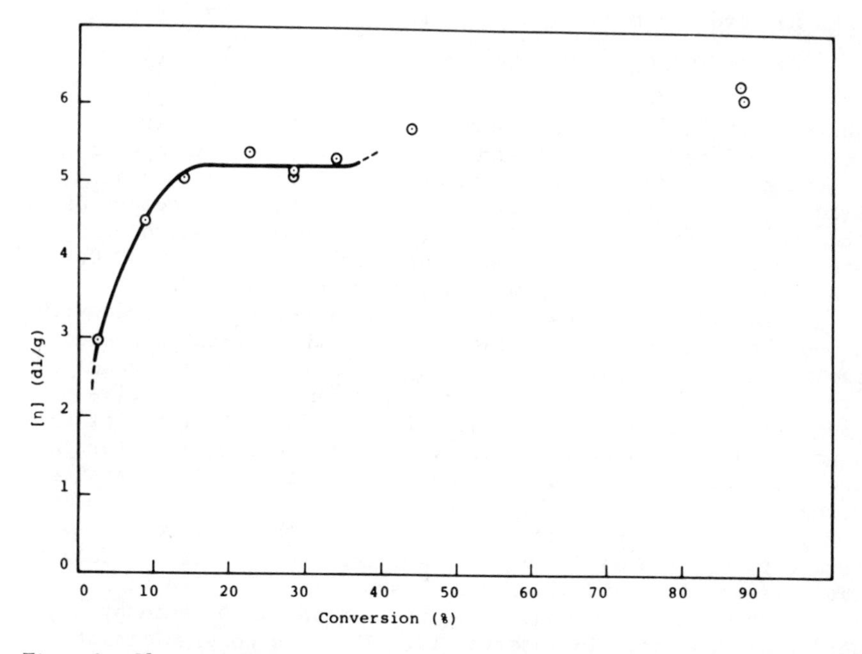

Figure 8. Change in intrinsic viscosity with conversion for poly(methyl methacrylate)

Several investigators have compared molecular weights, obtained from viscosity measurements and Mark-Houwink relationships, i.e., $[\eta] = KM^a$, with those predicted theoretically. This was not done in this investigation, for the following reasons:

1) the exponent a in the Mark-Houwink relationship has been shown to vary drastically for high molecular weight polymers in good solvents (12),

2) corrections for shear rate effect are often not applied to the experimentally observed intrinsic viscosities determined in capillary viscometers used to obtain the Mark-Houwink relationships, and

3) a large number of widely varying K and a values are observed in the literature for intrinsic viscosity data obtained for the same polymer under supposedly identical experimental conditions.

An example of the magnitude of error involved in using various Mark-Houwink relationships is shown in the table below. The intrinsic viscosity values, determined in acetone at 30°C,

Intrinsic Viscosity (dl/g)	$\bar{M}_v \times 10^{-6}$ a) (g/mole)	$\bar{M}_v \times 10^{-6}$ b) (g/mole
4.2	7.01	5.85
5.3	10.00	6.91

a) determined using: $[\eta] = 1.4 \times 10^{-4}\ \bar{M}_v^{\ 0.654}$ (16)

b) determined using: $[\eta] = 7.7 \times 10^{-5}\ \bar{M}_v^{\ 0.70}$ (4)

shown in the first column, were obtained for two different poly(methyl methacrylate) samples. The values of $\bar{M}_v$ appearing in the second column were calculated using the equation of Krause et al. (16), and those appearing in the last column using the equation preferred by Gardon (4). Both of these equations are for intrinsic viscosities determined in acetone at 30°C. As can be seen, values of $\bar{M}_v$ may vary by as much as 40%.

Although the values of intrinsic viscosity determined with a low shear viscometer are the only ones which truly represent the intrinsic viscosity at high molecular weights, the results from the capillary viscometer are shown in Figure 7 to give an indication of the effect of shear in the viscosity range of the study. The values of intrinsic viscosity are different for the two types of viscometers, but the trend of intrinsic viscosity versus conversion is still the same.

By comparing Figures 5, 6, 7, and 8 with their respective rate curves, Figures 3 and 4, we see that the curves all follow the same trend. Initially, there is a large increase of molecular weight with conversion in interval I followed by a period of constancy in interval II. There is no significant increase in the curves in interval III for polystyrene, but quite a large increase in the case of poly(methyl methacrylate). The shape of molecular weight-conversion curve in interval III is determined by two processes whose effects oppose each other. The molecular weight would be expected to decrease continuously with increasing conversion if the rate of termination within the particles remains constant since the propagation rate is proportional to decreasing monomer concentration. However, if the particle size and/or the viscosity within the particle is large, the termination rate will decrease. This results in an increase in both polymerization rate and molecular weight.

From the intrinsic viscosity and GPC results, it is apparent that these two effects essentially cancel each other in the case of polystyrene. The molecular weight remains fairly constant to 85 percent conversion. The results for poly(MMA), on the other hand, are quite different. Here the effect of the decreased rate of termination is already quite prominent at approximately 40 percent conversion. It is proposed that this rather large increase in molecular weight in interval III is due to the viscosity or Trommsdorff effect and not to particle size.

Whereas Figures 5 and 8 show plots of cumulative values of molecular weight and intrinsic viscosity versus conversion, Figure 9 shows plots of both instantaneous and cumulative weight average molecular weights, determined by GPC for the polystyrene samples. The instantaneous molecular weights, $\bar{M}_w^p$'s, were estimated graphically using the equation

$$\bar{M}_w^{\,p} = \bar{M}_w^{\,o-p} + p\,\frac{d\,\bar{M}_w^{\,o-p}}{dp}$$

where $\bar{M}_w^{\,o-p}$ is the cumulative or experimentally measured $\bar{M}_w$ and p is the extent of reaction. Values of

$$\frac{d\,\bar{M}_w^{\,o-p}}{dp}$$

were obtained by taking slopes from the curves of cumulative $\bar{M}_w$ versus p (i.e., Figure 5).

An equation, similar to that shown above for $\bar{M}_w^p$, was used by Fitch et al. (19) for instantaneous viscosity average molecular weights and by Wall (20) and Robertson (21) for instantaneous intrinsic viscosities.

No serious attempt was made to use the GPC curves to obtain a molecular weight distribution, i.e., the heterogeneity index, M_w/M_n, because corrections for neither shear degradation nor instrument spreading were made for the GPC data. Although no attempt was made, it appears from the data shown in Figures 5 and 6 that the distribution is close to the most probable distribution. This is not in agreement with Gardon's theoretical

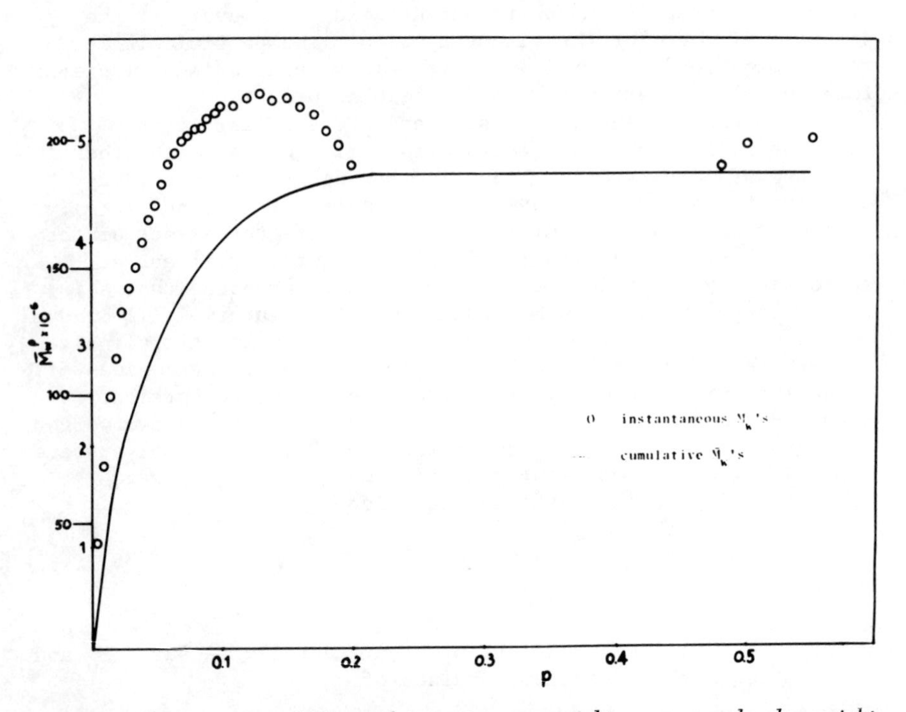

Figure 9. Change in cumulative and instantaneous weight average molecular weights with conversion

prediction (4) that the molecular weight distribution of polymers prepared by emulsion techniques would be $1 \leq \bar{M}_v/\bar{M}_n \leq 1.33$. The results of Grancio and Williams (3), Krackeler and Naidus (5, 22) and Kamath (2) all yielded values of $\bar{M}_w/\bar{M}_n$ in the range of 2.0 to 3.3.

Conclusions

In this investigation we have shown, with the use of dilute solution viscometry and gel permeation chromatography, that the molecular weight does remain constant for both polystyrene and poly(methyl methacrylate) during the period of constant rate in emulsion polymerization. It also appears that the ratio of $\bar{M}_w/\bar{M}_n$ is very close to the most probable distribution, i.e., $\bar{M}_w/\bar{M}_n \simeq 2$. The molecular weight was found to increase very rapidly during the particle formation period. In interval III of polymerization, the molecular weight of the polystyrene changes very little, as might be expected from kinetic considerations. However, in the case of the poly(methyl methacrylate), the molecular weight begins to increase, presumably due to a pronounced Trommsdorff effect for this monomer.

It is felt that the additional three items are needed for the study of molecular weight development of emulsion polymerization and they will therefore be carried out in future experimental studies: 1) absolute molecular weights should be obtained using light scattering and/or analytical untracentrifugation techniques. These experimental data would be helpful in determining the validity of theoretical molecular weight equations proposed by Smith and Ewart (1), Gardon (4, 23), Stockmayer (24), Katz and Saidel (25, 26), and Sundberg and Chassen (27); 2) corrections for the effect of shear rate on intrinsic viscosity should be made for the poly(methyl methacrylate) samples; and 3) further investigation into the effect of monomer solubility on the kinetics of emulsion polymerization may be gained by investigating the polymerization of monomers such as isoprene, vinyl acetate and acrylonitrile.

Acknowledgements

The authors are indebted to Waters Associates for supplying the GPC columns used in this investigation.

References

1. Smith, W. V. and R. H. Ewart, J. Chem. Phys. (1948) 16, 592.
2. Kamath, V. R., Ph. D. Thesis, The University of Akron, 1973.
3. Grancio, M. R. and D. J. Williams, J. Polymer Sci., Part A-1 (1970) 8, 2733.
4. Gardon, J. L., J. Polymer Sci., Part A-1 (1968) 6, 665, 687.
5. Krackeler, J. J. and H. Naidus, J. Polymer Sci., Part C (1969) No. 27, 207.
6. Roe, C. P., Ind. Chem. (1968) 60 (9), 20.
7. Wang, P. C., private communication.
8. Letchford, R. J., M. S. Thesis, The University of Akron, 1973.
9. Grubisic, Z., P. Rempp and H. Benoit, J. Polymer Sci. (1967) B5, 753.
10. Slagowski, E. L., L. J. Fetters and D. McIntyre, Macromolecules (1974) 7 (3), 394.
11. McIntyre, D., L. J. Fetters and E. Slagowski, Science (1972) 176, 1041.
12. Slagowski, E., Ph. D. Thesis, The University of Akron, 1972.
13. Berry, G. C., J. Chem. Phys. (1967) 46 (4), 1338.
14. Krigbaum, W. R. and P. J. Flory, J. Polymer Sci. (1953) 11, 17.
15. Cantow, H. J., J. Pouget and C. Wippler, Makromol. Chem. (1954) 14, 110.
16. Krause, S. and E. Cohn - Ginsberg, J. Polymer Sci., Part A (1964) 2, 1393.
17. Zimm, B. H. and D. M. Crothers, Proc. Natl. Acad. Sci., U.S. (1962) 48, 905.
18. Gardon, J. L., J. Polymer Sci., Part A-1 (1968) 6, 623.
19. Fitch, R. M., M. B. Prenosil and K. J. Sprick, J. Polymer Sci., Part C (1969) No. 27, 95.
20. Wall, F. T., J. Amer. Chem. Soc. (1945) 67, 1929.
21. Robertson, E. R., Trans. Faraday Soc. (1956) 52, 426.
22. Krackeler, J. J. and H. Naidus, J. Polymer Sci., Part C (1969) No. 27, 207.
23. Gardon, J. L., Rubber Chem. Technol. (1970) 43 (11), 74.
24. Stockmayer, W. H., J. Polymer Sci. (1957) 24, 314.
25. Katz, S. and G. M. Saidel, ACS Polymer Preprints (1966) 7(2), 737.
26. Saidel, G. M. and S. Katz, J. Polymer Sci., Part C (1969) No. 27, 149.
27. Sundberg, D. S. and J. D. Chassen, "Polymer Collids," edited by R. M. Fitch, Plenum Press, New York, 1971, p. 153.

14

Molecular Weight and Molecular Weight Distribution of Polystyrene Obtained with Suspension Polymerization

V. D. YENALYEV and V. I. MELNICHENKO

Donetsk State University, Donetsk, 340055 U.S.S.R.

The molecular weight (MW) and molecular weight distribution (MWD) influence the physical and mechanical, and melting properties of polymer to the great degree (1-3). Hence, for obtaining polymers with optimum physical property-fabrication relationship it is necessary to regulate their MW and MWD in the process of polymerization widely. Up to now this problem was solved only by means of experimental choosing of charges and conditions of polymerization.

The usage of methods of mathematical simulation of the process of polymerization may be a very effective way. To govern the process of radical free polymerization in obtaining the material with optimum values of MW and MWD is possible to be done in the following way:

1) by choosing the initial mixture of reagents allowing the optimum conduct of the process;

2) by supporting optimum composition of reagents introducing one or several components into the system during the process;

3) by"constructing" charges of polymerization mixture and conducting the process of polymerization according to the definite temperature-time regime.

Some researchers have already attempted to obtain the polymer with definite values of MW and MWD in the processes of block and solution polymerization (4,5). It's impossible to introduce these or other components into polymerization system of suspension polymerization, because their uniform distribution according to volume is impossible due to the limited mass exchange. In this case it's possible to govern the polymerization process only by changing temperature in reactor according to the definite program. This paper deals with obtaining polystyrene with definite MW and MWD of thermal and initiated polymerization of styrene by means of suspension method.

Experimental details.

Uninhibited styrene containing 99.7% of the main material being vacuum distillated in nitrogen three times was used for polymerization. The degree of conversion was calculated from the weight of the obtained dry precipitating polymer and by dilatometer method. Benzoyl peroxide (BP) and t-Butyl Perbenzoate (BPB) were used as initiators. The concentration of BP in the system of styrene-polystyrene was defined by polarographic method (6). MWD of polystyrene was defined by the method of precipitation chromatography (7) and with ultracentrifuge of the type G-120.

Thermal polymerization.

Some mechanisms were suggested to explain rather high rate of thermal initiating (8,9). The paper (10) is one of the first dealing with the attempt to obtain the mathematical model of thermal polymerization process of styrene. The results of the kinetic calculation of polymerization for isothermal conditions given by the authors agree with the experimental data quite sufficiently. However, if it is more reasonable to conduct the thermal polymerization of styrene according to the step temperature regime, it will be undoubtfully interesting to use the suggested by the authors mathematical model for kinetic calculation of polymerization and attempt to state the dependence of MW and MWD on the character of temperature regimes. With the help of the computer "Minsk-22" we have calculated kinetics of polymerization of styrene for several two-step temperature regimes under which obtaining polymer with the predicted MW and MWD was possible. Fig.1 deals with calculation dependence of styrene conversion on time for 2 temperature regimes. The values of MW and $\bar{M}_w/\bar{M}_n$ of the final product are given in table I.

The given results of calculation at the computer allow to predict the important characteristics of the obtained final product at the step temperature regime of polymerization. But for obtaining such information of the process for conditions which somewhat differ from the given above it is necessary to calculate kinetics of polymerization at the computer again. Generalized dependencies of $\bar{M}_n$, $\bar{M}_w$, $\bar{M}_w/\bar{M}_n$ of the final product on polymerization conditions could be more effective for technological calculations. For obtaining interpolar formulas for $\bar{M}_n$, $\bar{M}_w$, $\bar{M}_w/\bar{M}_n$ we have used the method of statistic planning of experiment (11).

Table I.

Theoretical and experimental values of $\bar{M}_w$ and $\bar{M}_w/\bar{M}_n$ of polystyrene; S_1 - final conversion of step 1; t_1 - temperature of step1; t_2 - temperature of step 2.

№№ of regime	S_1, %	t_1, °C	t_2, °C	Theory		Experiment	
				$\bar{M}_w \cdot 10^{-5}$	$\bar{M}_w/\bar{M}_n$	$\bar{M}_w \cdot 10^{-5}$	$\bar{M}_w/\bar{M}_n$
I	60	112	189	3.0	3.0	3.8	3.2
II	36	90	212	3.0	6.0	3.6	8.2
III	70	97	174	5.0	3.0	6.6	3.3
IV	60	90	200	5.0	6.0	5.8	7.4
V	75	80	164	7.0	3.0	8.0	3.3
VI	77	88	228	7.0	6.0	-	-

As variable values we have chosen: temperature of step 1 is x_1, final conversion of step 1 is x_2, temperature of step 2 is x_3. The final conversion of step 2 was equalled to 97%. The values of x_1, x_2, x_3 are given in table 2.

Table II.

The values of factors of x_1, x_2, x_3 in the plan points.

Factor level	Factor		
	x_1, °C	x_2, %	x_3, °C
- 1	100	30	170
0	120	50	190
+ 1	140	70	210
Δ x	20	20	20

For conditions corresponding to planning matrix the kinetic calculations of thermal polymerization of styrene were made. After calculating the following equations of regression were obtained:

$$\hat{Y}_1 \cdot 10^{-5} = 1.11 - 0.15\,x_1 + 0.26\,x_2 - 0.38\,x_3 - 0.84 \cdot 10^{-1}\,x_1 x_2 + 0.11\,x_1 x_3 - 0.13 \cdot 10^{-2}\,x_2 x_3 + 0.53 \cdot 10^{-1}\,x_1^2 + 0.11\,x_2^2 + 0.54 \cdot 10^{-1}\,x_3^2 \quad (I)$$

$$\hat{Y}_2 \cdot 10^{-5} = 3.06 - 0.86x_1 + 0.66x_2 - 0.36x_3 - 0.23x_1x_2 + 0.12x_1x_3 + 0.17x_2x_3 + 0.13x_1^2 + 0.39 \cdot 10^{-1}x_2^2 + 0.18x_3^2 \qquad (2)$$

$$\hat{Y}_3 = 2.88 - 0.43x_1 - 0.13 \cdot 10^{-1}x_2 + 0.55x_3 + 0.51 \cdot 10^{-1}x_1x_2 - 0.19x_1x_3 + 0.19 \cdot 10^{-1}x_2x_3 - 0.12x_1^2 - 0.29x_2^2 + 0.19x_3^2 \qquad (3)$$

In the region of changing of variables of x_1, x_2, x_3 one can study influence of one of the factors on $\bar{M}_n$, $\bar{M}_w$ or $\bar{M}_w/\bar{M}_n$ by means of obtained equations of regression fixing the rest factors at the definite level.

In fig. 2 dependencies on initial stage temperature t_1 are represented for fixed values of S_1 and t_2. Dependencies of this type can be obtained for other values of S and t as well. Fig. 3 shows dependencies of $\bar{M}_w/\bar{M}_n$ on final stage temperature of polymerization at fixed t_1 and S_1. The equations (1), (2) and (3) make easier to a great extent the search of optimum regimes of polymerization which would allow obtaining polystyrene with the predicted values of $\bar{M}_n$, $\bar{M}_w$ and $\bar{M}_w/\bar{M}_n$, i.e. solving practical purposes. For example, in table III. the temperature regimes of thermal polymerization of styrene are given, they allow to obtain polymer with $\bar{M}_n$ approximately at $1.5 \cdot 10^5$ and with various values of the ratio of $\bar{M}_w/\bar{M}_n$.

In the table IV temperature regimes of polymerization are given, which allow to obtain polymer with $\bar{M}_n$ approximately at $1.0 \cdot 10^5$ and with the ratio of $\bar{M}_w/\bar{M}_n$ approximately at 3. These tables also give time necessary for polymerization for each temperature regime.

Initiated polymerization of styrene.

Kinetics of initiated radical polymerization of styrene is well studied at the initial stages. At the high-conversion of polymerization the growth of viscosity of media- influences the mechanism and kinetics of process greatly (12). In our research we have attempted to obtain the empirical dependence of some rate constants of initiated polymerization of styrene on conversion with the aim of using them at mathematical simulation of the process.

Investigation of kinetics of styrene polymerization was conducted the temperature interval of

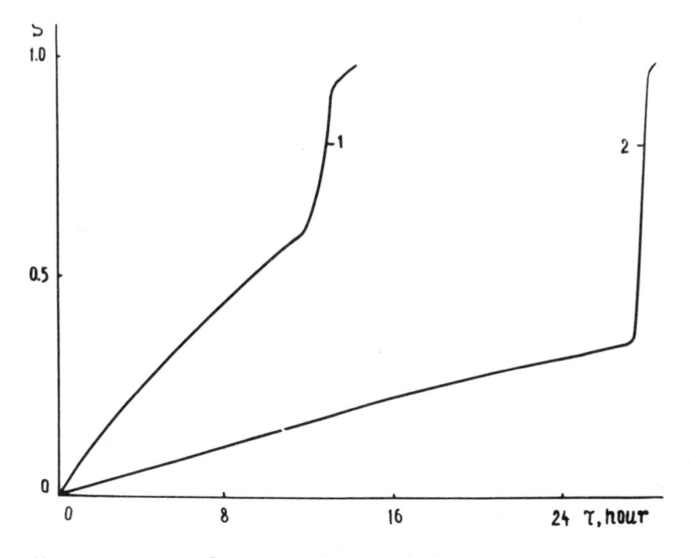

Figure 1. Dependence of styrene conversion on time. Step temperature regimes. 1, regime I; 2, regime II (Table I).

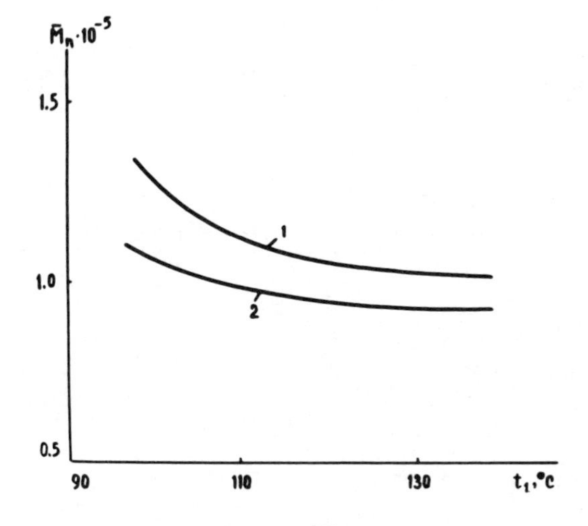

Figure 2. Dependence of $\overline{M}n$ on initial stage temperature. 1 — t_2 = 180°C, S_1 = 30%; 2 — t_2 = 100°C, S_1 = 40%.

Table III.

Temperature regimes of styrene polymerization: $\bar{M}_n = 1.5 \cdot 10^5$

№№	$t_1, ^oC$	$S_1, \%$	$t_2, ^oC$	$\bar{M}n \cdot 10^{-5}$	$\bar{M}_w/\bar{M}_n$	τ, hour
1.	90	30	190	1.5	3.0	27.0
2.	90	40	190	1.6	3.2	33.1
3.	90	60	210	1.4	4.2	44.7
4.	90	70	220	1.5	5.6	51.3
5.	100	40	190	1.4	3.1	21.0
6.	100	50	190	1.6	3.2	26.0
7.	110	70	210	1.4	3.6	22.6
8.	110	80	220	1.5	3.8	26.8
8.	120	40	180	1.4	2.6	7.3
9.	90	50	200	1.5	4.0	38.7
10	100	60	200	1.5	3.5	30.8

Table IV.

Temperature regimes of styrene polymerization $\bar{M}n = 1 \cdot 10^5$, $\bar{M}w/\bar{M}n = 3$

№№	$t_1, ^oC$	$S_1, \%$	$t_2, ^oC$	$\bar{M}n \cdot 10^{-5}$	$\bar{M}w/\bar{M}n$	τ, hour
1.	90	30	200	1.2	3.0	26.4
2.	100	30	200	1.0	3.0	15.4
3.	100	40	200	1.2	3.1	20.3
4.	110	50	200	1.1	3.1	16.1
5.	130	50	210	0.9	2.9	3.8
6.	130	70	220	1.1	3.0	6.7
7.	140	80	160	1.0	3.0	9.4

60-100 °C, i.e. in conditions at which initiators are not consumed too fast. As initiators benzoyl peroxide (Bz_2O_2) and t-Butyl Perbenzoate (TBPB) were chosen, they are most widely used in practice. Initial concentrations were chosen so that MW of obtained polymers were within the range of $10^4 - 10^6$.

In fig. 2 and 3 kinetic curves of polymerization of styrene are represented in presence of Bz_2O_2 and TBPB at temperatures of 75 and 80° C. Analogous curves for other temperatures have also been obtained. The given data shows that on all kinetic curves of polymerization one can distinguish three parts, which differ greatly from one another. Up to conversion of ~40% polymerization proceeds approximately at constant rate. But for one and the same initiator at this part the rate of polymerization is no longer constant as the polymerization temperature grows.

As the result of fast decomposition of initiator the rate of initiation decreases sharply and this leads to decrease of polymerization rate at the degree of monomer transfer of~20%. In consequence of decrease of initiator concentration the sharp decrease of initiation rate is not compensated by the decrease of rate constants of chain termination for at this stage of the process molecular weight of obtained polymer is not high and viscosity of polymerization system grows not too fast.

The second stage of reaction is characterized by considerable growth of polymerization rate as a result of sharp decrease of rate constant of termination of macromolecular radicals, acceleration, with respect to initial rate, growing with the growth of MW of polymer being obtained (fig.4).

The degree of conversion being 85-90% the sharp decrease of polymerization rate takes place, and at conversion of~95% the rate of reaction practically equals zero.

Monomer does not fully transform into polymer, maximal degree of polymerization decreasing with the decrease of temperature and initial concentration of initiator. In our opinion sharp decrease of polymerization rate is connected not only with the decrease of monomer concentration, but also with sharp decrease of constant of growth rate.

The rate of polymerization is described by the known thesis:

$$\frac{d\ [M]}{d\ \tau} = \frac{Kp}{Kt^{1/2}} \left(2f\ Kd\right)^{1/2} [I]_o^{1/2} \exp\left(-Kd\cdot\tau/2\right) [M] + [M]\ /W\cdot dW/\ d\tau \qquad (4)$$

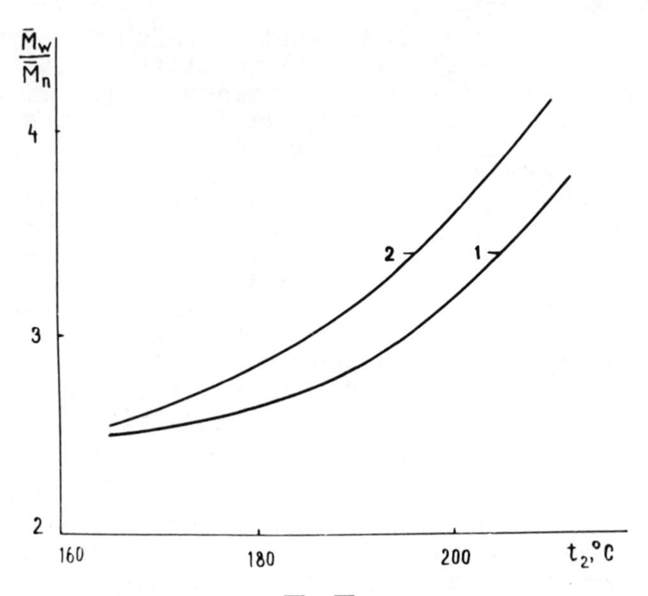

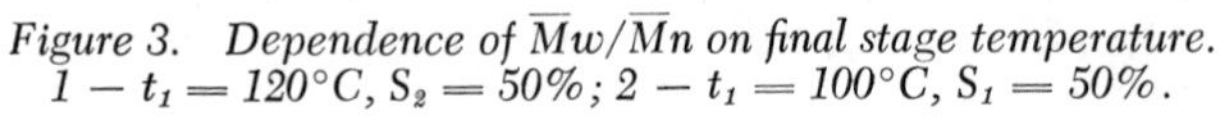

Figure 3. Dependence of $\overline{M}w/\overline{M}n$ on final stage temperature. 1 — $t_1 = 120°C$, $S_2 = 50\%$; 2 — $t_1 = 100°C$, $S_1 = 50\%$.

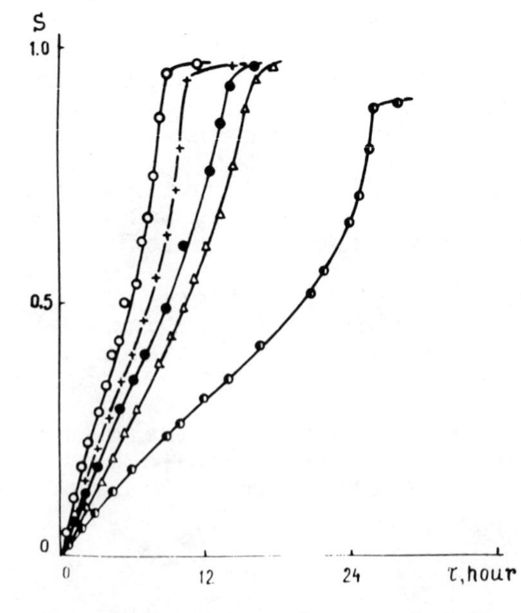

Figure 4. Kinetic curves of styrene polymerization. $t = 75°C$, Bz_2O_2 concentration, mol/l. · 10^2: +, 1.2; ●, 0.9; △, 0.7; ○, 0.2; ⭘, 2.0.

where W is volume of mass being polymerized.

The values of polymerization rate and initiation rate being known, from equation (4) we may calculate the ratio of constants of propagation and chain termination rate $K_p/K_t^{1/2}$ for any degree of monomer conversion. Dependences of $K_p/K_t^{1/2}$ on polymerization degree obtained in this way could be described by the function of definite type. But at the high stages of polymerization change of viscosity of media can considerably change not only the rate of propagation and termination chain, but also the initiation rate in consequence of change of efficiency of initiation f and rate constants of decomposition of initiator K_d. If this fact takes place, the calculated values of $K_p/K_t^{1/2}$ according to equation (4), f and Kd being constant, would differ greatly from real ones. In this case for different initiators would be "their own" values of $K_p/K_t^{1/2}$.

With the purpose of finding the viscosity of media influence on the constant of decomposition rate of Bz_2O_2 the kinetics of its decomposition during the process of styrene polymerization up to high stages was investigated. To define the concentrations of Bz_2O_2 in styrene-polystyrene system we applied the polarographic method, for application of other methods is difficult due to high viscosity of media.

In order to check if there is decrease of Bz_2O_2 concentration in solution in presence of polystyrene due to peroxide adsorption by the large surface of precipitated polymer , the calibration curves were built for Bz_2O_2 both in presence and absence of polymer. In fig.5 the calibration curve for Bz_2O_2 is presented, which had been obtained both in presence and absence of polystyrene. On the axis of ordinates the height of a wave is marked in mm, on the abscissa axis the concentration of Bz_2O_2 in solution obtained after the precipitation of polymer is marked. As seen in this figure all points lie on the straight line; hence, the precipitated polymer, does not adsorb peroxide from solution on its surface. In fig.6 the kinetic curve of Bz_2O_2 decomposition during the process of styrene polymerization up to the high stages at 75°C is given. On the axis of ordinates the weight part of peroxide in charge of polymerization mixture is marked. It is seen in this figure, that experimental and calculated from the equation

$$[I] = [I]_0 \exp(-K_d \tau) \qquad (5)$$

Curves of peroxide decomposition coinside within the limits of experiment error, which indicates the nonchanging of constant of Bz_2O_2 decomposition rate

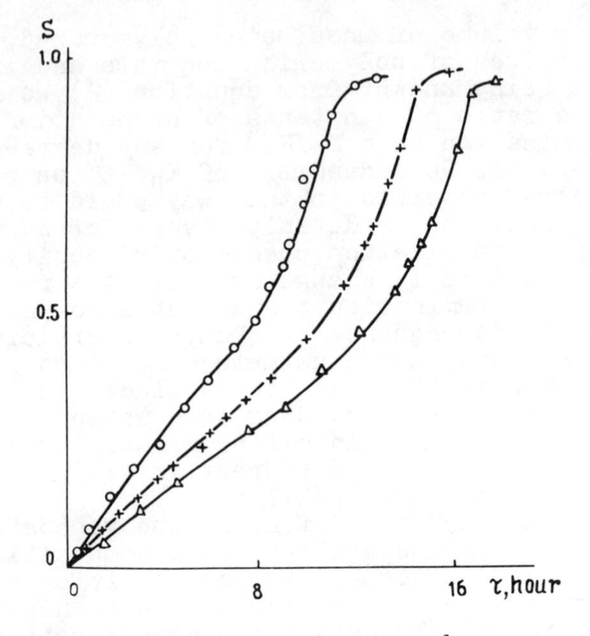

Figure 5. Kinetic curves of styrene polymerization. $t = 80°C$, TBPB concentration, mol/l. $\cdot$ 10^1: ○, 1.0; +, 0.75; △, 0.5.

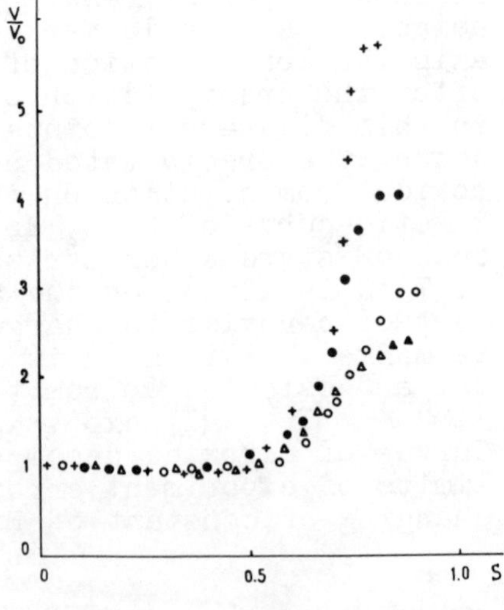

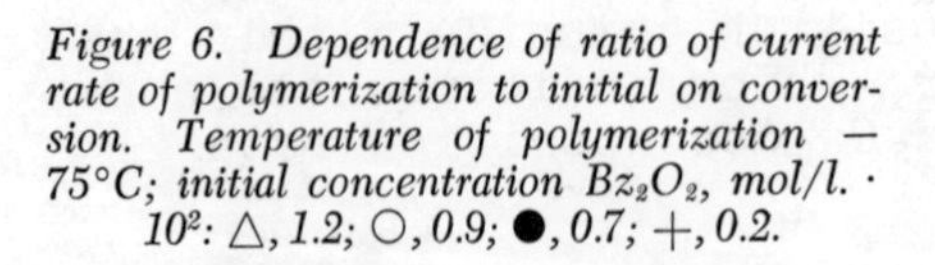

Figure 6. Dependence of ratio of current rate of polymerization to initial on conversion. Temperature of polymerization — 75°C; initial concentration Bz_2O_2, mol/l. $\cdot$ 10^2: △, 1.2; ○, 0.9; ●, 0.7; +, 0.2.

during the polymerization of styrene up to the high stages. Data given agree with the found results of the rate constants at Bz_2O_2 decomposition in model mixtures with different contents of polystyrene in CCl_4 (13,14).

Efficiency of initiation can be found on the basis of measuring the average degree of polymerization supposing that the reaction of chain transfer to the polymer may be disregarded (15). In this case we can write the following:

$$\bar{P} = \frac{[M]_o - [M]}{f\ \{[I]_o - [I]\}} \qquad (6)$$

or taking the change of volume during the polymerization into account:

$$\bar{P} = \frac{[M]_o \qquad S}{f\ [I]_o [1 - \alpha S - \exp(-Kd/\tau)]} \qquad (7)$$

To calculate the degree of polymerization measuring MW of polystyrene specimens was carried out, the specimens having been obtained at different degrees of polymerization. Using the equation (7) on the basis of data on the styrene polymerization the values of $\bar{f}$ have been calculated(average "accumulated") for different degrees of conversion of monomer, these values are represented in table IV. As seen from the obtained data there is no definite dependence of initiation efficiency by benzoyl peroxide on the degree of polymerization.

On the basis of the obtained results we have assumed, that in case of styrene polymerization in the presence of Bz_2O_2 the rate of initiation changes only due to change of concentration of peroxide. Possible change of initiation efficiency will, in this case, be included in the effective value of $Kp/Kt^{1/2}$.

To calculate the rates of styrene polymerization with the help of experimental data on conversion-time the method of the digital differentiation using five points (16) was applied. In table V the values of $Kp/Kt^{1/2}$ for different conversions are represented, the values have been obtained during the styrene polymerization in the presence of different initial concentrations of Bz_2O_2 at temperature of 70° C. Up to conversion of ~ 40% the ratio of $Kp/Kt^{1/2}$ somewhat increases and does not depend on the initial concentration of initiator. But at more high stages of polymerization the difference of MW of polymer being formed reveals itself and the ratio of $Kp/Kt^{1/2}$ increases as quickly as the initial concentration of initiator decreases. Dependences of $Kp/Kt^{1/2}$ on the con-

Table V.
Values of $\bar{f}$ for benzoyl peroxide at different degrees of polymerization.

$\bar{f}$	75 °C			70°C	
$[I]_0$ mol/l 10^2 / S ,%	1.2	0.9	0.7	0.9	0.7
10	0.72	0.66	0.75	0.70	0.80
15	0.83	0.64	0.73	0.73	0.78
20	0.85	0.67	0.75	0.73	0,79
25	0.71	0.67	0.79	0.74	0.73
30	0.83	0.66	0.63	0.74	0.78
40	0.80	0.69	0.72	0.74	0.69
50	0.79	0.70	0.69	0.74	0.72
55	0.77	0.70	0.62	0.75	0.66
60	0.75	0.70	0.60	0.66	0.69
65	0.70	0.70	0.59	0.73	0.71
70	0.70	0.66	0.65	0.76	0.70
75	0.70	0.64	0.60	0.75	0.75
80	0.70	0.63	0.65	0.73	0.73
90	0.68	0.66	0.58	0.67	0.74
95	0.69	0.68	0.60	0.66	0.70

Table VI.
Values of $Kp/Kt^{1/2}$ for different degrees of polymerization at T= 70°C

$[I]_0$ mol/l 10^2	1.5	0.9	0.7
S , %	$Kp/Kt^{1/2} \cdot 10^2$	$Kp/Kt^{1/2} \cdot 10^2$	$Kp/Kt^{1/2} \cdot 10^2$
5	2.92	2.90	2.93
10	3.11	3.00	2.96
20	3.52	3.48	3.54
30	4.22	4.26	4.51
45	6.41	6.51	7.43
50	7.19	9.52	10.4
60	12.5	15.0	20.1
65	18.2	21.5	24.3
70	26.4	27.2	34.1
75	36.1	44.6	48.1
80	50.3	64.1	71.2
85	68.1	93.9	114

version for other temperatures are of analogous character. This agrees completely with the data given in figure 4, concerning the rate of polymerization at high stages for different initial concentrations of Bz_2O_2. Thus for one and the same temperature the ratio of rate constants $Kp/Kt^{1/2}$ may be described by some surface of response with the coordinates of $Kp/Kt^{1/2}$ - S. To approximate the dependence of $Kp/Kt^{1/2}$ on conversions the following equation was chosen:

$$Kp/Kt^{1/2} = A_0 \exp\left(A_1 S + A_2 S^2 + A_3 S^3 + A_4 \bar{P} S \right), \quad (8)$$

where S is specific conversion; A_0, A_1, A_2, A_3, A_4 - coefficients which do not depend on conversion, A_0 - being the value of $Kp/Kt^{1/2}$ at zero conversion; $\bar{P}$ - the degree of polymerization.

The search of appropriate coefficients for the equation (8) was done by the method of regressive analysis. Dependences of coefficients on the temperature are given below:

$$\begin{aligned} A_0 &= 409 \exp(-6500/RT), \\ A_1 &= 31.0 - 9.56 \cdot 10^{-2}\, T, \\ A_2 &= -103 + 0.32T, \\ A_3 &= 109 - 0.32T, \\ A_4 &= 2.43 \cdot 10^{-3} - 5.22 \cdot 10^{-6}\, T \end{aligned} \quad (9)$$

According to mathematical model we have calculated the kinetics of the initiated polymerization of styrene up to the high stages using the equation (8). As seen in fig.7 calculated and experimental data agree with each other satisfactorily. As in case of the thermal styrene polymerization the kinetics of the initiated styrene polymerization for obtaining the polymer with predicted MW and MWD can be calculated. With the purpose of obtaining polymer with narrowing MWD at the initiated styrene polymerization we have calculated the temperature-time regimes of polymerization, which allow to maintain constant the average degree of polymerization during the whole process. Here we took into consideration the following: sizes of macromolecules of polymer are characterized by the average degree of polymerization $\bar{P}$, i.e. by the number of monomer molecules included into the macromolecule or by the average length of the kinetic chain:

$$\nu_1 = \frac{Kp\,[M]}{2\,\{fKd \cdot Kt \cdot [I]\}^{1/2}} \quad (10)$$

or taking into account the chain transfer to the monomer and initiator:

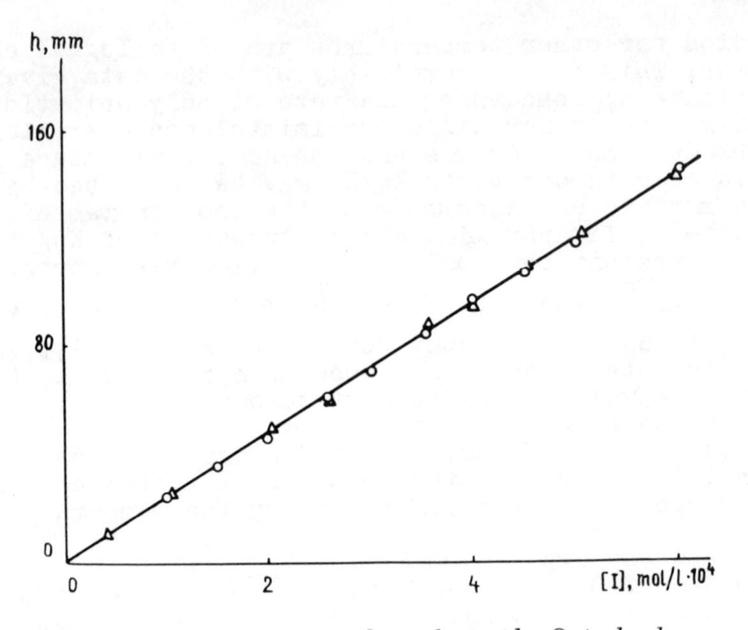

Figure 7. Calibrating diagram for benzoyl peroxide. ○, in the absence of polymer; △, in the presence of polymer.

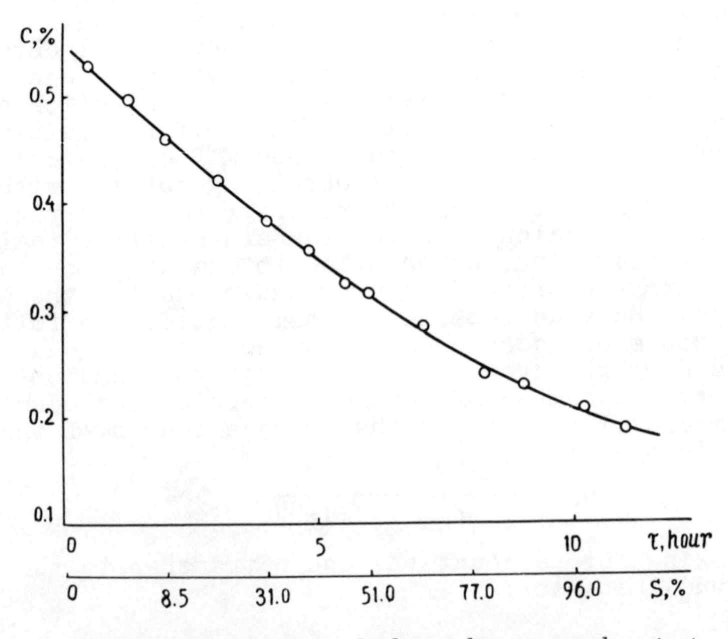

Figure 8. Decomposition of Bz_2O_2 during the styrene polymerization up to high stages. Temperature — 75°C; $[I]_0 = 2.0 \cdot 10^{-2}$ mol/l.; ○, experimental; —, calculated according to equation (5).

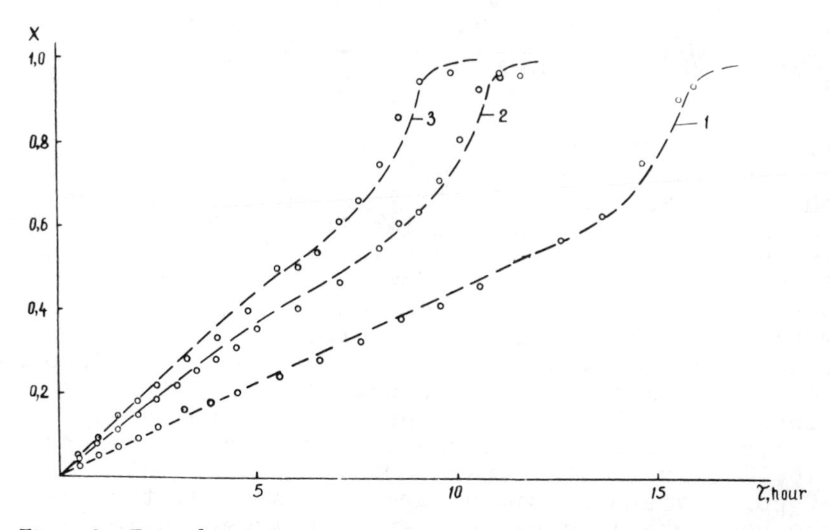

Figure 9. Dependence of styrene conversion on time. 1 — t = 70°C, $[I]_o$ = 1.2 · 10^{-2} mol/l.; 2 — t = 75°C, $[I]_o$ = 1.2 · 10^{-2} mol/l.; 3 — t = 75°C, [I] = 2.0 · 10^{-2} mol/l.; initiator — Bz_2O_2; – – –, calculated; ○, experimental.

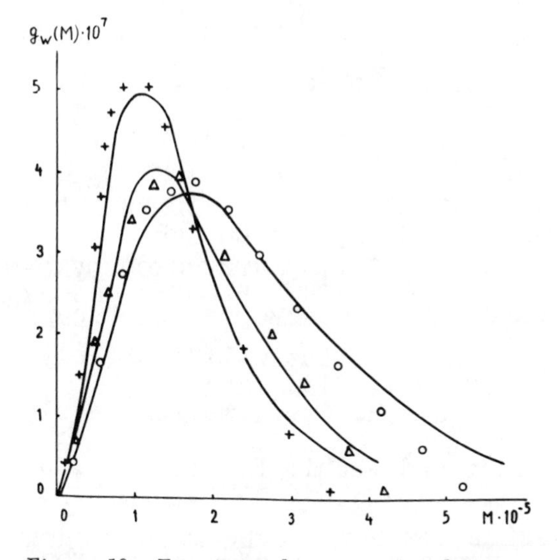

Figure 10. Experimental curves of polystyrene MWD. 1, S = 11.5%; 2, S = 39.9% (isothermal polymerization, t = 60°C, $[I]_o$ = 0.0142 mol/l.); 3, S = 42.1% (regime); —, calculated; ○, +, △, experimental.

$$\nu_2 = \frac{1}{\frac{1}{\nu_1} + 2\,\frac{K_f m}{K_p} + 2\,C_I\,\frac{[I]}{[M]}} \qquad (11)$$

i.e. by the number of elementary acts of chain propagation, which were takingplace from the moment of chain initiating up to its termination. Thus the sizes of the obtained macromolecules can be expressed by the length of the kinetic chain ν_2; which is the function of temperature (for the rate constants depend on temperature) and concentration of reacting substances. It is to be noted that the value of ν_2 (and, thus, of MW) can be predicted beforehand, proceeding from the initial conditions of polymerization. Algorythm of computing the temperature-time regimes of polymerization was described in papers (16,17). In fig.10 experimental and calculated MWD of polystyrene specimens are represented, the latter has been obtained by isothermic polymerization and also according to the calculated temperature-time regimes. Thus, the experimental data obtained prove the theoretical suppositions.

LITERATURE CITED:

1. T. Alfrev, Jn. Mechanical Bhaviour of High Polymers, Interscience, N.Y., 1948.
2. H.W. Mc Cormick, F.M. Brewer, L.Kin, J. Polymer Sci.,(1959), 39, 87.
3. Р.Хувинк,А.Ставерман,Химия и технология полимеров, т.2,изд."Химия",М.-Л.,I965.
4. R.F. Hoffman, S. Schreiber, G. Rosen, Jn. Eng. chem., (1964), 56 , 51.
5. F.Y. Teaney, R.G. Antony , J. Appl. Polymer Sci., (1970), 14, 147.
6. В.Д.Безуглый,Полярография в химии и технологии полимеров,изд."Химия",Л.,I968,I46.
7. А.И.Шатенштейн и др.,Практическое руководство по определению молекулярных весов и молекулярно-весового распределения полимеров,изд."Химия",М.-Л., I964.
8. W.A. Pryor and L.O. Lasswell, ACS Polymer Preprints, (1970), 11, 713.
9. W.A. Pryor and J.H. Coco, Macromolecules, (1970), 3, 500.
10. Albert W.Hui and Archie E. Hamielec, J. Appl. Polym. Sci., (1972), 16, 749.
11. Л.П.Рузинов,Статистические методы оптимизации химических процессов,изд."Химия",М.,I972.
12. Г.П.Гладышев,К.М.Гибов,Полимеризация при глубоких степенях превращения и методы ее исследования,

изд."Наука",Каз.ССР,А.-А.,1968.
13. Г.П.Гладышев,П.Е.Мессерле.Т.Т.Омаров,С.Р.Рафиков, Докл.АН СССР,(1966),168,1093.
14. С.С.Иванчев,Д.В.Скубилина,Е.Т.Денисов.Высокомолек. соед.(1967),Б9,706.
15. Nichimura N., Bull. Chem.Soc., (1958), 80, 5927.
16. В.Д.Енальев,Дж.Х.Харвуд,В.И.Мельниченко,Тезисы X Украинской республиканской конференции по физической химии,изд."Наукова думка",К.,1972,стр.21.
17. В.Д.Енальев,Доп.АН УРСР (1973),серIя Б,№ 7,623.

15

Redox Polymerization in Emulsion

HENRY WARSON

Solihull Chemical Services, 284 Warwick Rd., Solihull, England, B92 7AF

Redox polymerisation, the use of an oxidizing and a reducing agent simultaneously to ensure a rapid free radical polymerisation at relatively low temperatures, i.e. ambient or even sub-ambient, was discovered independently in the USA, Great Britain and Germany during the period 1940-5, but not generally published until 1946. Since work prior to 1955 has been reviewed in detail (1), this paper will be restricted to mention of a few newer systems, and to some ideas of the mechanisms of redox systems.

THEORETICAL TRENDS

The vast number of redox processes which have been disclosed and which will be the subject of a separate publication are not matched by theoretical developments in elucidating the systems. A number of general features are becoming slowly established. One is the pH sensitivity of many systems, particularly where there is a third component. Thus whereas the potassium persulfate - sodium bisulfite system is accelerated by Fe^{2+} in acidic media, acceleration of polymerisation occurs with Cu^{+} in an alkaline medium. Unusually with Cu^{+}, a second peak occurs at alkaline pH if both methyl methacrylate and vinyl acetate are added, one of the few examples where a vinyl acetate copolymer is formed in the presence of a copper compound. (2). Complex formation is indicated, producing active intermediates which form radicals.(3) Vinyl bromide is polymerised in presence of H_2O_2 and a chelate iron compound at 32°.(4) This has already been mentioned by Bacon(1), whilst coordination compounds and their effect on radical polymerisation have been discussed.(5)

Redox systems which have been the subject of recent examination include potassium permanganate - tartaric acid (6), and potassium persulfate - ascorbic acid.(7) Whilst experiments were with the water soluble acrylamide, they should be adaptable to emulsion conditions. The ascorbic acid reductant is of interest as it is not interfered with by air or monomer stabilisers. With $KMnO_4$ - tartaric acid, the termination rate seems to be bimolecular.

In the view of the current writer it is probable that in most cases the reducing moiety plays the major part in forming the radical which initiates polymerisation. This was suggested by Sully (8), but evidence, albeit indirect, comes from another source. When a persulfate - bisulfite redox initiation of vinyl acetate is performed, and the resultant polymer film, optionally pigmented applied to a steel substrate, it has been observed that this film is much more efficient in imparting corrosion resistance than one prepared with identical stabilisers, but with a persulfate initiator only. The effect is most marked with copolymers of vinyl acetate and the vinyl ester of a highly branched C_{10} acid, known as "Veova" (Europe) or "VV10" (USA - Shell Chemicals). This may be accounted for by assuming that the radical initiating polymerisation is such that there is a sulfur to carbon bond, $XO_3S.CH_2CH(OCOCH_3)$....... as an end group rather than an oxygen to carbon bond as $XO_3SOCH_2CH(OCOCH_3)$......... The latter would probably occur with a radical $^{\bullet}SO_4$ derived from a persulfate. Whilst a sulfate end group could be assumed to hydrolyse fairly readily, hence giving rise to corrosion of a substrate, hydrolysis is much less likely to occur where there is a sulfonate end group as would probably result from a radical derived from a sulfur reducing acid. It is also to be noted that in the hypophosphorous acid - diazonium salt system, to be mentioned later, there is strong evidence that there is bound phosphorus in the poly(acrylonitrile formed. (9) (10) (40)

Another interesting factor in emulsion polymerisation is varying solubilities of the initiators in the monomers. One suggestion is that the most active hydroperoxides are those which are the least soluble in the aqueous phase. Reductants are not indicated, the reference being to USSR rubber production. An explanation might well be that the radical - producing reaction is fundamentally a surface one, under which conditions byproducts and secondary reactions are at a minimum. The differential reactivities of the hydroperoxides might also be a factor, and careful comparisons would be necessary, operating at such temperatures that their half lives were equal. (11)

There seems little doubt that the most efficient way of operating a redox system is to add the components in stages as suggested in earlier publications, by gradual addition, or by adding 1 component, usually the per-compound, at the start, and adding the reductant gradually. (12) (13).

Some emulsifiers act directly as reductants. This applies particularly to quaternary ammonium compounds. Thus cetyl pyridinium bromide, with benzoyl eroxide or hydrogen peroxide will initiate the polymerisation of chloroprene, methyl methacrylate and styrene in emulsion, although the mechanism is not clear; possibly there may be a radical formed from the quaternary nitrogen. It is interesting to note that polystyrene, prepared in emulsion with cetyl pyridinium bromide and hydrogen peroxide , has a syndiotactic structure , suggesting that under some conditions these systems are stereo-directive. (14)(15) It is possible that under special conditions the hydroxyl groups in many nonionic surfactants may have suitable reducing properties for redox initiation. Polyvinyl alcohol may also function as a redox component under some circumstances. (see below)

Graft Copolymers

Graft copolymerisation on to an existing polymer backbone has proved to be one of the major interests with redox polymerisation systems. The main factor for efficiency is the removal of a hydrogen atom from a C atom in the polymer backbone to produce a radical on which a branch chain will form. The reductant probably functions by assisting in the rapid formation of hydroperoxide radicals in the first place, e.g., from cumene hydroperoxide, probably the most frequently encountered of graft initiators. It is used in conjunction with sodium sulfite, sodium formaldehyde sulfoxylate, dextrose and dextrin, with a range of polymers and copolymers. (16)(17)(18)(19). The very frequent simultaneous inclusion of tert-dodecyl or a similar mercaptan, not only with butadiene copolymers, suggests that it plays a direct part in the initiation, rather than act as only a chain transfer agent. (20)

Polyvinyl alcohol, when used as an emulsifier, undoubtedly acts as a graft base to some extent, and on to which the monomer, usually vinyl acetate, forms branches during polymerisation. Its water solubility assists the effciency of grafting, which may be best with the partially hydrolysed polyvinyl acetate normally used as an emulsion stabiliser. The type of polymer, either formed on a "seed" of about 20 - 25% of itself, or by a "continuous" process is probably so highly branched as a result of multiple grafts that it is de facto crosslinked, as shown by the lack of solubility in ethyl alcohol. In the standard preparation, with either hydrogen peroxide or tert-butyl hydroperoxide together with Na (or Zn) formaldehyde sulfoxylate, these are added continuously with monomer. These conditions produce the maximum graft, and a stable emulsion, of particle size about 1μm, although with a rather wide scatter of sizes, is formed, and the latex does not gel on addition of ethyl alcohol. It is not possible to hydrolyse these modified polyvinyl acetates completely to polyvinyl alcohol. The reducing groups in this case may provide charged end groups to the polymer chains, improving latex stability. A recent paper discusses further details of graft copolymerisation in emulsion. (21)

Chloroformic Ester with a Peroxide or a Persulfate

As early as 1946 the current author, working in the laboratories of Vinyl Products, Carshalton, England, was endeavouring to reduce the MW of polymethyl methacrylate formed in emulsion, (22)(23) with various halogen - containing compounds, one of them being ethyl chloroformate $Cl.COOC_2H_5$. Experiments were of a simple character, a pre-emulsion being heated in a water bath in a 2 oz. sealed bottle, hydrogen peroxide being the initiator.

An extremely rapid reaction took place on heating, the polymer formed precipitating, the liquid having a pH about 2.

In further examination of the hydrogen peroxide - ethyl chloroformate initiation system, with emulsions as known at the time, it was found that polymerisation of methyl methacrylate or vinyl acetate could be controlled readily at ambient temperature, there being some inhibition period. This tended to be prolonged when emulsions were stirred under open reflux, but vinyl acetate emulsions commenced to polymerise with extreme vigour at the azeotropic reflux temperature (66.5°). These experiments, performed before the symposium in London in which the nature of reduction activation was disclosed, were not pursued further at the time.

A disclosure was made in 1961 of polymerisation in emulsion with a chloroformic ester, a buffer such as $NaHCO_3$, and, rather unusually, permonosulfuric acid. (24). This system operates best at acid pH, and the presence of heavy metals, e.g., 0.001 - 0.5% of iron, cobalt, copper or silver is desirable. Chloroformic ester is added continuously during the polymerisation. In a typical example vinyl chloride is polymerised with 0.25% of potassium permonosulfate and 0.2% of methyl chloroformate, the inclusion of the latter improving the yield in 8 hours at 52°from 20.1% to 77.9%. The process is also suitable for the polymerisation of styrene. The specification is associated with 2 others, describing, inter alia, the reaction products of cyclohexyl chloroformate and potassium permonosulfate, yielding in situ cyclo-C_6H_{11}-O-C(O)-O-O-SO_2OK as a free radical initiator, operating in emulsion (25). These specifications assume that the sole function of the chloroformic ester is to form the reactive per-compound in situ.

Subsequently there have been a number of specifications claiming the use of chloroformic ester with H_2O_2 for initiation, principally for vinyl chloride in suspension. Whilst percarbonate is known to give fast copolymerisation, it is difficult to account for some of the earlier phenomena by the assumption that chloroformic ester plus a persulfate or H_2O_2 form percarbonate in situ, especially as a multivalent metal ion seems desirable with a persulfate. (26 - 32).

Some earlier papers have suggested that during the fission of chloroformic esters there may be unimolecular fission. (33 - 36). Thus with isopropyl chloroformate there is alkyl - oxygen fission. Polymerisation with this initiator may occur as follows:

Scheme A. $ClCOOC_2H_5 = {}^{\bullet}Cl + {}^{\bullet}COOC_2H_5$ (medium speed) (1)

$${}^{\bullet}COOC_2H_5 + H_2O_2 = HOCOOC_2H_5 + HO^{\bullet} \text{ (fast)} \quad (2)$$

$${}^{\bullet}Cl + H_2O_2 = H^+ + Cl^- + HO^{\bullet} \text{ (fast)} \quad (3)$$

Equation (3) is followed by either (4) or (5)

$$HOCOOC_2H_5 = HOC_2H_5 + CO_2 \quad (4)$$

$$2HOCOOC_2H_5 = (C_2H_5)_2CO_3 + CO_2 + H_2O \quad (5)$$

Scheme B:

Alternative mechanism from isopropyl chloroformate.

$$Cl.COOC_3H_7 = ClCOO^{\bullet} + {}^{\bullet}C_3H_7 \quad (6)$$
$$^{\bullet}C_3H_7 + H_2O_2 = C_3H_7OH + {}^{\bullet}OH \quad (7)$$
$$ClCOO^{\bullet} = CO_2 + Cl^{\bullet} \quad (8)$$

(8) is followed by (3) and once again $HO^{\bullet}$ initiates polymerisation

The above equations assume that there is no metal single electron activator.

A transition metal ion is desirable under some conditions. A ferrous - ferric transformation such as (9) would imply that chloroformic ester has oxidizing properties and is improbable.

$$Fe^{2+} + ClCOOC_3H_7 = Fe^{3+} + Cl^- + {}^{\bullet}COOC_2H_5 \quad (9)$$

The following may be a possibility:

$$Fe^{2+} + H_2O_2 = Fe^{3+} + HO^{\bullet} + {}^-OH \text{ (Haber - Weiss)} \quad (10)$$
$$Fe^{3+} + ClCOOC_3H_7 = Fe^{2+} + ClCOO^{\bullet} + {}^+C_3H_7 \quad (11)$$

This is followed by (8), thence (3)

$$ClCOO^{\bullet} = CO_2 + {}^{\bullet}Cl \quad (8)$$
$$^{\bullet}Cl + H_2O = H^+ + Cl^- + HO^{\bullet} \quad (3)$$

and also by

$$^+C_3H_7 + H_2O = H^+ + C_3H_7OH \quad (12)$$

$HO^{\bullet}$, formed as in (3) initiates the polymerisation, but this scheme is very speculative

DIAZONIUM SALTS IN REDOX SYSTEMS

Kornblum et al. (36)(37) have establised that the reaction between diazonium salts and hypophosphorous acid is unimolecular. Nonhebel and Waters (38) have shown that the first stage of the Sandmeyer reaction

$$ArN_2^+ + Cu^+ = Ar^{\bullet} + N_2 + Cu^{2+} \quad (13)$$

where Ar is an aryl radical, can be markedly accelerated by hypophosphorous acid, particularly in presence of a cupric salt. Some quantitative studies od diazonium salts as initiators of polymerisation have been made. (39)

A systematic study of the diazonium salt/hypophosphorous acid redox initiation is available. (40) Whilst this involves an aqueous solution - precipitation system, it would be interesting to apply the results to emulsions. Under the acid conditions used there is evidence that the hypophosphite radical($H_2PO_2^{\bullet}$) is that which actually initiates polymerisation. Electrophilic substituents in the diazonium salt, e.g., p-nitro, accelerate the polymerisation rate. Several schemes have been suggested to account for the reaction. (41) One involves the formation of an ion radical:

$$(ArN_2)^+ + (H_2PO_2)^- = (Ar\dot{N}_2H)^+ + (H\dot{P}O_2)^- \quad (14)$$

$$(ArN_2H) + (H_2PO_2)^- = ArH + N_2 + (H_2PO_2)^\bullet \quad (15)$$

$$CH_2{:}CHCN + (H_2PO_2)^\bullet = H_2PO_2CH_2CHCN^\bullet \quad (16)$$

Polymerisation then continues.

The cupric salt probably functions by being first reduced to the cuprous state, providing a hypophosphorous radical which causes initiation, and at the same time superimposed is the first stage of the Sandmeyer reaction providing aryl radicals. There is evidence that copper is chemically combined with the polymer, and copper compounds may also function as chain - terminating groups.

The diazonium salt - sodium bisulfite system also initiates polymerisation in the absence of hypophosphorous acid. (42) There is an optimum molar ratio of 10 : 1.75 respectively for the maximum R_p. The decrease in R_p when the bisulfite ratio is increased may be due to the formation of an addition compound between acrylonitrile and sodium bisulfite.

Amongst other compounds which function as accelerators with diazonium salts are thiourea dioxide (formamidine sulfinic acid), and p-tolylhydrazine. Whilst copper is the most efficient of the metal ions, catalytic quantities giving a large increase in R_p, some other transition metals have a marked effect. These include titanic sulfate and vanadyl sulfate.

In the most recent developments, a peroxy group and an aliphatic azo group have been used in conjunction in the same molecule. (43). This is of major interest in the formation of block copolymers because of the active end groups.

It seems probable that diazonium salts might be efficient initiators in the presence of radiation, in which respect they would have some semblance to the azo-bis initiators. (44) An extensive review id available of free radical reactions of diazonium salts, with limited reference to polymerisation. (45)

ADDENDUM. MIXED SULFUR REDUCING ACIDS

It has recently been shown that ammonium persulfate (2 millimoles) together with sodium pyrosulfite ($Na_2S_2O_5$) (1.5 millimoles) AND SODIUM THIOSULFITE(0.5 millimoles), together with copper sulfate (0.002 millimoles), buffered with sodium bicarbonate, the quantities being in 1 litre of water, form an exceedingly effective redox system for vinyl acetate emulsion polymerisation, the reaction, commencing at 25°, being virtually complete in 30 minutes with an exotherm rising to 70°. (46)

Literature Cited

1. Bacon, R.G.R., Quart. Reviews, (1955), 9, 287-310
2. Belyakova, A.V. et al., Volokna Sin. Polim.(1970), 62-7
3. Brown, C.W., Hobson, D.B., J. Appl. Pol. Sci.(1974), 18(5) 1269-77

4. Tsuchida, H.(Moriroku Shoji), Japan Kokai 73-20,884
5. Tazuke, S., Progress in Polymer Sci.(Japan), (1971), 1, 69-148
6. Misra, G.S. Ravello, J.J. Makromol. Chem.(1974),175, 3117-31
7. Mehta, P.C. Nair, G.P. India P. 131,842, 1974
8. Sully, R.D. J. Chem. Soc., 1950, 1498
9. Warson, H. et al.(Dunlop), BP 1,220,937, 1971
10. Warson, H. Paint Mfre. 1970, 40(11), 52-4
11. Shalatov, V.P. et al. Tr. Voronezh. Gos. Univ.(1969) 73(2), 104-5
12. Lindemann, M.K., Wacome, D.M.(C.S. Tanner), USP 3,732,184,1973
13. Roll, H. et al.(Huels), GP 2,208,442, 1973
14. Lebedeva, Y.V. et al. Vysokomol Soedin. B(1973), 15, 612-5
15. Trubitsyana, S.N. Lebedeva, V. Askarov, M.A. Sin. Vysokomol Soedin. (1972), 35-40
16. Cusano, C.M. et al.(Texaco), USP 3,691,261, 1972
17. Ono, T. et al.(Teijin), Japan Kokai 73 - 04,533
17A Nagai, I. et al.(Toray), Japan P. 71 - 26,864
17B Nomiya, Y. et al.(Kanegafuchi), BP 1,298,895, 1972
18. Kishimoto, A. et al.(Toray), Japan P. 73 - 07,848
19. Nishide, M. et al.(Dainippon Ink), Japan P. 73 - 06,170
20. Kolthoff, I.M., O'Connor, P.R., Hansem, J.L., J. Polym. Sci. (1955), 15, 459-73
21. Warson, H., Polymer, Paint, Col.J. (1974), 164(3879),758-60
22. Mayne, J.E.O., Reichard, H., Warson, H.(Vinyl Products), BP 607,704, 1948
23. Mayne, J.E.O., Warson, H.(Vinyl Products), BP 648,001, 1950
24. Hoechst, BP 1,009,304, 1965
25. Hoechst, BP 1,009,302-3, 1965
26. Dixon, D.R., Cunningham, J.A.(ICI), GP 1,920,974, 1969; BP 1,203,120, 1970
27. Porrvik, S.E., Kolacny, F.(Stockholms Superfosfat), Swed. P. 315,404, 1969
28. Thomas, C.M. et al.(BP Chemical), BP 1,254,349, 1971
29. Brown-Benton, J.L.(BP Chemical), BP 1,262,706, 1972
30. Goodyear, BP 1,282,125, 1972
31. Smith, E.S., Glazer, E.J.(Goodyear), USP 3,780,008, 1973
32. Langsam, M.(Air Products), GP 2,322,843, 1974
33. Crunden, E.W., Hudson, R.F. J. Chem. Soc. (1961), 3748-55
34. Queen, A., Canad. J. Chem.(1967), 45(14), 1619-29
35. Baxendale, J.H. Advances in Catalysis, 1952, 4, 31
36. Kornblum, N. Iffland, D.C. J. Amer. Chem. Soc.(1949)71, 2137
37. Kornblum, N. et al. ibid,(1950), 72, 3013 - 21
38. Nonhebel, D.C., Waters, W.A. Proc. Roy. Soc(London) Ser A. (1957), 242
39. Cooper, W., Chem. & Ind. 1953, 17, 503
40. Warson, H., Makromolek. Chem. (1967), 105, 228 - 45
41. Warson, H., Thesis, University of London, 1963
42. Warson, H. Makromolek. Chem. 1967, 105, 246 - 50
43. Sheppard, C.S. et al.(Wallace & Tiernan), GP 1,905,915, 1969
44. Lewis, E.S., Holliday, R.E., Hartnung, L.D., J. Amer. Chem. S

Soc. 1969, 91(2), 430-3
45. Ruechardt, C. et al. Chem. Soc.(London), Special Publ. (1970) 24, 51-70
46. Edelhauser, H., Nato Advanced Study Institute, Polymer Colloids, Trondheim, 1975

16

Functional Colloidal Particles for Immunoresearch

S. P. S. YEN and A. REMBAUM
Jet Propulsion Laboratory, California Institute of Technology, Pasadena, Calif. 91103

R. W. MOLDAY and W. DREYER
Biology Division, California Institute of Technology, Pasadena, Calif. 91125

The biological applications of polystyrene latex particles were summarized by Vanderhoff[1] in 1964. One of the most important medical applications of these particles has been in serologic tests for the diagnosis of rheumatoid arthritis and other disorders. The principles of a large variety of diagnostic tests by means of polystyrene particles were reviewed by Singer.[2] The majority of the above tests relied on the physical adsorbtion of antibodies onto the latex particle surface. The latter when mixed with the patient's blood serum agglutinated if the serum contained the corresponding antigen. These agglutination tests were used for the diagnosis of rheumatic[2,3] and kidney[4] diseases, gonorrhea, syphilis[6], pregnancy[5,7], etc. However, to date little work has been reported on the use of very small polymeric particles as microscopic markers for antigens or antibodies on the surface of cells. The application of latex particles for the detection and localization of cell surface molecules by scanning electron or optical microscopy constitutes a relatively new approach.[8]

Polystyrene latex particles, 0.2 μ in diameter, have recently been used as immunochemical markers for scanning electron microscopy[9] (SEM). But applications of such a reagent are limited because the hydrophobic surface of the polystyrene particles makes them stick nonspecifically to many surfaces and molecules. The same disadvantage applies to agglutination tests. Furthermore, reliance on weak adsorption forces to hold the antibodies on the particles is not always satisfactory[10] and chemical bonding of antibodies to polystyrene particles is virtually

This paper represents one phase of research performed by the Jet Propulsion Laboratory, California Institute of Technology, sponsored by the National Aeronautics and Space Administration, Contract NAS7-100.

impossible. We have designed new hydrophilic particles based on 2-hydroxyethyl methacrylate and containing functional groups for covalent bonding to antibodies and have shown that these are capable of specific bonding and can be used to label a variety of living cells.[8]

Although an emulsion technique was found to be satisfactory to synthesize small functional particles for the study of labeled cells in SEM, it was, however, necessary to increase particle size for observations under the ordinary light microscope. For this purpose, a Coγ irradiation technique was developed to polymerize 2-hydroxy ethylmethacrylate in absence or in presence of a variety of comonomers.[11]

In the present paper we report:

(1) The design and synthesis of spherical particles containing hydroxyl and carboxyl groups on their surface in a wide range of sizes (30 to 340 nm in diameter) by emulsion copolymerization.

(2) The preparation of spherical particles in the range of 300 nm to 3 μ containing a variety of functional groups by means of ionizing radiation (Coγ).

(3) The experimental conditions for the covalent bonding of fluorescent molecules and antibodies to the spheres by means of the cyanogen bromide, carbodiimide and glutaraldehyde methods.

(4) The successful application of antibody tagged particles as markers for antigens on the surface of red blood cells and lymphocytes.

(5) The identification of mouse spleen lymphocytes (B & T cells) by SEM and laboratory light microscopy using immunolatex spheres as specific markers.

(6) The labeling of cells with fluorescent particles visible in the light microscope.

(7) The labeling of cells bearing Conconavalin A receptors.

Experimental

Materials. The following monomers were obtained from Rohm & Haas and purified as indicated: 2-hydroxyethyl methacrylate (HEMA) distilled in the presence of 0.5% hydroquinone at 95°C, 1 mm Hg pressure; methyl methacrylate (MMA) distilled at 63°C, 200 mm Hg pressure; methacrylic acid (MA) distilled at 60°C, 10 mm Hg pressure and ethylene glycol dimethacrylate (EGD) distilled at 98°C, 4 mm Hg pressure. Allyl amine was distilled at atmospheric pressure (b.p. 56.5°C). Fluorescein isothiocyanate (Nutritional Biochem Co. Cleveland, Ohio), was used as received. Polyethylene oxide (PEO) was acquired from Union Carbide (sample WSR 35). Commercial HEMA is quoted by the manufacturer to be

94% pure and contains higher boiling homologues of 2-hydroxyethyl methacrylate as well as MA (3.5%) and EGD (1.5%).

Methods.

A. Emulsion Polymerization. The polymerization was carried out in an aqueous medium containing an emulsifying agent and a water soluble free radical initiator. A typical procedure and composition designed to form 340 nm spheres is as follows: 19.9 g of methyl methacrylate, 3.5 g of methacrylic acid, 10.5 g of hydroxyethyl methacrylate, 1.1 g of ethylene glycol dimethacrylate, 0.1 g of the emulsifier, sodium dodecyl sulfate (SDS) and 0.01 g of the free-radical initiator, ammonium persulfate, are added to 64.9 g of distilled water. This results in a final concentration of 35% total monomer and 0.1% emulsifier. The reaction mixture is purged of oxygen by saturating the system with argon. Polymerization is carried out for 1 hour at 98°C in sealed tumbling containers. Under these conditions the polymer yield is 99-100%. Any large aggregates which may have formed during the reaction can be separated out by low-speed centrifugation. SDS and residual ionic impurities are removed by titrating the latex suspension to pH 7 and passing it through a mixed-bed ion-exchange column consisting of Biorad AG 1 x 10 and AG 50 W x 12 resins.

Latex spheres with the same composition can be synthesized in a wide range of sizes by a similar procedure. Under the conditions described, the size of the spheres is dependent on the concentration of total monomer. As shown in Figure 1, the diameter of the spheres can be decreased from 340 nm to 35 nm by reducing the total monomer concentration from 35% to 3%. A linear relation between these parameters is observed.

A scanning electron micrograph of 340 nm latex spheres consisting of 57% MMA, 30% HEMA, 10% MA, and 3% EGD is illustrated in Figure 2. Smaller latex spheres have a similar appearance. The spheres are quite uniform in size with a standard deviation generally less than ±1.5% of their average diameter. The diameter of the spheres measured by scanning electron microscopy has been found to be as much as 20 nm greater than that measured by transmission electron microscopy. This difference is primarily due to the conductive gold coating used in the preparation of samples for SEM. A density of 1.23 ± 0.01 g/cc has been measured for these copolymer methacrylate spheres by centrifugation on a continuous sucrose gradient.

Since polymerization is carried out in an aqueous emulsion system, a high concentration of hydrophilic hydroxyl and carboxyl groups is present on the surface of the spheres. At neutral and alkaline pH, the spheres are negatively charged due to the ionization of the carboxyl groups. At pH values below 5.5, when the extent of ionization of the carboxylic groups is decreased, aggregation of the particles is observed. Hydrogen

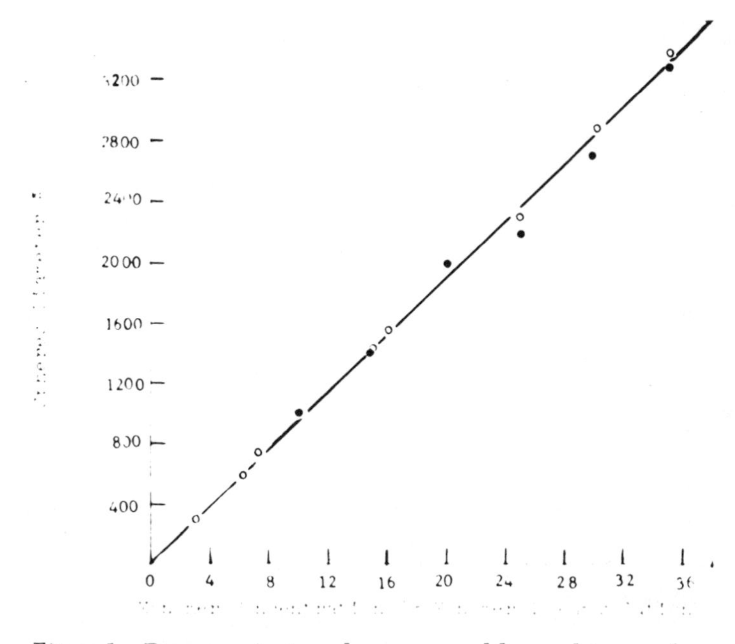

Figure 1. Diameter of microspheres prepared by emulsion copolymerization (determined by means of SEM) as a function of total monomer concentration

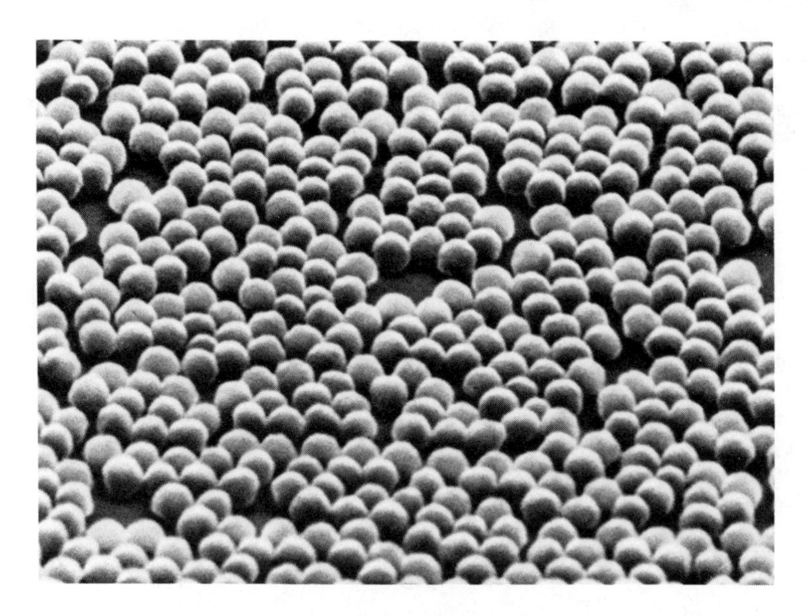

Figure 2. SEM photo of microspheres prepared by emulsion copolymerization of 340 nm in diameter

ion titration measurements indicate that copolymer latex spheres with an average diameter of 60 nm contain approximately 4200 titratable carboxyl groups per sphere. The hydrophilic, negatively-charged surface not only prevents the spheres from aggregating in the absence of an emulsifying agent, but also prevents the spheres from binding nonspecifically to the negatively-charged cell surfaces.

The size and shape of the particles is maintained in aqueous and organic solvents commonly used in the preparation of biological samples for electron microscopy. This stability is largely a result of the cross-linked nature of the polymeric matrix.

B. Polymerization by Means of Ionizing Radiation. The design of microspheres of various sizes and containing functional groups was based on a preliminary observation that Coγ irradiation of HEMA dissolved in water yielded water insoluble spherical particles. Subsequent studies[11] showed that the size and properties of these particles vary considerably as a function of monomer concentration, the functional groups of comonomer, concentration of crosslinking agent, etc. Most of the described results were obtained with copolymer systems in presence of PEO which acts as a steric stabilizer and permits the use of relatively high concentration of monomers (up to about 20%). Highly fluorescent and stable microspheres could be obtained by addition to the comonomer mixture of a fluorescent monomer such as dansyl allyl amine or an adduct of allylamine and fluorescein isothiocyanate. Typical examples of compositions used are shown in Table I. These mixtures, when irradiated in a Coγ source at room temperature in the absence of air for one hour (0.8 mr), yielded fluorescent particles the diameter of which was 1.7 (Table I composition a) and 0.7 μ (Table I composition b). The diameter was determined in presence of water by means of a hemacytometer and photographic enlargements of microscope pictures.

TABLE I

	Composition a* weight g	Composition b* weight g
HEMA	9.0	7.0
MA	-	2.0
BAM	1.0	1.0
Dansylallylamine	0.1	-
Allylamine	-	0.5
Fluorescein isothiocyanate	-	0.05
PEO	0.8	0.8

*made up to 200 cc of H_2O.

After irradiation in order to remove impurities and PEO the products were centrifuged several times in distilled water either in an international clinical centrifuge or an ultracentrifuge depending on the size of the particles. Parameters affecting size, aggregational properties and concentration of functional groups were previously described.[11]

C. Preparation of Immunolatex Conjugates. The presence of hydroxyl and carboxyl groups on the latex spheres enable proteins and other molecules containing primary amino groups to be covalently bonded to the spheres. Many of the same chemical procedures used in the preparation of absorbents for affinity chromatography[12] and in the immobilization of enzymes on solid supports can be used to prepare antibody-latex conjugates. In our laboratories, antibodies purified by immunoabsorbent chromatography have been coupled to the latex spheres by either the cyanogen bromide, carbodiimide, or glutaraldehyde method. The reactions are carried out under conditions which maintain the antigen binding activity of the antibodies, but minimize aggregation of the latex particles. The number of antibody molecules bound per sphere can be increased by increasing the concentration of antibody in the reaction. Immunolatex conjugates having only a few antibody molecules per sphere (1-2 antibody molecules per sphere 60 nm in diameter) have been routinely used in cell surface labeling studies.[8]

Cyanogen Bromide Method. In this reaction the hydroxyl groups on the latex spheres are first activated with cyanogen bromide at alkaline pH and then coupled to amino groups on proteins between pH 7 and 10. It has been suggested by Porath and co-workers from studies on dextrans and agarose derivatives that stable iminocarbonate, isourea, and N-substituted carbamate derivatives are the major stable products. They proposed a reaction mechanism shown in Figure 3. The coupling reaction is dependent on the pH, proceeding more efficiently at higher pH.[14]

Procedure: Immunolatex conjugates are prepared by a method adapted from that used by Cuatrecasas.[15] An aqueous suspension of latex spheres (20-50 mg/ml) adjusted to pH 10.5 with NaOH is activated at 25°C with 10 mg of CNBr per ml of suspension. The reaction mixture is maintained between pH 10 and 11 by the slow addition of 2N NaOH. After 15 minutes, the suspension is diluted with an equal volume of cold 0.1M borate buffer at pH 8.5 and equilibrated at 4°C. A solution of purified antibody in 0.1M borate buffer pH 8.5 is added to the latex suspension resulting in a final protein concentration of 0.5 mg/ml. The reaction is allowed to proceed at 4°C for at least 4 hours after which it is terminated by the addition of an equal volume of 0.1M glycine buffer at pH 8.5.

Immunolatex conjugates prepared by this procedure have been shown to be active in their capacity to bind antigens. However, some aggregation of the conjugates has been observed during the coupling reaction, even when conditions are carefully controlled.

Carbodiimide Method. In this reaction, a water-soluble carbodiimide derivative is used to couple carboxyl groups on the latex spheres to amino groups on the antibody molecules via a peptide bond.[16] The reaction mechanism is shown in Figure 4. The reaction is carried out between pH 6 and 7 in the absence of an added buffer or in the presence of a buffer which does not interfere with the reaction, i.e., N-2-hydroxyethyl piperazine-N-ethane sulfonic acid (HEPES).

Bonding of glycine to the latex spheres by the carbodiimide reaction has been found to be highly sensitive to the concentration of glycine but less sensitive to the carbodiimide concentration or pH.

Procedure: Antibodies are conjugated to the methacrylate spheres as follows: 1 mg of purified antibody is added to 50 mg of latex spheres in 2 ml of 0.1M NaCl. The solution is adjusted to pH 6.5 and equilibrated at 4°C. Ten milligrams of 1 ethyl-3-(3-dimethyl amino propyl) carbodiimide are then added to the solution with stirring. After 2 hours the reaction is terminated by the addition of excess glycine buffer.

The advantage of this coupling procedure is its simplicity. However, since proteins contain both amino and carboxyl groups, intermolecular and intramolecular crosslinking of antibody molecules can occur. This leads to the formation of aggregates and a decrease in immunological activity.

Glutaraldehyde Method. Glutaraldehyde has been extensively used as a bifunctional coupling reagent to prepare immunoferritin[17] and immunoenzyme conjugates[18] for localizing antigens by transmission electron microscopy. The reaction is thought to involve the addition of amino groups on proteins to α,β-unsaturated aldehyde polymers which are present in aqueous glutaraldehyde solutions. Figure 5 represents a reaction mechanism proposed by Richards and Knowles.[19]

In order to use this reaction to bond antibodies to the methacrylate latex spheres, diamino compounds such as diaminoheptane must first be derivatized onto the latex spheres. This can be done by either the carbodiimode or cyanogen bromide reaction. Only a small number of free amino groups (approximately 100) are bound, relative to the number of carboxyl groups on the spheres (approximately 4000), so as to maintain the net-negative charge on the spheres. Antibodies are then conjugated to the glutaraldehyde derivatized latex particles in either a one-step or two-step reaction.[20] In the one-step procedure, the coupling of antibodies to the latex spheres takes place in the presence of excess glutaraldehyde. In the two-step method, derivatized spheres are first activated with glutaraldehyde; antibodies are then coupled to the spheres after the excess glutaraldehyde has been removed. Recently, a comparative study of immunoferritin conjugates prepared by a one-step and two-step

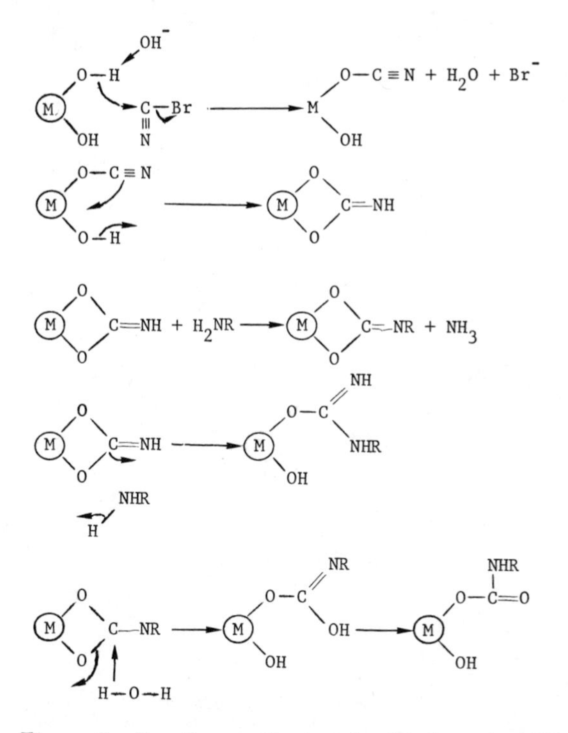

Figure 3. Reaction mechanism for binding vic. OH groups with amino groups by means of cyanogen bromide

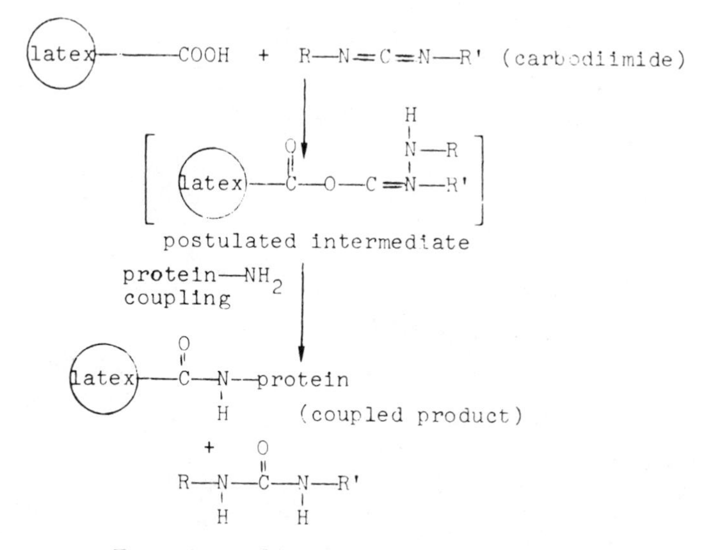

Figure 4. Mechanism of the carbodiimide reaction

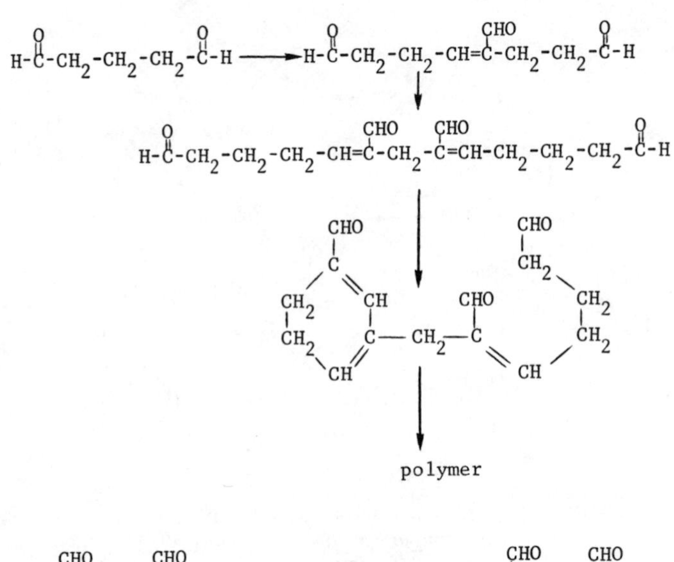

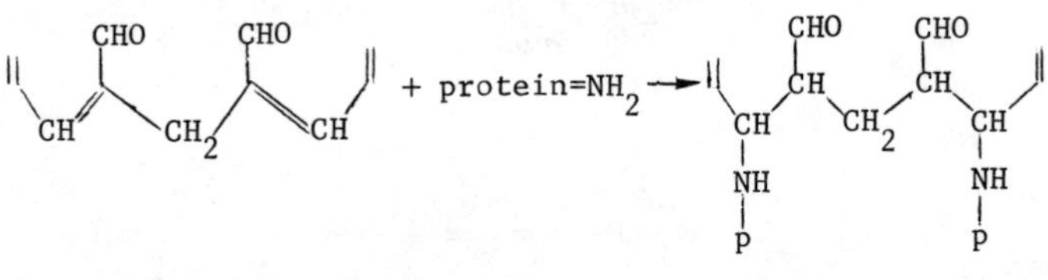

Figure 5. Mechanism of the glutaraldehyde reaction

glutaraldehyde procedure has been reported by Otto et al.[20] Their results indicate that conjugates prepared by the latter procedure have higher immunological activity.

The following procedure has been routinely used in our laboratories to prepare immunolatex conjugates of high immunochemical activity for applications in cell surface labeling.

Procedure: The latex spheres are derivatized with diaminoheptane by adding ten mg of 1-ethyl-3-(3 dimethyl amino propyl) carbodiimide to 5 ml of latex spheres (20-30 mg/ml) in 0.01M diaminoheptane at pH 7. After 2 hours at 4°C the solution is exhaustively dialyzed first against 0.1M NaCl and then against 0.01M phosphate buffer pH 7.0. The two-step coupling procedure is carried out by adding one volume of 25% glutaraldehyde to twenty volumes of derivatized latex spheres. After 1 hour at 25°C, the excess glutaraldehyde is removed by dialysis at 4°C for 24 hours against several changes of 0.01M phosphate - 0.01M NaCl buffer, pH 7.0. Finally, 2.5 mg of purified antibody is added to 5 ml of the glutaraldehyde-activated latex spheres at 25°C. After stirring for 5 hours at 25°C, an equal volume of 0.1M glycine pH 7.0 is added to react with any remaining glutaraldehyde-activated groups on the latex spheres.

The two-step glutaraldehyde coupling reaction has the advantage over the one-step reaction and the carbodiimide coupling reaction in that intramolecular and intermolecular crosslinking of the antibody molecules is avoided. Furthermore, under these conditions, aggregation of the latex particles does not occur.

D. Purification and Analysis of Immunolatex Conjugates. Antibody-latex conjugates can be separated from unbound antibody molecules by such physical techniques as column chromatography, filtration, centrifugation and electrophoresis. Differential centrifugation is a convenient separation procedure for latex spheres of diameter greater than 30 nm. In this procedure, the reaction mixture is layered on a discontinuous gradient consisting of a 10-20% (w/w) sucrose solution overlaying a dense 58% (w/w) sucrose solution. After high speed centrifugation (100,000 g) in a swirling bucket rotor, the immunolatex conjugate forms a narrow band at the interface between the two sucrose solutions. Since the density of the immunolatex conjugate (1.24 g/cc) is less than that of the dense sucrose solution (1.27 g/cc), the conjugate does not penetrate this layer. The time required for sedimenting the conjugate depends on both the size of the immunolatex spheres and on the concentration of the sucrose solution, i.e. viscosity, in the upper layer. For example, conjugates consisting of antibodies bound to latex spheres 110 nm in diameter sediment through a 10% sucrose solution at 100,000 g in less than 1 hour. Unbound antibody remains near the top of the tube. The conjugates are centrifuged onto a dense sucrose cushion in order to facilitate redispersion of the conjugate. To ensure complete removal of unbound antibody, this centrifugation procedure is repeated 2-3 times. In addition to separating the conjugate from the unbound antibody, this procedure also can be used to concentrate the conjugate.

The conjugate which is collected from the gradient is then dialyzed against phosphate buffered saline solution pH 7.40 (PBS) to remove the sucrose. Any large aggregates which may be present can be removed by low speed centrifugation. Conjugates are stored in small vials at a concentration of 10-20 mg per ml at 4°C. In the presence of 10^{-4}M NaN_3, conjugates prepared by the two-step glutaraldehyde procedure showed no significant loss in immunological activity after 3 months.

The immunochemical activity of the immunolatex conjugates can be determined by a latex agglutination assay. In the case in which goat antibodies directed against rabbit immunoglobulin (Ig), i.e., goat antirabbit Ig antibodies, are coupled to the latex spheres, immunolatex particles of high activity are precipitated by rabbit serum or rabbit Ig over a wide range of concentrations. Serum or immunoglobulins which show no cross reactivity with goat anti-rabbit Ig antibodies should not aggregate the immunolatex conjugates over the same range of concentration.

Alternately, the activity of conjugates to be used with the indirect labeling method can be tested by determining the conjugate concentration required to saturate antibody-sensitized cells with latex spheres. The number of markers bound per cells can be determined by SEM or by radioactivity if the spheres have been labeled with a radioactive isotope.

E. Labeling of Cell Surface Antigens with Immunolatex Spheres. The indirect method has been routinely used to label cell surface antigens with immunolatex markers for visualization by SEM. In this method cells are first sensitized with antibodies against specific cell surface antigens. These immunoglobulin (Ig) molecules are then tagged with immunolatex conjugates consisting of latex spheres coupled to anti-Ig antibody molecules. This method has the advantages that 1) antiserum containing anti-Ig antibodies can be prepared easily or obtained commercially in large quantities; 2) the anti-Ig antibodies can be purified on Ig immunoabsorbents according to well established procedures; and 3) a given preparation of anti-Ig-latex conjugate can be used to localize many different cell surface antigens for which antisera from the appropriate species is available.

In a typical procedure, a suspension of cells are washed in PBS by centrifugation. If prefixed cells are to be used, the washed cells are resuspended in 0.25% glutaraldehyde in PBS, and incubated at 25°C for 15 minutes. The cells are then sedimented at about 150-300 g, washed once in PBS to remove excess glutaraldehyde, and incubated in PBS containing 0.01M glycine for at least 30 minutes at room temperature. The glycine is used to react with any remaining glutaraldehyde-activated groups on the cell surface. Approximately 10^6 cells are resuspended in rabbit antiserum containing antibodies against the specific cell surface antigen. Routinely, the antiserum is centrifuged at 20,000 g for 20 minutes to remove protein aggregates. After 15 minutes at 25°C or 1 hour at 4°C, the sensitized cells are washed 3-4 times in PBS buffer, and finally, resuspended in 0.2 ml of goat anti-rabbit Ig antibody-latex conjugates. The concentration of the conjugate should be sufficient to saturate accessible sites. After 15 minutes at 25°C or 1 hour at 4°C, the cells are separated from unbound immunolatex reagent by repeated centrifugation at about 150 g.

In the control experiment to test for nonspecific binding, the cells are treated with nonspecific rabbit serum prior to incubating in goat anti-rabbit Ig antibody-latex conjugate.

F. Preparation of Specimens. For observation by SEM, tissue and cell samples must be carefully dried under conditions which preserve the cell surface features. This is generally accomplished by fixing, dehydrating, and critical point drying the specimens.

Cells must be attached to a support to prevent the loss of sample during critical point drying. Several techniques have been developed. Polliack[23] et al. have described a technique for aspirating cells onto silver "Flotronic" membranes; Wetzel[24] and co-workers have filtered cells onto Nucleopore membranes. Alternatively, cells can be centrifuged onto 12 mm glass coverslips inserted into filter holders. Adhesion of glutaraldehyde-fixed cells to glass coverslips is increased by pretreating the coverslips with the positively-charged polymer, 3,3 ionene chloride, which forms a monolayer on glass surfaces.[25] It should be noted, however, that when any of these techniques are used with a mixed population of cells, selective attachment of specific cell types may occur due to differences in cell surface adhesive properties.

Results

A. Red Blood Cells: Use of Spheres of Different Sizes. Immunolatex reagents consisting of goat anti-rabbit Ig antibodies coupled to latex spheres can serve as SEM markers to detect and map the distribution of a variety of different cell surface antigens by the indirect labeling method. The activity and specificity of these conjugates can be readily determined on red blood cells.[8] In this test system, unfixed or glutaraldehyde-fixed human red blood cells (Figure 6b) are treated with rabbit anti-human RBC antiserum. This commercially available antiserum (Cappel) contains antibodies against a variety of surface components on RBC. After the excess antiserum is removed by washing, the sensitized cells are incubated with goat anti-rabbit IG antibody-latex conjugates and washed in buffer. The immunochemical activity of the antibody-latex conjugate can be determined by titering out the conjugate and quantitating

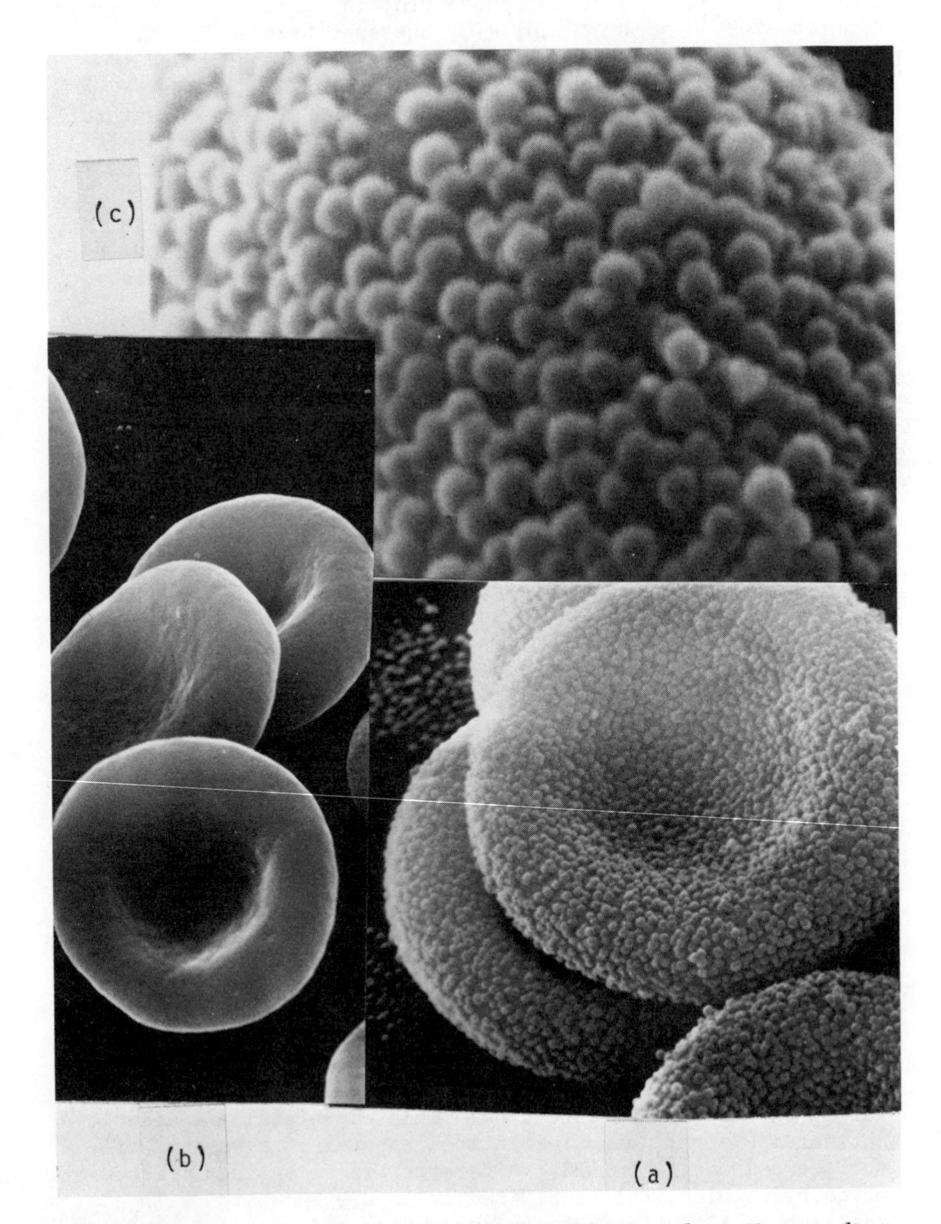

Figure 6. SEM photo of (a) human RBC labeled with microspheres 60 nm in diameter (11,560×); (b) control: nonspecific antiserum bonded to microspheres was used (7140×); (c) higher magnification of (a)

the number of spheres bound per RBC by SEM.

A scanning electron micrograph of human RBC labeled with immunolatex spheres of diameter 60 nm (600 Å) and at a concentration of 20 mg/ml is shown in Figure 6a. The same dense distribution of spheres was observed when the conjugate concentration was reduced ten fold. Markers of this size are readily seen in the SEM and can be used to identify cells exhibiting specific surface antigens in either mixed cell populations or tissue specimens.

Red blood cells indirectly labeled with markers 35-40 nm in diameter (25-30 nm by TEM) are displayed in Figure 7a. Particles of this size also coat the surface of the cells. Some loss in resolution is apparent due largely to the conductive layer of gold used in the preparation of the cells for SEM. In conjunction with high resolution SEM, latex markers in this size range can be used to obtain information about the topographical distribution and mobility of specific cell surface components.

The specificity of the immunolatex markers can be demonstrated in control experiments designed to measure the amount of nonspecific binding to cell surfaces. RBC, which have not been sensitized with rabbit antihuman RBC antibodies, are treated with goat anti-rabbit Ig antibody-latex conjugates (20 mg/ml) and washed in the usual manner. As shown in Figure 7b, only a few spheres adhere to the surface of the red blood cells.

B. Thymocytes: Labeling of the Theta Antigen. The application of immunolatex particles as SEM markers for specific cell surface antigens can be demonstrated in the case of mouse thymus cells. Thymocytes from different strains of mice possess either the theta AKR or theta C_3H antigen on their surface.[26] Alloantisera directed against these specificities can be readily prepared by immunizing C_3H mice with AKR thymocytes to yield C_3H anti θAKR antiserum and by immunizing AKR mice with C_3H thymocytes to yield AKR anti θC_3H antiserum. Primary antisera obtained in this way can be used with goat anti-mouse Ig antibody-latex conjugates to detect theta antigens on thymocytes by the indirect method.

Thymocytes from AKR/J and C_3H mice (Jackson Laboratory, Bar Harbor, Me.) were obtained by gently teasing apart thymus tissue in PBS. The cells were washed in buffer and lightly-fixed in 0.125% glutaraldehyde-PBS for 20 minutes. Thymocytes from each strain were incubated with C_3H anti θAKR antiserum, washed in buffer, and then treated with goat anti-mouse Ig antibody-latex conjugates.

As shown in Figure 8, AKR/J thymocytes are densely labeled with immunolatex spheres 110 nm in diameter. Over 95% of the cells were labeled. C_3H thymocytes, as well as thymocytes from a number of other strains of mice which do not display the AKR specificity, were not labeled with the spheres. In the reverse

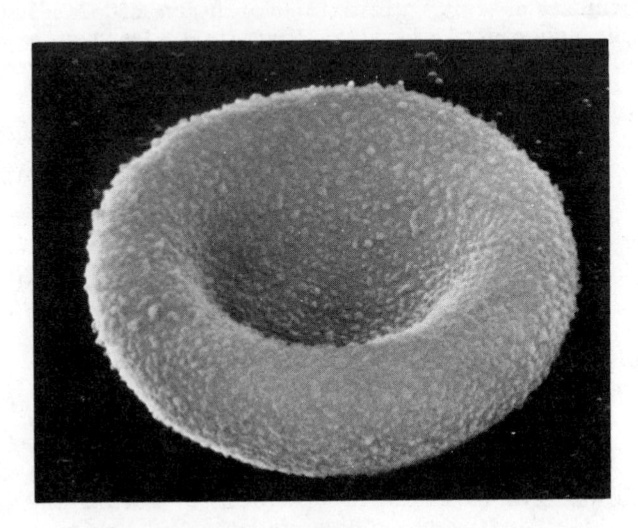

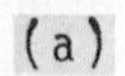

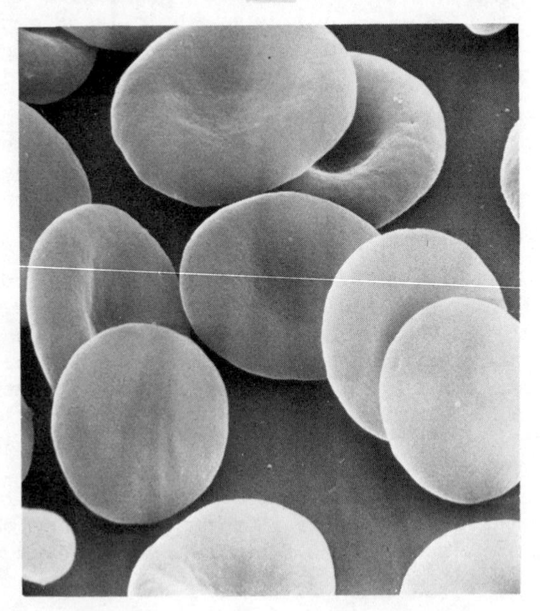

(b)

Figure 7. SEM photo of (a) human RBC labeled with microspheres of 35–40 nm in diameter (11,900×) and (b) control (5950×)

experiment in which thymocytes are first treated with AKR anti θC_3H antiserum prior to the addition of the immunolatex markers, C_3H thymocytes were found to be heavily labeled, whereas AKR/J thymocytes were not.

C. Lymphocytes: Identification of Cells Bearing Immunoglobulin Receptors (B-cells). Many studies have been directed toward distinguishing B-lymphocytes or thymus-independent cells from T-lymphocytes or thymus-dependent cells by differences in either cell surface antigens or cell surface morphology. Results obtained using immunological markers for fluorescent and transmission electron microscopy, indicate that B-cells display immunoglobulin (Ig) molecules on their surface, whereas T-cells do not. SEM observations on human lymphoid cells derived from various sources suggest that B-cells and T-cells differ in cell surface features.[27] Direct identification of B-lymphocytes by SEM can be achieved using immunolatex spheres as markers for cell surface Ig molecules.[8]

Mouse spleen lymphocytes were separated from blood cells and dead cells by centrifugation on a Ficoll-isopaque gradient. The cells were lightly fixed in 0.25% glutaraldehyde-PBS buffer in order to preserve their surface morphology and restrict the movement of cell surface receptors. Rabbit anti-mouse Ig antiserum was added to an equal volume of 10^6 cells suspended in PBS buffer. The cells were incubated at 25°C for 15 minutes, and subsequently, washed with buffer by repeated centrifugation. After resuspending the lymphocytes in 0.2 ml of goat anti-rabbit Ig antibody-latex conjugates for 20 minutes at 25°C, the cells were washed and prepared for observation by SEM. Control samples were carried through the same steps, but nonspecific rabbit serum was substituted for rabbit anti-mouse Ig antiserum.

Scanning electron micrographs of mouse spleen lymphocytes labeled for surface Ig molecules with latex spheres are illustrated in Figure 9. Many of the labeled cells (B-cells) were found to have microvilli-like structures; latex markers were densely distributed over their cell surface and microvilli. Recent SEM studies indicate that human T-lymphocytes, which form rosettes with sheep blood cells, also exhibit numerous microvilli on their surface.[28]

D. Direct Labeling of Photoreceptors. It is known that the plant lectin conconavalin A (Con A) binds to the carbohydrate unit of rhodopsin. This binding site is exposed on the surface of photoreceptor cells. The Con A receptors could be very simply labeled by the use of latex spheres (40 to 50 nm in diameter) Con A conjugates prepared by the glutaraldehyde procedure as described in experimental. Photoreceptor cells isolated from retinas of cattle eyes were then suspended in presence of latex Con A conjugates for 0.5 to 1 hour at 25°C. Excess reagent was removed by rinsing or centrifugation. The

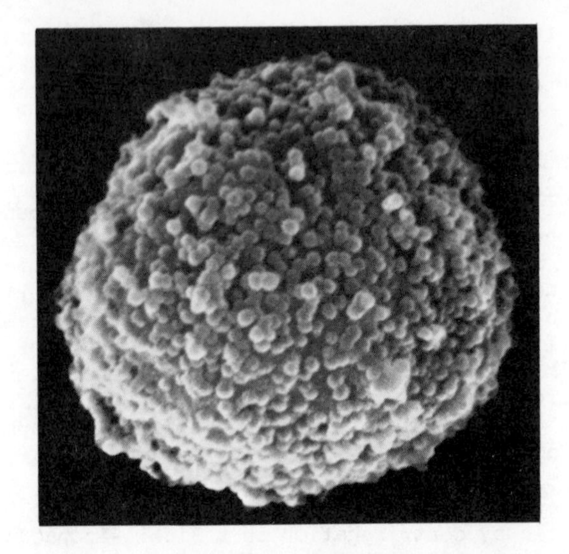

Figure 8. A SEM photo of AKR/J thymocytes labeled with microspheres 110 nm in diameter (15,300×)

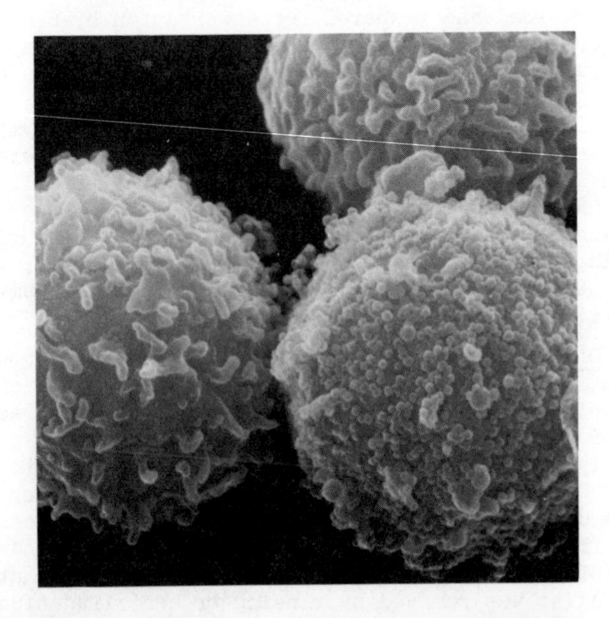

Figure 9. SEM photo of a labeled murine B cell. Unlabeled cells are believed to be T cells (12,750×)

distribution of Con A receptors along the entire length of photoreceptor cells is shown in Figure 10.

E. Labeling of Cells for Light Microscope Studies. The procedure for labeling cells with microspheres larger than 0.4 μ was identical to that described above except (1) a low speed centrifuge could be used, and (2) the cells could be separated from unreacted particles by means of a discontinuous density gradient. The cells remained at the interface of the gradient and the particles sedimented to the bottom of the container. Human red blood cells (RBC) and lymphocytes were labeled with fluorescent or nonfluorescent functional particles of various sizes by the indirect or direct immunological technique as described in the experimental part. Figure 11 shows a fluorescent microscope photograph of RBC sensitized with rabbit anti-human RBC antiserum and labeled with 340 nm goat antirabbit IgG antibody-fluorescent latex conjugate. In Figure 12 murine lymphocytes were labeled with 0.5 micron microspheres synthesized by means of Coγ irradiation and conjugated with goat anti-mouse IgG antibody. Since T cells do not have Ig receptors, the labeled cells are believed to be B cells. Figure 13 represents human lymphocytes labeled with 0.7 μ microsphere-goat antihuman conjugates. The assumption that the labeled cells are human B cells was confirmed by independent experiments.

Conclusions

A new class of immunochemical reagents consisting of antibodies covalently bonded to polymeric microspheres has been prepared and shown to serve as convenient markers for the detection of cell surface antigens by scanning electron and light microscopy. These reagents have been used to locate antigens on red blood cells, on mouse and human lymphocytes and on the surfaces of photoreceptors. They offer a number of advantages and applications for the study of cell surfaces for immunodiagnosis.

(1) Latex spheres can be synthesized in a wide range of sizes and compositions to suit particular requirements and can be stored indefinitely.

(2) Biological molecules such as antibodies or lectins can be bound to the latex spheres by any of a variety of standard chemical procedures for use in the identification of specific populations of cells, as well as in the detection and localization of specific cell surface receptors. This suggests that hormones and toxins can similarly be bound.

(3) Different sizes of spheres can be used in multiple labeling experiments and in conjunction with different types of microscopy. For example, acrylic spheres the size of ferritin and hemocyanin, i.e., 150-350 Å in diameter, can serve as

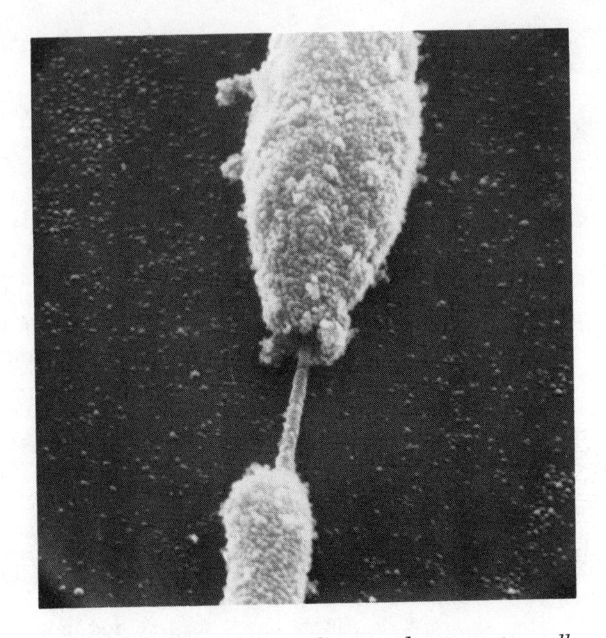

Figure 10. SEM photo of bovine photoreceptor cells labeled with Con A microsphere conjugates. No labeling was observed in presence of α-methylmannoside (14,450×).

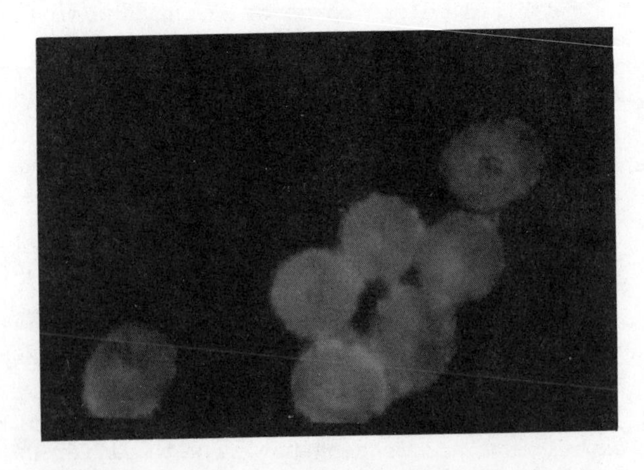

Figure 11. Microscope photo of human RBC sensitized with rabbit antihuman antiserum labeled with fluorescent microspheres (340 nm in diameter) bonded to goat anti-rabbit IgG

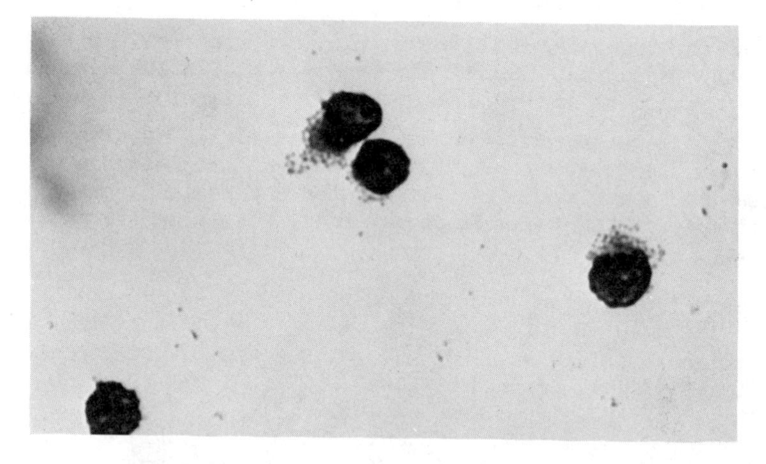

Figure 12. Microscope photo of murine spleen cells labeled with 0.5 μ microsphere conjugates. The unlabeled cells are believed to be T cells.

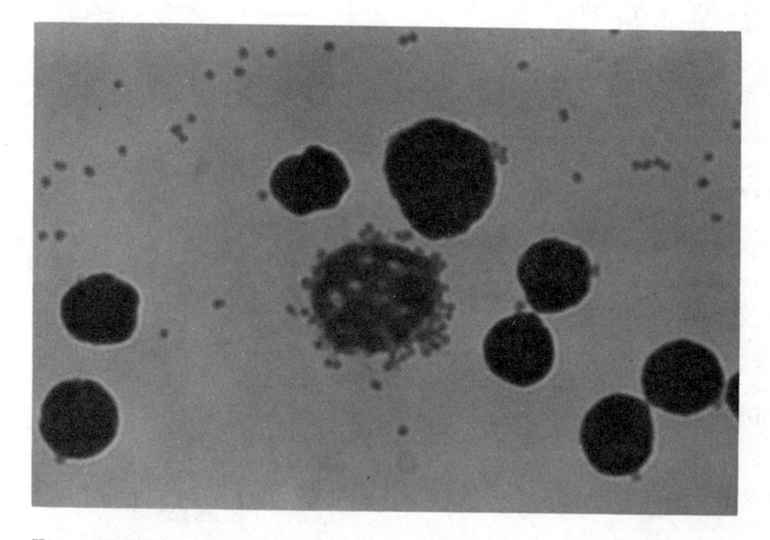

Figure 13. Microscope photo of human lymphocytes labeled by means of the direct method (the lymphocytes were not sensitized prior to labeling) with 0.7 μ microspheres goat antihuman antibody conjugates

markers for transmission electron microscopy as well as in high resolution scanning electron microscopy; spheres larger than 0.2 μ in diameter can be used with ordinary light microscopy.

(4) These microspheres can serve as highly sensitive fluorescent probes and quantitative reagents for biochemical and immunological studies. Binding the antibodies permits a high degree of tagging without adversely affecting the antibody activity.

References

1. Vanderhoff, J. W. and Kennedy, C. C., ACS Div. Org. Coating Plastics Chem. Preprints, (1964), 24 (2), 223-32.
2. Singer, J. M., Am. J. Med., (1961), 31, 766.
3. Singer, J. M. and Plotz, C. M., Am. J. Med., (1956), 21, 888.
4. Kramer, N. C., Watt, M. F., Howe, J. H. and Parrish, A. E., Am. J. Med., (1961), 30, 39.
5. Robbins, J. L., Hill, G. A., Carle, B. N., Carlquist, J. H. and Marcus, S., Soc. Exp. Biol. and Med (proceedings), (1962), 109, 321.
6. Stevens, R. W., Am. J. Clin. Pathol., (1965), 43, 490.
7. Pollack, W., U. S. 3,234,096 (to Ortho Pharm. Co.), Feb. 8, 1966.
8. Molday, R. W., Dreyer, W., Rembaum, A. and Yen, S. P. S., Nature, (1974), 249, 81.
9. Lo Buglio, A., Rinehart, J. and Balcerzak, S., "Scanning Electron Microscopy Part II," 313, IIT Research Institute, Chicago (1972).
10. Milgram, A. and Goldstein, R., VOX Sang, (1962), 7, 86.
11. Rembaum, A., Yen, S. P. S., Cheong, E., Wallace, J., Molday, R. and Dreyer, W., (in press).
12. Zaborsky, O., "Immobilized Enzymes," CRC Press, Cleveland (1973).
13. Porath, J., Nature, (1968), 218, 834.
14. Axen, R., Porath, J. and Eynback, S., Nature, (1967), 214, 1302.
15. Cuatrecasas, P., J. Biol. Chem., (1970), 245, 3059.
16. Goodfriend, T., Levine, L. and Fasman, G., Science, (1964), 144, 1344.
17. Neauport-Santes, C. and Silvestre, O., Transplant, (1972), 13, 536.
18. Sternberger, L., "Electron Microscopy of Enzymes: Principles and Methods," (M. Hayat, Ed. 1973).
19. Richards, F. and Knowles, J., J. Mol. Biol., (1968), 37, 231.
20. Otto, H., Takamiya, H. and Vogt, A., J. Immunol. Methods, (1973), 3, 137.
21. Boyde, A., "Scanning Electron Microscopy," (O. Johari and I. Corrin eds.), 257, IIT Res. Inst., Chicago (1972).

22. Cohen, A., "Principles and Techniques of Scanning Electron Microscopy," (M. Hyat, ed.) Van Nostrand-Reinhold Co., New York and London (1974).
23. Polliack, A., Lampen, N., Clarkson, B. and de Harren, E., "Scanning Electron Microscopy," Part III, 529, IIT Res. Inst., Chicago (1973).
24. Wetzell, B., Cannon, G., Alexander, E., Erikson, B. and Westbrook, E., "In Scanning Electron Microscopy," Part III, 581, IIT Res. Inst., Chicago (1974).
25. Rembaum, A., Appl. Polymer Symp., (1973), 22, 299.
26. Reif, A. and Allen, J., Nature, (1966), 209, 521.
27. Polliack, A., Lampen, N., Clarkson, B. and de Harren, E., J. Exp. Med., (1973), 138, 607.
28. Kay, M., Belohradsky, B., Yee, K., Vogel, J., Butcher, D, Wybran, J. and Fudenberg, H., Cl. Immunol. Immunopath., (1974), 2, 301.
29. Heller, J. and Lawrence, M., Biochemistry, (1970), 9, 864.

17

Mechanical Stability of Vinyl Chloride Homopolymer and Copolymer Latices

O. PALMGREN

Research Centre, Norsk Hydro a.s., 3900 Porsgrunn, Norway

The object of this study was to clarify some aspects of the mechanism of shear-induced flocculation in colloidal dispersions. Vinyl chloride homopolymer and copolymer latices were prepared by emulsion polymerization using sodium dodecyl sulphate as emulsifier. Agglomeration behavior in these latices was studied by measuring the mechanical stability using a high speed stirring test. The latex particle size was measured by an analytical centrifuge. Molecular areas of emulsifier in the saturated adsorption layer at the surface of homopolymer and copolymer latex particles were estimated from adsorption titration data.

Experimental

Materials. All monomers were of commercial quality: vinyl chloride (VCM), vinyl acetate and VeoVa 10. The latter is a vinyl ester of a saturated monocarboxylic acid with a highly branched structure containing 10 carbon atoms. This monomer is produced by Shell Chemicals. The emulsifier, sodium dodecyl sulphate (SDS), was the commercial material Berol 474 from Berol Kemi, Sweden. $Na_2S_2O_8$ supplied by Noury & Van der Lande, was used as initiator. The $NaHCO_3$ used as buffer and for adjustment of the electrolyte concentration, was Merck p.a. grade.

Polymerization. The polymerizations were carried out in a 40 liter reactor. Water, emulsifier, initiator and buffer were charged, de-aeration followed, and monomer was added. Then the reaction was started by raising the temperature of the batch to 50°C. The remaining emulsifier solution was added continuously during the polymerization, following a procedure that took into account particle total surface. The latices were in this way prepared at a low level of stability, with good shelf stability, but practically zero stirring stability.

Particle size measurement. The particle size was measured by an analytical centrifuge (1). In this technique the sedimentation and separation of particles is followed through an optical/electronic system based on the scattering and light-absorbing properties of particles. The monodisperse latices obtained are characterized by the particle size corresponding to the maximum light absorption, approximately equal to the weight average particle size.

In some cases particle size measurements were also performed by electron microscopy and nitrogen adsorption.

Determination of residual VCM. This determination was done by gas chromatograph.

Surface tension measurement. Adsorption titration, also called soap titration, (2,3) was carried out by the drop volume method at different polymer concentrations. The equivalent concentration of salt was held constant. The amount of emulsifier necessary to reach the critical micelle concentration (CMC) in the latex was determined by each titration. The total weight of emulsifier present in the latex is the weight of emulsifier in the water plus the weight of emulsifier adsorbed. The linear plot of emulsifier concentration (total amount of emulsifier corresponding to the end-point of each titration) versus polymer concentration gives the CMC as the intercept and the slope determines the amount of emulsifier adsorbed on the polymer surface in equilibrium with emulsifier in solution at the CMC (E_m).

E_m is related to the particle diameter by

$$E_m = \frac{9.96 \cdot M}{\rho \cdot D_p \cdot A_m}$$

Here M is the molecular weight of the emulsifier, ρ is the polymer density, D_p is the volume-to-surface average particle diameter in Ångström units, and A_m is the area per emulsifier molecule on the polymer surface in $Å^2$.

In some cases the surface tension also was measured by the ring method using a du Noüy tensiometer.

Latex stability measurement. Latex stability was measured by a high speed stirring test (4-9). The stability of a latex is given by the length of time required to produce complete flocculation by high speed stirring of 150 grams of the latex in a 600 ml glass beaker. The stirring speed was 14000 rpm with a circular disk 21 mm in diameter.

Although the absolute significance of the stability test is not clear, the test is suitable for comparison of stability levels under different test conditions.

Usually there was no trouble with foaming. In some cases antifoaming agent was added to the latex prior to stability measurements. This essentially eliminated the foaming without affecting the numerical value of the stability of these latices.

Both ASTM and BSI have dealt with the standardisation of the high speed stirring test (10,11).

Various methods for testing the mechanical stability of latices have been described by other authors (12-15).

Results and discussion

Polymerization. Figure 1. shows how the formation of particles during the polymerization depends on the initial concentration of SDS in the aqueous phase. In order to achieve a monodisperse latex, the particle nucleation must be confined to the initial stage of the polymerization. No new particles can be permitted to form during the reaction, and agglomeration of latex particles must be prevented. Thus both a too high and a too low emulsifier concentration must be avoided (16). This was accomplished by careful postaddition of emulsifier.

The linear relationship between the logarithm of the resulting number of particles and the logarithm of the initial emulsifier concentration, either in the presence or in the absence of micelles, has also been reported by Ugelstad, Mörk, Dahl and Rangnes (17). Nucleation of latex particles has recently been discussed by Goodwin, Hearn, Ho and Ottewill (18) and by Fitch (19).

Latex stability. Effect of particle size and emulsifier level. Latex stability data for three latices with different particle size, are plotted in Figure 2. At a given emulsifier level, expressed as weight per cent of polymer, the stability increases with increasing particle size. The logarithm of the stability is a linear function of the emulsifier concentration (20).

In Figure 3. the same stability data are plotted versus the surface concentration of emulsifier. The surface concentration is given as per cent of total coverage of the particle surface, as determined by adsorption titration.

The adsorption isoterms of SDS on polyvinyl chloride (PVC) latex particles in the presence of various concentrations of sodium ions in the bulk solution, have been determined by Bibeau and Matijevic (21). In the latices which were examined by us, the surface concentration of emulsifier was confined to the region well below the saturation level. Under these circumstances only a negligible fraction of the total amount of emulsifier will be dissolved in the aqueous phase.

At a given surface concentration of emulsifier the stability decreases with increasing particle size. The logarithm of the stability is a linear function of the surface concentration of emulsifier with a slope which is independent of the particle size. The stability level could be raised by reducing the polymer concentration or it could be lowered by increasing the salt concentration. The slope remains unaltered.

In Figure 4. the same stability data are plotted versus the surface concentration of emulsifier. But this time the surface concentration is expressed as number of molecules per square cm.

Van den Hul and Vanderhoff (22) have shown that polystyrene latices polymerized with a persulphate initiator have permanently charged groups present on the particle surface, resulting from the ionic free radical initiator. Surface charges on polystyrene latex particles have also been studied by Hearn, Ottewill and Shaw (23). A permanent particle surface charge was also found for PVC latex particles by Bibeau and Matijevic. The particle size was 4800 Å. $1.5 \cdot 10^{13}$ end groups per square cm were found on these particles. This corresponds to approximately one end group at the particle surface per each of seven polymer molecules. For a particle size of 2000 Å this

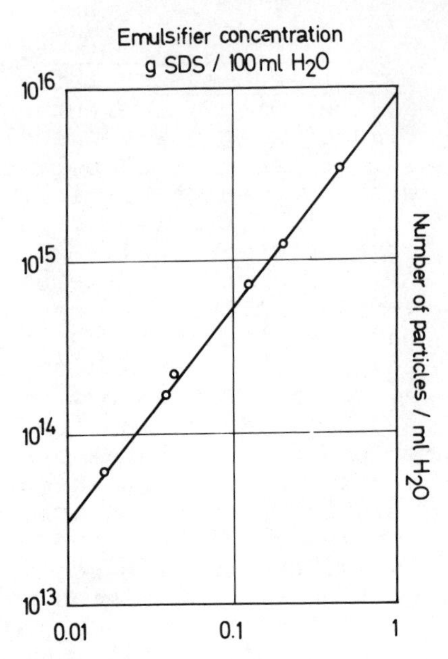

Figure 1. Polymerization—formation of particles. Effect of initial emulsifier concentration. Na^+: 0.01 mol/l.

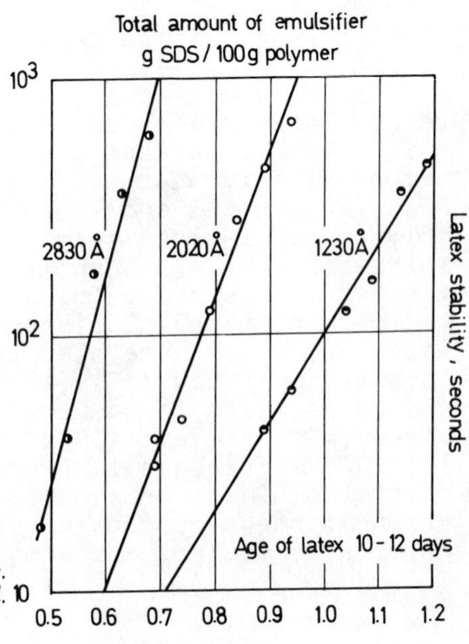

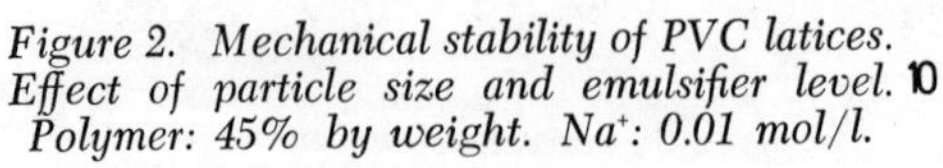

Figure 2. Mechanical stability of PVC latices. Effect of particle size and emulsifier level. Polymer: 45% by weight. Na^+: 0.01 mol/l.

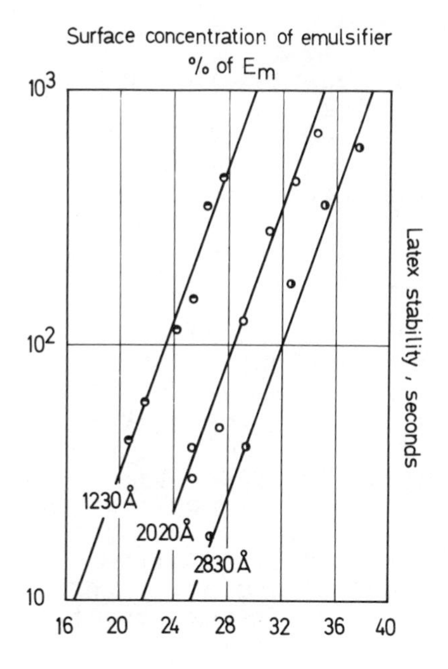

Figure 3. Mechanical stability of PVC latices. Effect of particle size and emulsifier level. Polymer: 45% by weight. Na^+: 0.01 mol/l.

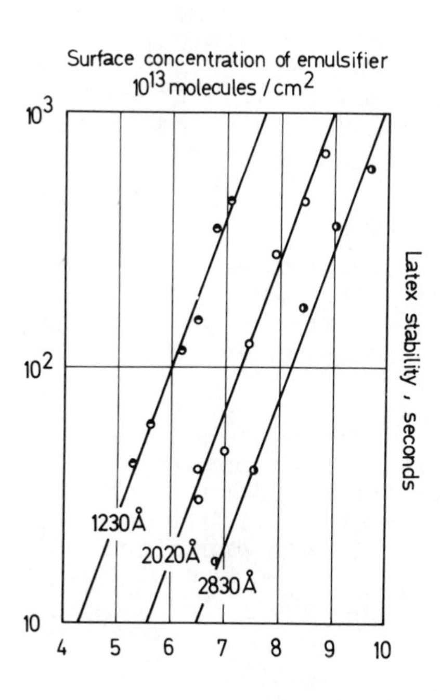

Figure 4. Mechanical stability of PVC latices. Effect of particle size and emulsifier level. Polymer: 45% by weight. Na^+: 0.01 mol/l.

will represent $6 \cdot 10^{12}$ charges per square cm. The surface concentration of emulsifier recorded in Figure 4. is approximately 10 times this value.

Similar results to those obtained here by the stability measurements have been reported by Roe and Brass (7,8). They studied polystyrene latex stabilized by potassium palmitate. The analysis supplied by these authors shows that the order of magnitude of the slope of the stability curves can be accounted for as an entropic effect of crowding of adsorbed molecules during an encounter between two particles. They pointed this out as a possible explanation as the amount of emulsifier adsorbed strongly affects the stability without altering the electrophoretically derived double-layer potential.

Bibeau and Matijevic studied the stability of a PVC latex by addition of electrolytes. They also found surface ion concentrations derived by electrophoresis to be poor indicators of latex stability. Their stability results were found to compare favorably with DLVO theory predictions, using the actual surface concentration of potential-determining species as the basis for interpretation. That means taking into account both fixed charges and adsorbed emulsifier.

Latex stability will be determined by the combined effect of two factors: the probability of collision between particles and the fraction of the encounters between particles which lead to permanent contact. Tha first factor, the collision frequency, will increase with increasing particle size and particle number. It will also increase with increasing shear rate. The influence of various test conditions on the second factor ought to be discussed on the basis of the DLVO theory of colloid stability.

In the theory developed by Derjaguin and Landau (24) and Verwey and Overbeek (25) the stability of colloidal dispersions is treated in terms of the energy changes which take place when particles approach one another. The theory involves estimations of the energy of attraction (London-van der Walls forces) and the energy of repulsion (overlapping of electric double layers) in terms of inter-particle distance. But in addition to electrostatic interaction, steric repulsion has also to be considered. The subject of flocculation kinetics and the stabilization of dispersions has been dealt with in many recently published papers. Some of them are cited here (26-34).

The adsorbed layer of emulsifier on the particle surface can affect the stability of latices in three main ways:

1. By increasing the charge on the particles, which will increase the repulsive forces.
2. By altering the value of the effective Hamaker constant, which means modifying the inherent attractive forces.
3. By sterically hindering convergence of the particles.

Bibeau and Matijevic used a fixed value for the Hamaker constant and interpreted the increased stability by addition of emulsifier as being exclusively an effect of increased surface charge. But the various stabilizing mechanisms are not mutually exclusive and may function co-operatively. The effect of adsorbed layers on the energy of attraction between two particles has been considered by Vold (35) and by Vincent (36).

Excellent reviews on colloid stability are given by Napper and Hunter (37) and by Ottewill (38).

Latex stability. Effect of residual monomer. Stability depends on the age of the latex, as shown in figure 5. Presumably the raising of the stability level by storage can be attributed to the decreasing content of residual monomer. The data in this figure were obtained some years ago before special attention was focused on the residual monomer because of the health hazard arising from exposure to it. Initially the monomer content in the particles of this latex was 6-8 per cent. Only a small fraction of the monomer is dissolved in the aqueous phase.

Experimental polymerizations are now run to higher conversion than previously, and the monomer is stripped from the latex particles by the end of polymerization to a content of 0.25 per cent by weight of polymer. This raises the stability level by a factor of more than 100 as compared to a latex with approximately 3 per cent VCM, as shown in Figure 6.

A short evacuation of this 3 per cent VCM latex has the same effect as several days of storage. The process can also be reversed by addition of VCM, making the latex more unstable. Storage has no effect on the low monomer latex, and further lowering of the monomer content by additional stripping has no effect on the stability level.

The residual monomer content may change the inherent attractive forces between the particles. These forces depend on the nature of the polymer material

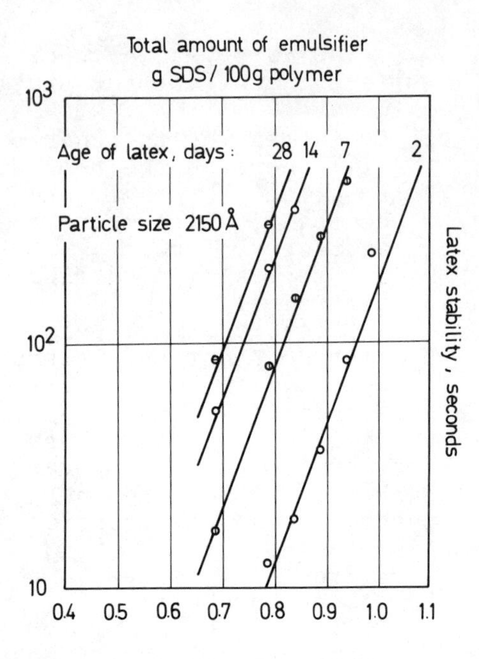

Figure 5. Mechanical stability of PVC latices. Effect of age of latex. Polymer: 45% by weight. Na⁺: 0.01 mol/l.

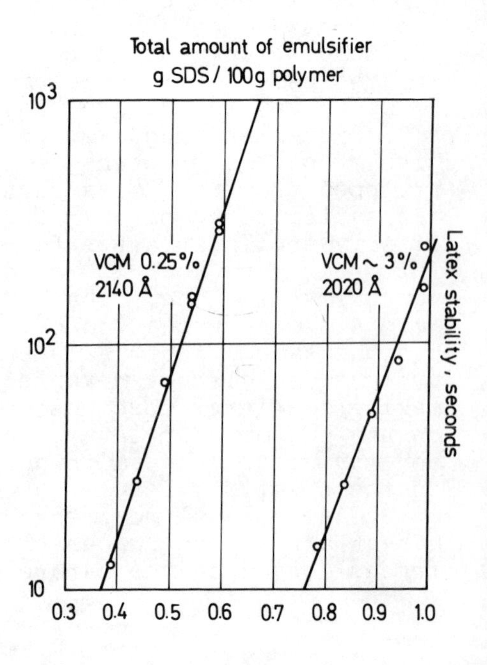

Figure 6. Mechanical stability of PVC latices. Effect of residual VCM in the latex particles. Polymer: 45% by weight. Na⁺: 0.05 moles/l.

of which the particles consist, particularly the number of atoms per unit volume and the polarisability. A residual monomer content of 3 per cent will presumably change these properties very little.

The residual monomer content may change the repulsive forces between the particles. To some extent migration of monomer through the polymer-water interface will take place during the stability test. This migration could disturb the double-layer potential. The migration will be more pronounced the more monomer-swollen the particles are. Monomer may be found in the locality of the charged surface, adjoining the adsorbed emulsifier molecules, giving a shielding effect of surface charge.

The residual monomer content will by external plasticization cause a considerable lowering of the polymer glass transition temperature. A correlation between stability and softness of the polymer particles may exist. The hydrophobic part of the emulsifier molecules may partly penetrate the particle surface and thus be anchored to the surface to some extent. The resistance to deformation of such a stabilizing layer, when subjected to mechanical shear, is assumed to be dependent on the polymer particle softness. With soft particles polymer chain entanglement may also occur on particle-to-particle contact, making redispersion of agglomerates more unlikely.

Latex stability. Effect of copolymerization. When copolymerizing VCM with vinyl esters it appears to be the combination of two competing effects which determines the latex stability. A stability increasing effect seems to arise from increasing the polarity of the polymer particle surface, and a stability decreasing effect from increasing the softness of the polymer particles by internal plasticization.

Copolymerization with vinyl acetate has a strong effect on the nature of the surface of the polymer particles, but the plasticization effect is comparatively weak. With increasing content of vinyl acetate in the copolymer the latex stability will pass through a distinct maximum before decreasing below the stability level of the homopolymer.

Copolymerization with VeoVa has little effect on the nature of the particle surface, but the plasticization effect is somewhat more pronounced than with vinyl acetate. With increasing content of VeoVa in the copolymer the latex stability will pass through a small maximum before decreasing far below the stability level of the homopolymer.

The stability of the copolymer latices shown in Figure 7. are near the maximum level both for the vinyl acetate and the VeoVa copolymer.

Vinyl acetate units in the copolymer are more hydrophilic than the vinyl chloride units, and they will to some extent hydrolyse, introducing hydroxyl groups on the particle surface. The branched, bulky structure of VeoVa makes its ester group difficult to hydrolyse.

The specific density were 1.39, 1.36 and 1.33 gram per cubic cm for homopolymer, copolymer with 10 per cent vinyl acetate, and copolymer with 10 per cent VeoVa respectively. The latices were of low viscosity.

The outlined relationship between stability and polymer characteristics has to be confirmed by further investigations.

Latex stability. Effect of pH. The pH of the latices were adjusted to about 8 before doing any of the previously described stability measurements. The effect of pH on the stability is described in Figure 8. A sharp drop in stability occurs below pH 2. This is possibly due to deionization of the sulphate group of the emulsifier, indicating an electrostatic repulsion to have been operative.

Latex stability. Effect of temperature. Usually no temperature control was imposed. The rise in temperature during the test was 1-2 °C. For purposes of temperature control the bottle containing the sample was provided with a water jacket through which water at specified temperature was circulated. The effect of temperature on the stability is described in Figure 9. To avoid confusion the experimental points are not indicated on the figure, except those obtained at room temperature. The temperature dependence was found to obey the Arrhenius equation. At any given emulsifier level a linear plot of the logarithm of the stability versus 1/T could be obtained.

Latex stability. Effect of stirring speed and spindle disk diameter. Figures 10. and 11. show how sensitive the stability test is with regard to the speed of stirring and the diameter of spindle disk.

Characterization of particle size. In Figure 12. a comparison is given between particle size determination by the analytical centrifuge, by electron microscope and by nitrogen adsorption on the carefully

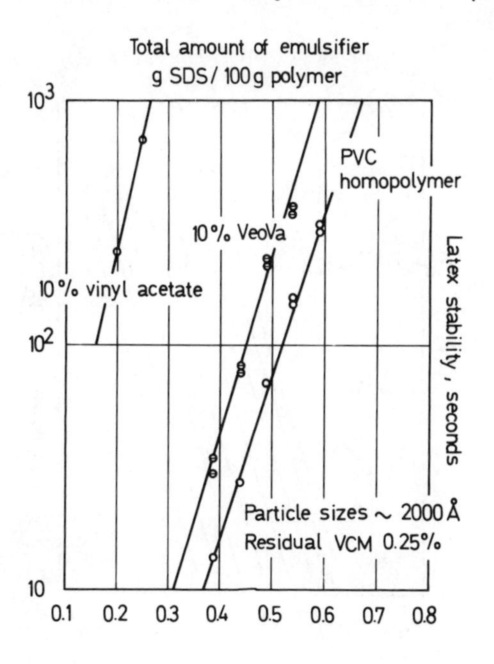

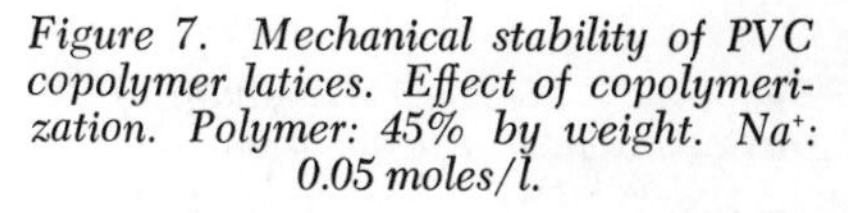
Figure 7. Mechanical stability of PVC copolymer latices. Effect of copolymerization. Polymer: 45% by weight. Na^+: 0.05 moles/l.

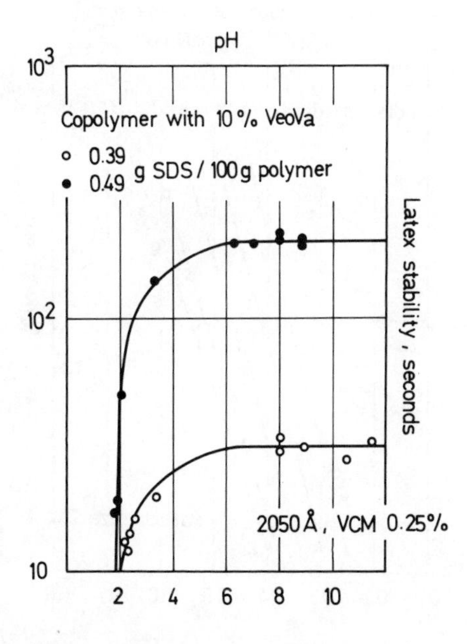

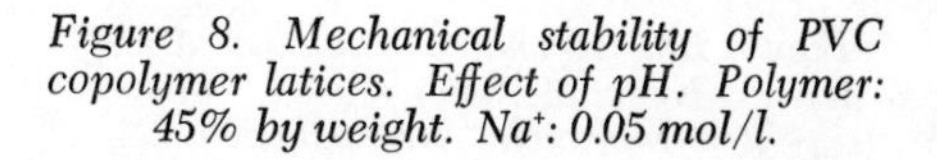
Figure 8. Mechanical stability of PVC copolymer latices. Effect of pH. Polymer: 45% by weight. Na^+: 0.05 mol/l.

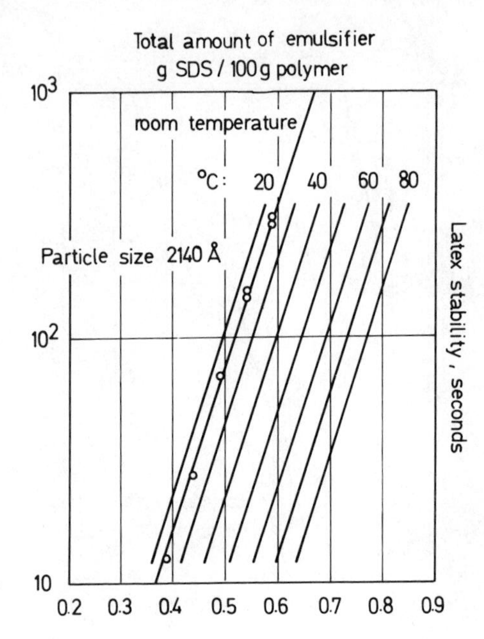

Figure 9. Mechanical stability of PVC latices. Effect of temperature. Polymer: 45% by weight. Na^+: 0.05 mol/l.

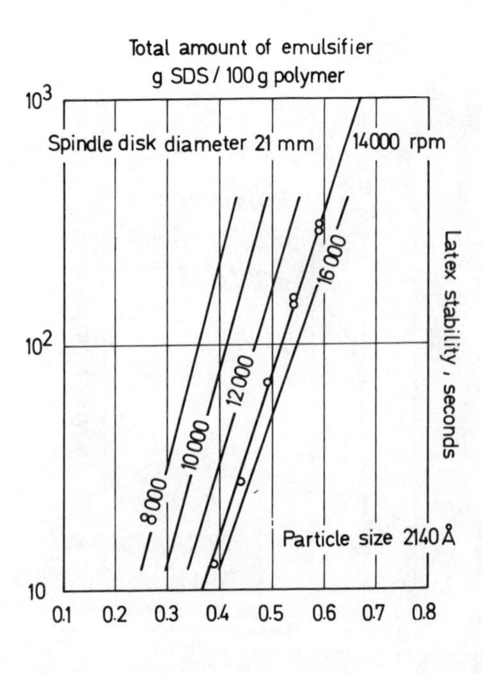

Figure 10. Mechanical stability of PVC latices. Effect of stirring speed. Polymer: 45% by weight. Na^+: 0.05 mol/l.

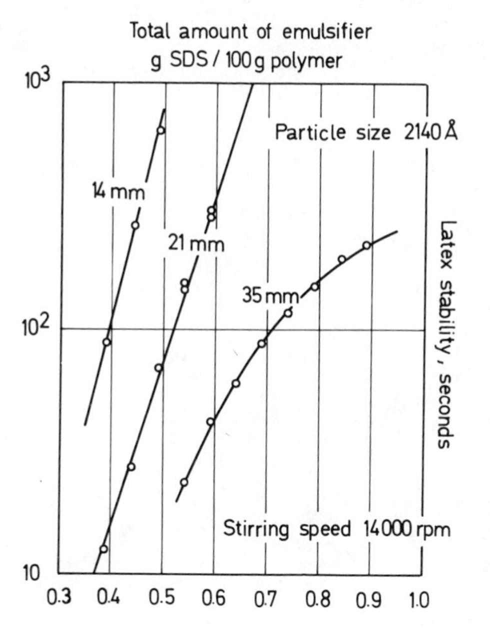

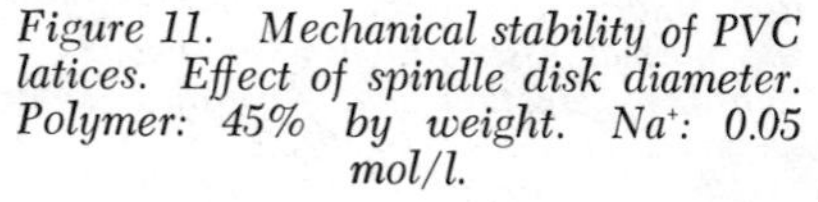

Figure 11. Mechanical stability of PVC latices. Effect of spindle disk diameter. Polymer: 45% by weight. Na^+: 0.05 mol/l.

Figure 12. Particle size measurements by different methods

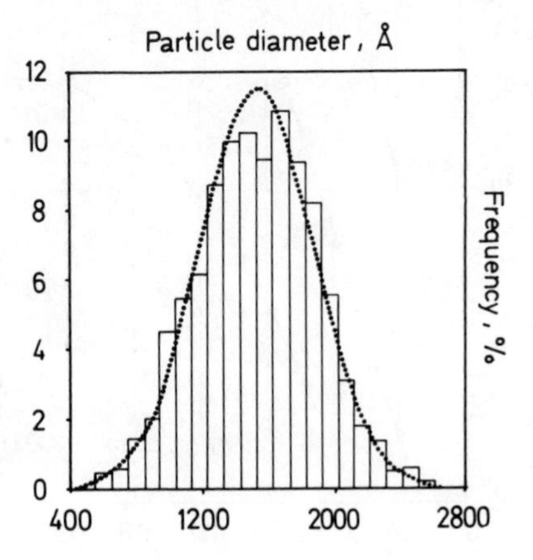

Figure 13. Histogram: particle size distribution obtained by electron microscopy. Curve: the normal distribution with mean 1530 and standard deviation 340. Analytical centrifuge: 1960 Å.

	$Å^2$/molecule
Residual VCM ~3%	
PVC	36.4 - 39.9
Residual VCM 0.25%	
PVC	37.2 - 42.0
Copolymer, 10% VeoVa	39.1 - 43.5
Copolymer, 10% vinyl acetate	66.4 - 68.3

Figure 14. D_p measured by analytical centrifuge. E_m determined by soap titration. Estimation of A_m from these data.

dried powder obtained by freezing latex. The results obtained by electron microscope are corrected for a shrinking of 20 per cent. Some sintering of the powder during drying is a possible explanation of the somewhat high values obtained by nitrogen adsorption.

Vinyl chloride is built into the copolymer somewhat faster than the vinyl esters. The latices described in Figure 12. were prepared with the same initial concentration of emulsifier. The difference in particle size between the homopolymer and the copolymer latices are considered to be within experimental error.

In Figure 13. the histogram shows a particle size distribution obtained by electron microscopy. This distribution may be described by a normal distribution, illustrated by the dotted curve.

Characterization of adsorption of emulsifier. The area occupied by each adsorbed emulsifier molecule at the polymer-water interface (A_m) was estimated from the mean particle size of the latices, determined by the analytical centrifuge, and the amount of emulsifier adsorbed at the interface corresponding to full coverage, determined by adsorption titration.

The results obtained after having examined several latices are summarized in Figure 14. The values found for A_m do not differ significantly for the homopolymer and the VeoVa copolymer. For the vinyl acetate copolymer the area is distinctly increased as a result of increased interface polarity (39).

The molecular areas recorded in Figure 14. has to be adjusted upwards by approximately 15 per cent for not using the surface average diameter.

Acknowledgment

The author is grateful to Mr. K.Gill, Mr. A.Hansen, Mrs. K.Mürer and Mr. A.Talmoen for carrying out the polymerization, stability and adsorption experiments, and to Norsk Hydro for permission to publish this work.

Literature cited

1. Rangnes P. and Palmgren O., J.Polymer Sci. C 33 (1971), 181
2. Maron S.H., Elder M.E. and Ulevitch I.N. J.Colloid Sci. 9 (1954), 89
3. Paxton T.R., J.Col.Interface Sci. 31 (1969), 19

4. Jordan H.F., Brass P.D. and Roe C.P., Ind.Eng.Chem.Anal.Ed. 9 (1937), 182
5. Dawson H.G., Anal.Chem. 21 (1949), 1066
6. Madge E.W., Collier H.M. and Duckworth I.H., Trans.Inst.Rubber Ind. 28 (1952), 15
7. Roe C.P. and Brass P.D., J.Colloid Sci. 10 (1955), 194
8. Roe C.P., Ind.Eng.Chem. 60 (1968), 20
9. Roe C.P., J.Col.Interface Sci. 37 (1971), 93
10. ASTM specification D 1076-71
11. BSI specification 1672-72 and 3397-70
12. Maron S.H. and Ulevitch I.N., Anal.Chem. 25 (1953), 1087
13. Stamberger P., J.Colloid Sci. 17 (1962), 146
14. Greene B.W. and Sheetz D.P., J.Col.Interface Sci. 32 (1970), 96
15. Utracki L.A., J.Col.Interface Sci. 42 (1973), 185
16. Vanderhoff J.W., Vitkuske J.F., Bradford E.B. and Alfrey T., J.Polymer Sci. 20 (1956), 225
17. Ugelstad J., Mörk P.C., Dahl P. and Rangnes P., J.Polymer Sci. C 27 (1969), 49
18. Goodwin J.W., Hearn J., Ho C. and Ottewill R.H., Br.Polym.J. 5 (1973), 347
19. Fitch R.M., Br.Polym.J. 5 (1973), 467
20. Palmgren O., IUPAC Symp. on Macromolecules, Helsinki 1972, Priprint Vol.4, III-49
21. Bibeau A.A. and Matijevic E., J.Col.Interface Sci. 43 (1973), 330
22. Van den Hul H.J. and Vanderhoff J.W., Br.Polym.J. 2 (1970), 121
23. Hearn J., Ottewill R.H.and Shaw J.N., Br.Polym.J. 2 (1970), 116
24. Derjaguin B.V. and Landau L., Acta phys.- chim. URSS 14 (1941), 633
25. Verwey B.V. and Overbeek J.T.G., "Theory of the Stability of Lyophobic Colloids", Elsevier, 1948
26. Watillon A. and Joseph-Petit A.M., Disc.Faraday Soc. 42 (1966), 143
27. Ottewill R.H. and Shaw J.N., Disc.Faraday Soc. 42 (1966), 154
28. Napper D.H., Ind.Eng.Chem.Prod.Res.Develop. 9 (1970), No4, 467
29. Dunn A.S. and Chong L.C.H., Br.Polym.J. 2 (1970), 49
30. Napper D.H. and Netschey A., J.Col.Interface Sci. 37 (1971), 528
31. Hesselink F.T., Vrij A. and Overbeek J.T.G., J.Phys.Chem. 75 (1971), No14, 2094

32. Hatton W. and McFadyen P., J.Chem.Soc., Faraday Trans. 70 (1970), No4, 655
33. Ottewill R.H. and Walker T., J.Chem.Soc., Faraday Trans. 70 (1970), No5, 917
34. Khanna R.K., J.Oil.Col.Chem.Assoc. 57 (1974), 161
35. Vold M.J., J.Colloid Sci. 16 (1961), 1
36. Vincent B., J.Col.Interface Sci. 42 (1973), 270
37. Napper D.H. and Hunter R.J., MTP Int.Rev.Sci., Series I 7 (1972), Chap.8
38. Ottewill R.H., Specialist Periodical Reports, Col.Sci. 1, Chap.5, The Chemical Soc., London 1973
39. Yeliseyeva V.I. and Zuikov A.V., Polymer Preprints 16 (1975), No1, 143

18

Effect of Surfactant Structure on the Electrodeposition of Cationic Latexes

E. H. WAGENER, S. R. KUROWSKY, D. S. GIBBS, and R. A. WESSLING

Physical Research Laboratory, The Dow Chemical Co., Midland, Mich. 48640

The effect of emulsifier structure on the electrodeposition of cationic latexes has been investigated. The electrocoating performance of these systems can be correlated with the reducibility of the emulsifier. This phenomenon has no counterpart in conventional electrodeposition processes.

The electrophoretic deposition of aqueous colloids has been known for many years. An extensive technology was developed to fabricate articles from natural rubber latex. Noble (1) traces its origins back to 1908.

The use of electrodeposition to apply paint is a comparatively recent innovation. The history of this development has been detailed by Brewer (2,3). The electrodeposition of paint, or electrocoating as it is commonly labeled, is derived from the old rubber latex technology and has many features in common; but it differs in one important respect-- it provides rapid and complete current cutoff.

The modern process of electrodeposition can thus be described as a combination of three basic elements: (a) Electrophoresis - migration of charged polymer particles to metal surface; (b) Deposition - colloidal destabilization of particles at the metal-bath interface; and (c) Insulation - formation of an adherent, nonconductive layer of resin on the metal surface. The last named element is responsible for the high throwing power which can be achieved with the electrocoating process.

Commercial electrocoating formulations are made up with low molecular weight resins containing ionizable groups. Typically, they are converted to aqueous dispersions by first dissolving the resin in a water-miscible coupling solvent, then adding the appropriate solubilizing agent and finally adding water to form the dispersion.

Both anodic and cathodic electrocoating systems are in use. The anodic systems contain carboxylated polymers where the acid groups are neutralized with amines or KOH. The cathodic systems contain amino-functional resins where the amine groups are neutralized with an acid.

The colloidal state in these systems is poorly defined. It depends on the level of organic solvent, the number of ionizable groups in the resin and the degree of neutralization. In an anodic system, for example, the polymer may be completely soluble at high pH and change gradually to a hydrophobic colloid as the pH is decreased to 7. At lower pH, the system flocculates.

The uncertainty of the colloidal state makes it very difficult to study the process of electrodeposition in these systems quantitatively. However, the consensus is that deposition takes place primarily by a charge neutralization mechanism (3):

$$R(COO^-)_x(COOH)_y \quad + \quad xH^+ \rightleftharpoons R(COOH)_{x+y}\downarrow \qquad (1)$$

Anodic

$$R[R_2NH^+]_x[R_2N]_y \quad + \quad x\ OH^- \rightleftharpoons R(R_2N)_{x+y}\downarrow \qquad (2)$$

Cathodic

The H^+ and OH^- ions are supplied by the simultaneous electrolysis of water during electrodeposition. Electrode reactions involving the resins are of little significance. However, oxidation of the metal substrate does play a role in anodic electrocoating.

$$M^o \longrightarrow M^{+n} + ne^- \qquad (3)$$

$$M^{+n} + R(COO^-)_n \longrightarrow M^{+n}R(COO^-)_n\downarrow \qquad (4)$$

Latexes can also be electrodeposited but if they are stabilized by ionized groups such as carboxyl ions, the same problem with pH dependent particle charge and colloidal stability is encountered.

Latexes stabilized with fully ionized groups such as sulfonate ions are well known in the coatings field. But, they have not been utilized in electrodeposition because they do not cut off current. This is not the case for certain cationic latexes. An electrodeposition process based on sulfonium stabilized colloids has been reported (4). The sulfonium ion, like the analogous quaternary ammonium ion, is highly ionized over a wide pH range so that particle charge is independent of pH. Moreover, it has been shown that sulfonium systems can electrodeposit with current cutoff comparable to that of an amine-stabilized latex, whereas quaternary ammonium latex deposited conductive films. This condition holds over a wide range of coating conditions (5). A comparison of coating weight deposited at various pH's is shown in Figure 1. The quaternary ammonium stabilized latex deposited as a conductive gel over a pH range of 2-10. The sulfonium latex deposited under the same conditions yielded an insulating film. Not evident from the figure is the fact that deposition had essentially stopped in the sulfonium case after about 15 seconds; it was still increasing in the quaternary ammonium system when the experiment was terminated (2 minutes).

Since particle charge in these latexes is not pH dependent, the mechanism outlined above for the conventional electrocoating systems cannot apply. It was clear from the start that the difference in the behavior of sulfonium and quaternary ammonium stabilized latexes is related to the greater reactivity of the sulfonium ion. Though stable in dilute aqueous solutions, sulfonium ions might be expected to undergo rapid decomposition under the conditions obtained at the cathode surface while current is flowing. However, the specific reactions involved were not known.

The present study was undertaken to determine whether the electrochemical reducibility of the stabilizing cations was an important factor. In order to simplify the problem experimentally, a latex stabilized with adsorbed emulsifier was selected as the model system. This permits study of the cation independent of the polymer particle.

A study of reduction processes was a natural choice since we were dealing with cathodic electrodeposition; but the motivation to look at this aspect was generated by earlier work with benzylic sulfonium salts on mercury cathodes (6,7). It was shown that these salts could be reductively coupled in water whereas the corresponding quaternary ammonium salts were unreactive. Reductive coupling of sulfonium salts in water turns out to be a very general reaction. It can take place not only on mercury, but on hard metals like steel as well. This technique was used to develop unique electrocoating processes (8,9). The cathodic processes are illustrated schematically below:

$$R_1\text{-}S^{+}\begin{matrix} R_2 \\ R_3 \end{matrix} \xrightarrow{e^-} R_1\cdot + R_2\text{-}S\text{-}R_3 \quad (5)$$

$$2\ R_1\cdot \longrightarrow R_1\text{-}R_1 \quad (6)$$

$$R_1\cdot \xrightarrow{e^-} R_1\ominus \xrightarrow{H_2O} RH + OH^- \quad (7)$$

The reducibility of the cations can be influenced by changing the substituents on the hetero atom. We used this approach to synthesize surfactants with different reduction potentials. The electrodeposition of latexes stabilized by these surfactants was then studied and the results correlated with reduction potential.

Results

The data in Table I show that all of the sulfonium emulsifiers are easily reduced relative to water (∿-2.0 V in this experiment). The first wave usually appears at ∿-1.0 V, but an electron transfer process at very positive potentials is seen in the nitrobenzyl substituted sulfonium compound. It is believed that the nitro group itself is involved in this process though the sulfonium group is the one that ultimately gets reduced.

An aryl substituent on the sulfur leads to a slight loss in reducibility (more negative $E_{\frac{1}{2}}$), but the biggest change comes when no activating groups are present as in the alkyl sulfoniums. No significant steric effects are evident in any of the salts.

The effect of substituent structure is more pronounced in the nitrogen compounds. The "aromatic oniums" reduce as easily as sulfoniums, but the quaternary ammonium salts are reducible only when the nitrogen has activating substituents.

Electron with drawing groups like F, CF_3 and NO_2 on the benzyl substituent favor reducibility. The fluoro substituted compound #12 is reducible only competitively with water since its first wave appears as a shoulder on the solvent wave.

These data pertain to reductions on mercury only. On other metals, the reduction potential may be shifted to more negative values, and since the hydrogen overvoltage would be lower, the onium ions might not reduce preferentially. We speculate, however, that the order of reducibility would not be changed in comparisons on a given substrate. This does seem to be a reasonable approximation for sulfonium salts at least (11).

Properties of Emulsifiers

Table I-A - Sulfonium Salts, $R_1-\overset{+}{S}(R_2)-R_3\ A^-$

No.	R_1	R_2	R_3	A	$E_{1/2}$ V (SCE)
1	$C_{12}H_{25}-$	C_2H_5-	CH_3-	Cl^-	-1.63
2	$C_{12}H_{25}-$	C_6H_5-	CH_3-	Cl^-	-1.35
3	$C_{12}H_{25}-$	$C_6H_5-CH_2-$	CH_3-	Cl^-	-1.11
4	$C_{12}H_{25}-$	$O_2N-C_6H_4-CH_2-$	CH_3-	Cl^-	-0.64, -1.08, -1.38
5	$C_{12}H_{25}-C_6H_4-CH_2-$	CH_3-	CH_3-	Cl^-	-1.02, -1.13, -1.42
6	$C_{12}H_{25}-C_6H_4-CH_2-$	C_2H_5-	C_2H_5-	Cl^-	-1.00, -1.14, -1.44
7	$C_{12}H_{25}-C_6H_4-CH_2-$	C_3H_7-	C_3H_7-	Cl^-	-0.97, -1.11, -1.45
8	$C_{12}H_{25}-C_6H_4-CH_2-$	$HO-CH_2CH_2-$	CH_3-	Cl^-	-1.09
9	$C_{12}H_{25}-C_6H_4-CH_2-$	$HO-CH_2CH_2-$	$HO-CH_2CH_2-$	Cl^-	-0.88, -1.03, -1.36
10	$C_{12}H_{25}-C_6H_4-CH_2-$	C_4H_9-	C_4H_9-	Cl^-	-0.99

Table I-B - Quaternary Ammonium Salts, $R_1-N^+(R_2)(R_4)-R_3\ A^-$

No.	R_1	R_2	R_3	R_4	A^-	$E_{1/2}$ V (SCE)
11	$C_{12}H_{25}-$	$C_6H_5-CH_2-$	CH_3-	CH_3-	Cl^-	<-2.0
12	$C_{12}H_{25}-$	$F-C_6H_4-CH_2$	CH_3-	CH_3-	Cl^-	∿-1.90
13	$C_{12}H_{25}-$	$O_2N-C_6H_4-CH_2-$	CH_3-	CH_3-	Cl^-	-0.53, -1.33
14	$C_{12}H_{25}-$	$CF_3-C_6H_4-CH_2-$	CH_3-	CH_3-	Cl^-	-1.64
15	$C_{12}H_{25}-C_6H_4-CH_2-$	CH_3-	CH_3-	CH_3-	Cl^-	<-2.0
16	$C_{12}H_{25}-C_6H_4-CH_2-$	C_2H_5-	C_2H_5-	C_2H_5-	Cl^-	<-2.0
17	$C_{12}H_{25}-C_6H_4-CH_2-$	C_2H_5-	C_2H_5-	$HO-CH_2CH_2-$	Cl^-	<-2.0

Table I-C - Heterocyclic Salts, $R-Z^+\ A^-$

No.	R	Z	A	$E_{1/2}$ V (SCE)
18	$C_{12}H_{25}-$	pyridinium N-	Cl^-	-1.30, -1.65
19	$C_{12}H_{25}-C_6H_4-CH_2-$	pyridinium N-	Cl^-	-1.00
20	$C_{14}H_{27}-$	isoquinolinium N-	Br^-	-1.17

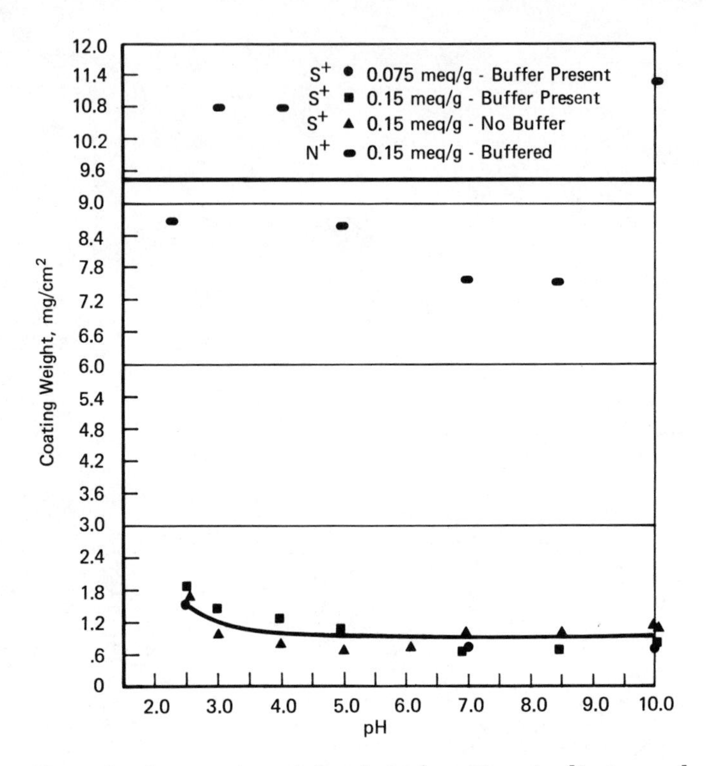

Figure 1. A comparison of the electrodeposition of sulfonium and quaternary ammonium latexes as a function of pH (5). Coating conditions 200 V, 2 min.

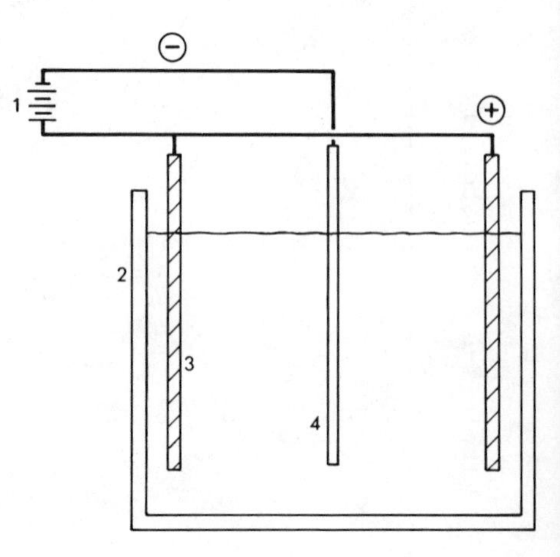

Figure 2. Schematic of the electrodeposition cell: (1) 500 V, 50 amp DC power supply, < 5% ripple at 200 V; (2) nylon cell; (3) graphite anodes; and (4) metal cathode. Circuit also includes integrator and recorder.

The surface activity of these compounds was not studied in detail. As mentioned in the Experimental Section, all were about equivalent in nucleating particles during emulsion polymerization. The resulting latexes when dialyzed to remove excess salt were stable against settling even at 10% solids over many months. Data on samples where both latex particle size and critical micelle concentration were measured is shown in Table II.

Table II

Comparison of Cationic Surfactants

Soap #	CMC at 25°C moles/l x 10^3	Latex Particle Size, Angstroms (by L.S.)
5	2.6	1000
6	3.1	1100
7	2.1	1190
8	2.5	1170
12	3.2	1190
14	2.5	1140

Vastly more data were collected on electrodeposition than can be covered in this paper. Our goal here is to present sufficient data to illustrate the trends and correlations.

An examination of the data suggests that the amount of polymer deposited varies both with emulsifier structure and metal type. In some cases the concentration of emulsifier has a significant effect. Since the amount of washoff did not usually vary significantly, the coating weight is a good indicator of the rate of current cutoff. This is illustrated further by the current-time curves in Figure 4. Two systems are compared: one formulated with an easily reducible (more positive value of $E_{\frac{1}{2}}$) emulsifier cut off the current very rapidly; the film was completely insulated in <10 seconds. The other latex formulated with an emulsifier having $E_{\frac{1}{2}}<-2.0$ V, cut off the current rather sluggishly; current continued to flow for the duration of the experiment. The high current flow is accompanied by heating of the bath, severe gassing, and continued film build. Thus, in the former case, the coating after rinsing with deionized water was velvety smooth and uniform in appearance. In the latter case, the coating was rough and bubbly with excessive build at the edges and corners of the specimen.

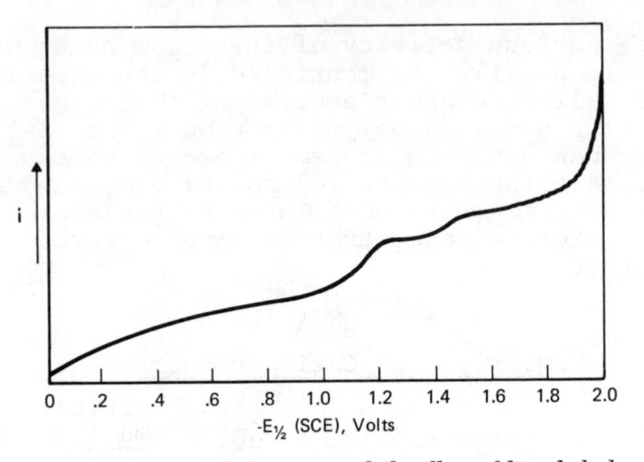

*Figure 3. Typical polarogram—dodecylbenzyldimethylsulfonium chloride, in 1.0*N *KCl, dropping mercury electrode; current full scale 5 μa. Halfwave potentials, −1.02* V*; −1.13* V*, major wave, one electron transfer; −1.44* V.

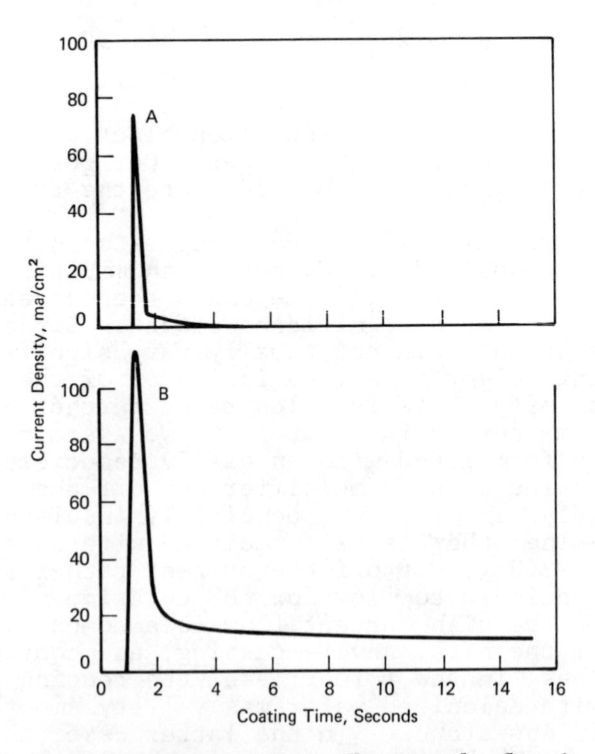

Figure 4. Current–time curves illustrating the effect of emulsifier structure on current cutoff. (A) p-*nitrobenzylmethyldodecylsulfonium chloride; (B)* p-*fluorobenzyldimethyldodecylammonium chloride.*

In general, a change in emulsifier from 0.06 meq/g to 0.1 meq/g had little effect. (This is a range of particle surface coverage of ∿50-80%.) This is illustrated by a comparison of some quaternary ammonium systems in Table III. On platinum they performed badly at both levels; on lead they looked equivalent at both levels though the efficiency decreased consistently at the higher charge. However on steel, a metal of intermediate character, concentration had a major effect in the case of the marginally reducible soap, #12.

Table III

Comparison of Electrodeposition Performance as a Function of Soap Concentration

Soap No.	Concentration meq/g solids	Metal	Coating Wt.*, mg	Efficiency mg/coul.
12	0.06	Platinum	311	35.4
12	0.1	Platinum	341	29.2
14	0.06	Platinum	270	32.7
14	0.1	Platinum	251	43.5
12	0.06	Steel	47.9	14.8
12	0.1	Steel	158.3	23.1
14	0.06	Steel	25.9	20.9
14	0.1	Steel	27.1	15.4
12	0.06	Lead	10.1	18.0
12	0.1	Lead	11.4	14.4
14	0.06	Lead	13.2	21.2
14	0.1	Lead	11.2	14.2

*Coated area, 7.3 cm^2.

This effect is illustrated even more vividly by comparing current-time curves obtained on different substrates with a latex stabilized with a marginally reducible soap. As shown in Figure 5, both the peak current and the residual current are highest on platinum and lowest on copper which like lead is a relatively high hydrogen overvoltage metal. It is worth noting that in the absence of any polarization of the cell, the peak current density would be of the order of 300 ma/cm^2. Since the amount of coating deposited is proportional to the number of coulombs passed, coating thickness must increase much more rapidly on platinum than on copper; but the freshly deposited film remains conductive. The difference is not related to the conductivity of the metal substrate, the order of which is Cu>Pt>Steel.

In general, all of the latexes stabilized by easily reducible emulsifiers deposited lower coating weights on all the metals tested except magnesium. Those latexes stabilized with difficultly reducible emulsifiers failed to coat any of the metals satisfactorily. Even where film build was not excessive because of washoff of the deposited polymer, coating appearance was poor.

Data on Bonderite® 37 treated steel are shown in Table IV. The samples are listed in order of reducibility. The difference in coating weights between reducible and non-reducible systems is clearly evident. The former deposit coating weights in the same range as for an amine-stabilized styrene/butyl acrylate latex. The latter was coated at pH 4 because of colloidal instability at a higher pH.

Table IV

Electrocoating Results on Bonderite® 37 Treated CRS Showing Correlation between Reduction Potential and Coating Weight

Emulsifier Number	$E_{1/2}$ (SCE) Volts	Conc. meq/g	pH	Conduc. µmho/cm	Coating Wt. mg/cm^2
4	-0.64	0.10	7.2	1,680	0.77
7	-0.97	0.10	7.0	1,460	1.42
6	-1.00	0.10	7.4	1,430	2.02
19	-1.00	0.10	7.4	1,020	0.85
5	-1.02	0.15	7.2	1,390	0.80*
20	-1.17	0.09	7.5	940	0.51
18	-1.30	0.10	7.4	1,070	1.30
14	-1.64	0.10	7.4	1,020	1.53
12	-1.90	0.10	7.8	1,150	5.18
16	<-2.0	0.10	7.5	1,110	4.08
17	<-2.0	0.10	7.5	1,100	5.74
15	<-2.0	0.15	6.8	1,480	7.50*
$C_{12}H_{25}NH_3^+C_2H_3O_2^-$	---	0.50	4.3	1,080	1.90*

*Cathode area was 100 cm^2 for these samples.

A finer differentiation between emulsifiers can be made by plotting coating weight on a given metal as a function of reduction potential as illustrated in Fig. 6. Lead appears to be the easiest metal to coat; magnesium and platinum are the most difficult. Leaving magnesium aside, there is a clear pattern in the electrodeposition performance involving both emulsifier reduction potential and metal activity or hydrogen over-voltage. The order of increasing over-voltage for the metals in question is approximately:

$$\xrightarrow{\text{Over-voltage increasing}} \\ \text{Pt/Fe/Al/Cu/Pb}$$

The zinc phosphate coating on steel probably shifts it in the direction of higher over-voltage. When the coating weight is plotted against type of metal on an arbitrary scale, as shown in Figure 7, a correlation emerges. Systems with easily reducible emulsifiers deposit thin films on all metals; those of intermediate reducibility deposit thin films on the high over-voltage metals but coat Pt and steel very poorly. The systems stabilized with difficult to reduce emulsifiers do not cut off current effectively on any substrate.

The argument is persuasive that electrodeposition performance is determined by emulsifier reducibility; but this has to be qualified by the fact that reduction potential is a measure of chemical reactivity, generally. Those salts which reduce most readily also hydrolyze more easily.

We have not established quantitatively that reduction of emulsifier does take place during electrodeposition. The only attempt to analyze the coating for decomposition products was made in a system stabilized with dodecylbenzyldimethylsulfonium chloride. The coatings were dissolved off the metal and the solution analyzed by mass spectrometry. The only product present that could be traced to the soap was p,p-bisdodecyldibenzyl. The most logical explanation for its formation is by reductive coupling. It is very possible that other more volatile products such as dodecyltoluene were lost during coating, drying and extraction.

Discussion and Conclusions

The model system was designed to isolate, to the extent possible, a single variable, the wet film conductivity. By using a latex stabilized with an adsorbed soap, and by keeping the soap concentration below saturation, we are able to adjust particle charge to a known value. The size of the particle is fixed

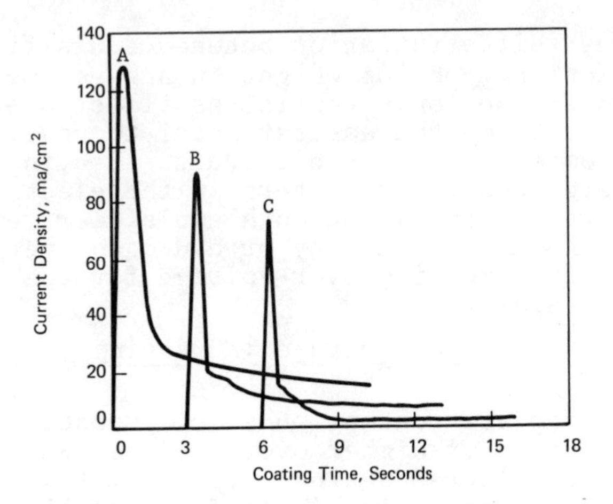

Figure 5. Effect of metal substrate on current cut-off. Latex stabilized by N(m-*trifluoromethylbenzyl)*-N-*dodecyl*-N,N-*dimethylammonium chloride.* $E_{1/2} = -1.64$ *V: (A) platinum; (B) cold rolled steel; (C) copper (soft sheet). Curves B and C displaced 3 and 6 sec, respectively.*

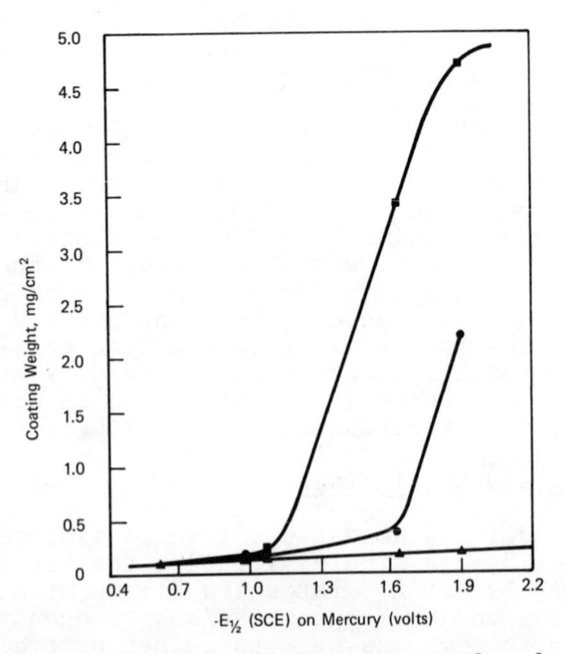

Figure 6. Correlation between coating weight and reduction potential. ■ *, platinum;* ● *, cold rolled steel;* ▲ *, copper.*

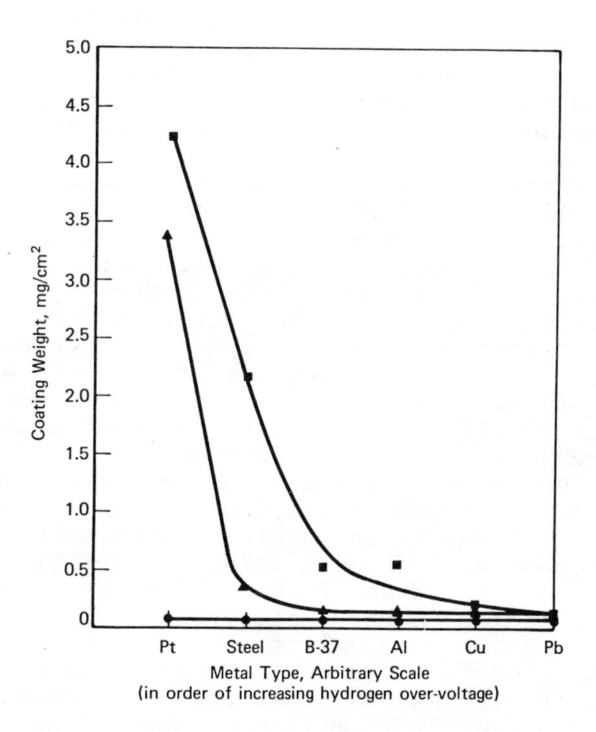

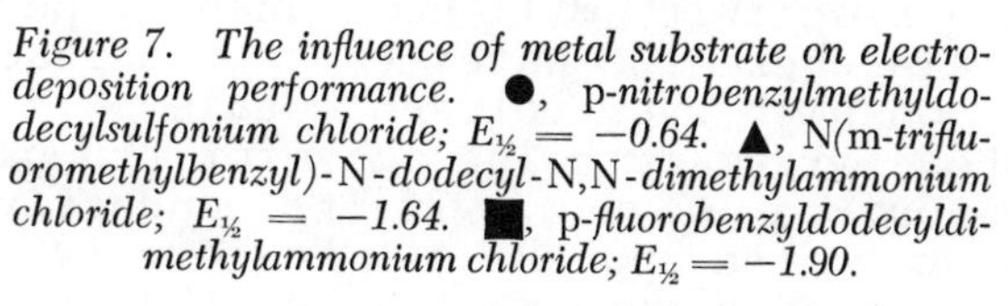

Figure 7. The influence of metal substrate on electrodeposition performance. ●, p-*nitrobenzylmethyldodecylsulfonium chloride;* $E_{1/2} = -0.64$. ▲, N(m-*trifluoromethylbenzyl*)-N-*dodecyl*-N,N-*dimethylammonium chloride;* $E_{1/2} = -1.64$. ■, p-*fluorobenzyldodecyldimethylammonium chloride;* $E_{1/2} = -1.90$.

and independent of charge density. Thus, when the conductivity and particle concentration are fixed, all systems should have approximately the same rate of migration of particles to the metal surface. Since this is a hydrophobic colloid, the particles can deposit by the mechanism of concentration coagulation even in the total absence of charge destruction, and the deposition process will be irreversible.

As the latex particles deposit, a certain amount of soap will be co-deposited and become buried in the film. The conductivity of the film, other factors being constant, will increase with concentration of ions and water.

If we assume that the order of reducibility of the emulsifiers determined on mercury also holds on other metals, then the experimental observations can be explained by the following argument: the rate of current cutoff is determined by the conductivity of the initial polymer deposit. The latter in turn depends on the concentration of ionic groups in the deposit. If we assume that electrophoretic migration of particles brings equivalent amounts of emulsifier to the metal surface, the differences in wet film conductivity must result from differences in the amount of charge destroyed electrochemically (either by primary reduction or secondary reaction with hydroxide ion and/or free radicals). It seems reasonable to assume that the number of cations destroyed is determined by the relative rates of reduction of the emulsifier versus water. If the emulsifier reduces easily, the system will show rapid current cutoff on all metals; if it reduces with difficulty, then the reduction of water to H_2 will be the dominant process. Systems employing emulsifiers of intermediate reactivity will consequently perform well only on metals where the water reduction reaction is slow, i.e., on high over-voltage metals.

The above argument holds only for the primary layer of deposited polymer. If its conductivity is low, gassing is prevented and the film conductivity continues to decrease even more rapidly as the film thickness increases. Thus, the process becomes self-limiting.

If the primary film is too conductive, however, an unstable condition is created. Further current flow at these high current densities increases the temperature at the interface, and therefore the conductivity of the wet film even more. Hydrogen evolution will continue at an increasing rate until the film ruptures or the electrode becomes insulated by a layer of gas bubbles. The gas polarization process will eventually lead to a drop in current flow whether or not the charged groups in the polymer are destroyed. However, it seems likely that some charge destruction occurs in all cases in the hot, alkaline environment which is created at the metal surface.

The above analysis pinpoints the importance of the processes occurring at the moment current begins to flow. Subsequent events will be determined primarily by the conductivity of the initial polymer deposit. This in turn, appears to be governed by the reducibility of the stabilizing cationic groups in systems where particle charge is independent of pH.

Experimental

Emulsifiers. Emulsifiers were prepared by alkylation of nucleophiles as illustrated by the preparation of p-dodecylbenzyltrimethylammonium chloride.

$$C_{12}H_{25}\text{-}C_6H_4\text{-}CH_2\text{-}Cl + (CH_3)_3N \xrightarrow{CH_3OH/H_2O} C_{12}H_{25}\text{-}C_6H_4\text{-}CH_2\text{-}\overset{+}{N}(CH_3)_3Cl^- \quad (8)$$

Alkylating Agent Nucleophile Surfactant

The reactions are 2nd order and are accelerated by H_2O. Therefore, reacting with an excess of nucleophile in aqueous media will produce good yields of the surfactant. The emulsifiers used in this study are listed in Tables I-A, I-B and I-C.

Most of the emulsifiers reported in this work were prepared in the same general recipe. In a few cases, alternate procedures were followed and these will be given in detail below. In general the soap was prepared by reacting the halide with the desired sulfide or amine in a methanol-water mixture (9% water by wt.). The mixture was allowed to react with stirring at ambient temperature. The reaction was followed by titrating for halide ion using 0.1N $AgNO_3$. To maximize the yield, additional deionized water was added incrementally after ∿80% conversion until upon further addition the reaction mixture remained clear. In the beginning stages of this program, the reaction mixture was worked up by extracting with ether; however, in many extractions emulsions were formed in the ether layer which could not be readily broken. A more efficient procedure was to extract the reaction mixture containing excess H_2O with a 50/50 (vol/vol) ether-n-hexane solution. An improved procedure was to extract the methanol/emulsifier mixture first with n-hexane prior to the addition of the excess water. After separation, the excess methanol and n-hexane were stripped off under vacuum and deionized water was back-added to the emulsifier to get the desired concentrations. Using this work-up procedure, aliquots of the emulsifier could be analyzed in the form of relatively pure concentrated solutions before the addition of water without resorting to additional purification techniques, e.g., freeze-drying, recrystallization, etc. It was necessary to handle the sulfonium soaps in solution because they tended to decompose in the dry state. Some representative preparations are listed below. Products were identified in some cases by NMR and IR. Concentration and purity were measured by titration with standard $AgNO_3$ and NaOH solutions.

Dodecylbenzyldiethylsulfonium Chloride. Into a 500 ml flask was added 76.6 g (0.200 moles) of practical grade p-dodecylbenzyl chloride, 26.5 g (0.294 moles) diethyl sulfide (Eastman White label), 200 g of methanol and 2 g deionized water. Stirring was started and after 218.5 hrs the reaction was terminated by extracting it 3 times with 125 ml portions of a 50/50 (vol/vol) n-hexane/ether mixture. The aqueous fraction was separated and concentrated under vacuum to give 429 g of solution which was 19.8% active. The conversion was 85%.

m-Trifluoromethylbenzyldimethyldodecylammonium Chloride. 47.7 g (0.245 moles) of m-trifluoromethylbenzyl chloride and methanol were mixed and 57.4 g (0.269 moles) of dimethyldodecylamine and 50 ml methanol added into the solution with rapid stirring. Then 350 ml of additional methanol and 100 ml of distilled water were added. After 300 hrs the solution was washed 3 times with excess ether. The aqueous phase was concentrated under vacuum to yield 500 g of solution which was 17.25% active. The conversion was 85.3%.

p-Nitrobenzylmethyldodecylsulfonium Chloride. Into a 500 ml flask equipped with condenser and heating mantel was added 40 g (0.233 moles) of p-nitrobenzyl chloride, 50 g (0.231 moles) dodecylmethyl sulfide, 200 g of methanol and 2 g of deionized water. Stirring was started and the temperature was brought up to 50°C. After 480 hrs the reaction was terminated by extracting 3 times with 125 ml portions of a 50/50 (vol/vol) n-hexane/ether mixture. The aqueous portion was separated and concentrated under vacuum. During evaporation, yellow needle-like crystals formed in the flask. Upon gentle warming the crystals dissolved. After stripping, 366.7 g of aqueous solution was recovered which was 13% active. The conversion was 53%. (A second preparation carried out at room temperature gave a conversion of only 17.3% after 380 hrs.) Upon standing at room temperature, crystals reappeared in the aqueous solution. The crystals were filtered and NMR confirmed the structure was the title compound: NMR ($CHCl_3$), δ = 0.87 (multiplet, methyl hydrogens of dodecyl group), 1.71 (multiplet, methylene hydrogens β to the nitrogen), 3.12-3.80 (broad multiplet, methylene hydrogens on the dodecyl group α to nitrogen), 3.29 (singlet, methyl hydrogens bonded to sulfur), 4.94-5.52 (multiplet, benzyl methylene hydrogens) and 8.16 (quartet, aromatic hydrogens). The aromatic sulfonium salt, #2, was prepared by a different procedure shown below.

Dodecylphenylmethylsulfonium Methylsulfate. Into a 500 ml flask equipped with a thermometer and a condenser with drying tube was placed 100 g (0.36 moles) of phenyldodecyl sulfide and 45.4 g (0.36 moles) of dimethyl sulfate. Stirring was started and the reaction mixture was heated to 70°C. After 17 hrs, a clear, amber-colored, viscous liquid was obtained. The solution was allowed to come to ambient temperature and 80 ml of n-hexane was added. After mixing, 200 ml of dry methanol was added to the reaction mixture and mixed thoroughly. Two phases separated immediately; a bottom amber-colored layer and a top pale yellow organic layer. The phases were separated and the organic layer discarded. The methanol layer was washed 3 times with 80 ml portions of n-hexane. The methanol and other volatiles were stripped off under vacuum to give 127.0 g of title compound (87.3% conversion). Methyl sulfate ion was then replaced with chloride using Dowex 1 beads. Dodecylamine acetate, dodecylethylmethylsulfonium chloride and tetradecylisoquinolinium bromide were obtained from commercial sources.

Latexes. Latexes were made in a monomer addition recipe described earlier (10). This is a seeded continuous monomer addition recipe using t-butylhydroperoxide/hydroxylamine hydrochloride redox couple as initiator. Polymerizations were carried out in stirred glass reactors at 50°C. The only variation in the original recipe was in the surfactants. In the present procedure, $\sim$1/3 of the soap ($\sim$1.5% based on total monomer) was used in the seed and the remainder fed to the reactor during polymerization. The monomer feed contained styrene and butylacrylate in a 40/60 ratio. This composition was selected because it is readily film-forming and is not affected chemically by the electrodeposition process. The polymer remains soluble and thermoplastic even after coating and baking 20 minutes at 175°C. This recipe typically yielded a fluid latex at 40-45% solids with a particle size in the range 1000-1500 Angstroms by light scattering.

Latexes were successfully prepared with all soaps except those containing nitro-substituents; conversions were lower in the latter case and for soap #13, no polymerization occurred. A latex stabilized by this soap was prepared with soap #15; it was then dialyzed exhaustively until the residual soap concentration reached 0.02 meq/g solid. Then soap #13 was back-added to the desired level.

In all cases, the latexes were dialyzed to remove water-soluble impurities and salts. This process also removed part of the soap. The dialyzed latexes were then analyzed and reformulated to the desired level of charge with the appropriate emulsifier.

Electrodeposition. The basic formulation used for electrocoating was:

Polymer Solids	-	10%
Conductivity	-	900-1500 μmho/cm
pH	-	6.8 to 7.5
Buffer	-	Diammonium hydrogen phosphate

Electrodeposition experiments were carried out in the apparatus described earlier (4). The equipment is illustrated schematically in Figure 2. The circuit also included a strip chart recorder and integrator. A variety of metals were coated at the cathode using the standard condition of 200 V for 2 min. The cathode area was held constant at 7.3 cm^2, unless otherwise specified. After deposition, the metal test pieces were rinsed with deionized water, dried and weighed to determine the amount of polymer deposited. The number of coulombs was measured simultaneously in order to calculate current efficiencies. Results were analyzed and compiled in a simple computer program. A typical printout is shown below for a system with poor current cutoff (Table V), and good cutoff (Table VI). The set of metals used as substrate are listed below.

Platinum
Copper
Cold Rolled Steel
Lead
Aluminum
Magnesium
Conversion Coated Steel (Parker Bonderite® 37)

Table V - Surfactant - #16, Dodecylbenzyltriethylammonium Chloride

Variable	Metal Type						
	Mg	Cu	Pb	UCS	PCS	Pt	Al
Composition	60 BA 40 STY	60 BA 40 STY	60 BA 40 STY	60 BA 40 STY	60 BA 40 STY	60 BA 40 STY	60 BA 40 STY
Efficiency, mg/coul.	30.347	18.640	18.026	28.173	24.418	34.189	25.550
Coating Weight, mg	172.866	43.533	21.599	217.966	54.566	229.300	104.599
Coulombs	5.697	2.320	1.201	7.728	2.242	6.706	4.098
Voltage	200	200	200	200	200	200	200
pH Range	6.9-5.9	6.9-5.9	6.9-5.9	6.9-5.9	6.9-5.9	6.9-5.9	6.9-5.9
Conductivity - Before After	810 780	810 780	810 780	810 780	810 780	810 780	810 780
Soap Conc., meq/g pol	0.06	0.06	0.06	0.06	0.06	0.06	0.06
Particle Size, Å	1390	1390	1390	1390	1390	1390	1390
% Solids	10	10	10	10	10	10	10
Coating, minutes	2.0	2.0	2.0	2.0	2.0	2.0	2.0
Thickness, mils	10.3	2.0	1.3	15.1	2.15	16.0	8.0
Residual Current, ma.	45.0	25.0	23.3	47.5	35.0	60.0	42.5
Electrode Area, sq.mm	729	729	729	729	729	729	729

Table VI - Surfactant - #4, p-Nitrobenzylmethyldodecylsulfonium Chloride

Variable	Mg	Cu	Pb	CRS	B-37	Pt	Al
Composition	60 BA 40 STY	60 BA 40 STY	60 BA 40 STY	60 BA 40 STY	60 BA 40 STY	60 BA 40 STY	60 BA 40 STY
Efficiency, mg/coul	13.502	18.882	15.151	20.998	17.499	12.368	14.147
Coating Weight, mg	8.333	7.433	6.566	7.700	5.633	6.799	6.599
Coulombs	0.623	0.393	0.432	0.367	0.322	0.549	0.464
Voltage	200	200	200	200	200	200	200
pH Range	7.2-7.2	7.2-7.0	7.2-7.0	7.2-7.0	7.2-7.0	7.2-7.0	7.2-7.2
Conductivity - Before After	1680 1820	1680 1730	1680 1730	1680 1830	1680 1730	1680 1830	1680 1820
Soap Conc., meq/g pol	0.1	0.1	0.1	0.1	0.1	0.1	0.1
Particle Size, Å	1050	1050	1050	1050	1050	1050	1050
% Solids	10	10	10	10	10	10	10
Coating Time, minutes	2.0	2.0	2.0	2.0	2.0	2.0	2.0
Thickness, mils	0.43	0.53	0.53	0.27	0.27	----	0.37
Residual Current, ma.	31.666	30.0	37.0	38.666	28.333	22.0	30.666
Electrode Area, sq.mm	729	729	729	729	729	729	729

We quickly discovered that magnesium was a difficult substrate. Results were erratic and most latexes coated this metal poorly. The problem appeared to be related to the variable oxide layer on the metal. No attempt was made to control this so the magnesium results are not indicated in the analysis.

Polarographic Measurements. The reduction potentials were measured with a Leeds and Northrup polarograph using a dropping mercury electrode. Reduction potentials, $E_{\frac{1}{2}}$, were measured with respect to a saturated calomel reference electrode. Into a cell provided with a snug-fitting Teflon cover containing a salt bridge leading to a saturated calomel electrode (SCE), dropping mercury electrode, and a glass tube used as a nitrogen inlet, was pipetted 10 ml of a 1 N KCl solution (supporting electrolyte). The nitrogen inlet tube was inserted into the solution such that its outlet was at the bottom of the cell, and prepurified nitrogen was allowed to bubble through the solution. The solution was deoxygenated for 20 min., after which the tube was lifted above the solution such that it was covered constantly with a blanket of nitrogen. The mercury level in the dropping electrode was raised to a height of 38 cm above the surface of the solution and the mercury droplets allowed to drop at a constant rate. The current range was set at 5 milliamperes and the polarogram of the KCl solution was run from 0 to -2.0 volts (limit of this supporting electrolyte) to ensure that there were no interfering impurities in the supporting electrolyte. Next 5 μl of a 0.01 N soap solution was rapidly syringed into the KCl solution by quickly removing the nitrogen gas tube, syringing in the solution and rapidly replacing the tube in such a way that the solution was always covered with nitrogen. The cell was tapped a few times to ensure adequate mixing and a polarogram was run as above. Next, another 5 μl of the 0.01 N soap solution was added to make a total of 10 μl in the KCl solution and the polarogram run. The concentration of the cell was increased in increments of 5 μl and polarograms taken, respectively, until a maximum started to appear on the primary half-wave potential (using the concentration of 0.01 N approximately 40 μl or greater was needed to reach this point). A typical polarogram is shown in Figure 3. Half-wave potentials were measured at the steps in the curve by the usual graphical technique. The values are listed in Table I.

In many cases, multiple waves were observed indicating more than one electron transfer process was occurring. All values are listed. At high concentrations, prewaves resulting from adsorption processes were usually observed at $\sim$-0.06 V. These are not listed.

Acknowledgments

The experimental work was carried out with the assistance of R. Goodenough, L. D. Yats, and B. W. Miller. The contributions of W. J. Settineri and T. Alfrey, Jr. in the formulation of the mechanism are also acknowledged.

References

1. Noble, R. J., "Latex in Industry", Palmerton Pub. Co., New York (1953), Chapter 16.
2. Brewer, G. E. F., in Encyclopedia of Polymer Science and Technology, Vol. 15, Interscience, New York (1971), p. 178.
3. Brewer, G. E. F., J. Paint Tech. 45, 37 (1973).
4. Wessling, R. A., D. S. Gibbs, W. J. Settineri and E. H. Wagener in G. E. F. Brewer (ed.), Electrodeposition of Coatings, Adv. Chem. Ser. 119, ACS, Washington, D.C. (1973) Chapter 8.
5. Wagener, E. H., L. D. Yats, H. D. Clarey and R. A. Wessling, Paper presented at Fall Scientific Meeting, Midland Section ACS (1973).
6. Wessling, R. A. and W. J. Settineri, U.S. 3,480,525 (1969).
7. Settineri, W. J. and R. A. Wessling, U.S. 3,681,419 (1972).
8. Wessling, R. A. and W. J. Settineri, U.S. 3,697,398 (1972).
9. Settineri, W. J. and R. A. Wessling, U.S. 3,852,174 (1974).
10. Gibbs, D. S., R. A. Wessling and E. H. Wagener, U.S. 3,637,432 (1972).
11. Settineri, W. J., The Dow Chemical Company, Midland, Michigan (Unpublished results).

19

Effect of Viscosity and Solubility Parameter of a Nonreactive Liquid Additive on the Emulsion Polymerization of Styrene

DONALD R. OWEN, DONALD McLEMORE and WAN-LI LIU
Department of Polymer Science, University of Southern Mississippi, Hattiesburg, Miss. 39401

R. B. SEYMOUR and WILLIAM N. TINNERMAN
Department of Chemistry, University of Houston, Houston, Tex. 77004

Monomers, such as styrene which are good solvents for their polymers do not retard the bulk polymerization rate. However, this rate does not increase in a viscous good solvent medium that is present toward the end of the polymerization. Heterogeneous solution polymerization in nonviscous poor solvents (1) and in viscous poor solvents is faster than rates observed in good solvents.

Polymerization of styrene in an emulsion polymerization has been shown to follow a kinetics scheme as first described by Smith and Ewart. When the vinyl monomer is not a good solvent for the polymer (i.e. acrylonitrile or vinyl acetate) large deviations from Smith-Ewart Theory kinetic predictions are observed.

It is the purpose of this investigation to attempt to modify a conventional styrene emulsion by the addition of a nonreactive oil soluble additive with different combinations of viscosity and Hildebrand solubility parameters. It was anticipated these additives would induce the same heterogeneous condition as in a monomer system with poor polymer solubility.

Thus, this study was undertaken to study the effect of various good and poor viscous and nonviscous solvents on the rate of polymerization and the molecular weight of the polymer obtained in these emulsion systems.

Experimental

All polymerizations were conducted in a 500 ml three-necked round bottom flask equipped with stirrer, reflux condenser, and zero grade nitrogen inlet. The flask submerged in a constant temperature bath at 50 ±

1 C. Styrene monomer was washed with 10% sodium hydroxide solution and vacuum distilled. All ingredients were oxygen-free and the system was continuously purged with nitrogen. The liquid additive and monomer were added to the aqueous surfactant solution prior to the addition of the initiation system. The formulation which consisted of a high surfactant concentration was as follows:

distilled water	320 ml
monomer	40.0 g
Triton X-405	2.0 g
sodium lauryl sulfate	0.60 g
potassium persulfate	0.100 g
sodium bisulfite	0.030 g
liquid additive	varied

1-2 ml aliquot samples were withdrawn from the polymerization system periodically and the polymerization sample was quenched by the addition of 5 ml of 2 percent aqueous hydroquinone. The emulsion was creamed by the addition of a saturated sodium chloride solution and broken via addition of 10 ml of 0.1 N sulfuric acid. The coagulated polymer was subjected to a series of washings with water, methanol and finally petroleum ether. The sample was then dried to constant weight in a heated vacuum desiccator.

Intrinsic viscosities were obtained using a Cannon-Fenske viscometer and the usual method of extrapolation to zero concentration by measuring the reduced viscosities at 0.1%, 0.05% and 0.025% by weight of polymer in benzene. The mean viscosity average molecular weights were determined using the Mark-Houwink equation and an "a" value of 0.72 and a "K" value of 12.3×10^{-5} (2).

The gel permeation chromatography data was obtained using a Model 501 Waters Gel Permeation Chromatograph at 75% pump capacity with a total column composed of 6 feet of 5×10^4 Styragel, 4 feet of 700°A pore size CPG and 2 feet of 3000°A pore size CPG.

Results and Discussions

As shown in Figure 1, the addition of 1 part benzene to 4 parts of styrene in the emulsion formulation caused a large reduction in the rate of polymerization. This effect was increased as the concentration of the benzene was increased. The chain transfer constant of benzene is not significant enough to account for this effect but rather the simple fact that the benzene is a nonreactant.

Benzene is termed a good solvent for polystyrene since its solubility parameter (δ=9.2H) is within a previously established range of ± 1.8 (3) for polystyrene (δ=9.2H). When hexane (δ=7.3H) was used at the same concentration, very little polymerization retardation was observed. The intrinsic viscosity and GPC elution times of the polymer resulting from the hexane modified emulsion indicated it was substantially lower in molecular weight than the control.

Figure 1 also shows the result of an additive that is a viscous good solvent, dibutyl phthalate (δ=9.3H). Very little effect on the rate of polymerization was noted. The δ value of this additive is within 0.1H of that of polystyrene and hence the difference in rates between this solvent and benzene must be attributable to the viscosity of this additive (4).

As shown in Figure 1, the rate of polymerization of styrene in the presence of diisoctyl phthalate (δ=7.4H) was much faster than the control. The small concentration necessary to cause an observable rate increase is difficult to explain by a viscosity increase or solubility parameter change of the entire emulsion system. However, if the viscous poor solvent additive is concentrated on the surface of the monomer-polymer emulsion particle, very small amounts of an additive could exhibit a profound effect on polymerization rate and molecular weight of the resulting polymer. One can postulate a mechanism whereby the viscous poor solvent at low concentrations plays a minor role in the initial stage of the emulsion polymerization, since its contribution to the total solubility parameter and viscosity of the system is slight. As the polymerization reaches the swollen monomer-polymer particle stage, higher surface concentrations of the viscous poor solvent arise due to the immiscibility of additive with polystyrene rich core. This high surface concentration of the additive can then cause the propagating macroradical to possess a lower translational diffusion rate due to the increased viscosity contribution of the additive. Also a lower segmental diffusion rate can be expected because of the increased solubility parameter difference between the reaction media and the propagating polymer species. This enhancement of polymerization rate in the presence of viscous poor solvents has also been observed in the solution polymerization of styrene (5) but at much higher concentrations.

As can be seen from Table 1, the diisoctyl phthalate systems produced polymers with a lower $\overline{M}_v$

and $\bar{M}_n$ than the control. As expected the faster (1/0.25) system produced the lowest molecular weight product.

TABLE I

ADDITIVE	CONCENTRATION		$\bar{M}_v x10^{-6}$	$\bar{M}_n x10^{-6}$
none	---	7.2	4.2	3.5
Nujol	1: 0.025	11.0	7.5	4.5
Nujol	1: 0.25	4.5	2.1	2.0
Nujol	1: 0.50	4.8	2.4	1.8
Nujol	1: 1	6.5	3.6	2.1
diisoctyl phthalate	1: 0.025	5.8	3.1	2.5
diisoctyl phthalate	1: 0.25	4.2	2.0	1.9
dibutyl phthalate	1: 0.25	6.8	3.8	3.5

As shown in Figure 2, rapid polymerization was also observed when Nujol (δ=6.4H) was used as the viscous poor solvent.

As in the case of diisoctyl phthalate, this mineral oil demonstrated an ability to increase polymerization rates and either increase or decrease molecular weights depending on the concentration as seen in Table 1.

With low concentrations of Nujol the resulting increased macroradical lifetimes should result in a larger percentage of high molecular weight species. The (1/0.025) Nujol system definitely demonstrated an increased high molecular weight fraction as indicated by Table 1 by the substantial increase in $\bar{M}_v$ over the control. At higher concentrations of Nujol, increased rates were observed as well as lower polymer molecular weights. One can postulate that at higher concentrations of Nujol a significant increase in the number of propagating species per particle results in more active sites competing for the same quantity of monomer. Therefore, a greater number of macromolecules result with a corresponding lower molecular weight. Similar results with good and poor solvents were also noted for the polymerization of methyl methacrylate, acrylonitrile, and vinyl acetate.

If indeed this viscous poor solvent additive effect is a result of a surface polymerization phenomena, one would expect to see kinetic deviations from Smith-Ewart Theory. The control styrene emulsion

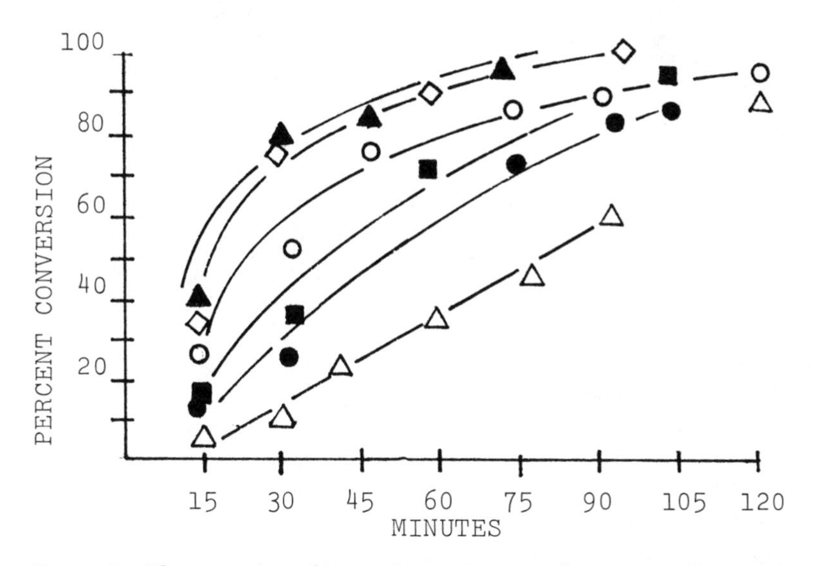

Figure 1. The rate of emulsion polymerization in the presence of various additives at a styrene-to-additive ratio of 1/0.25 by weight. ○, *styrene standard;* △, *benzene;* ●, *hexane;* ■, *dibutyl phthalate;* ◇, *Nujol;* ▲, *diisoctyl phthalate.*

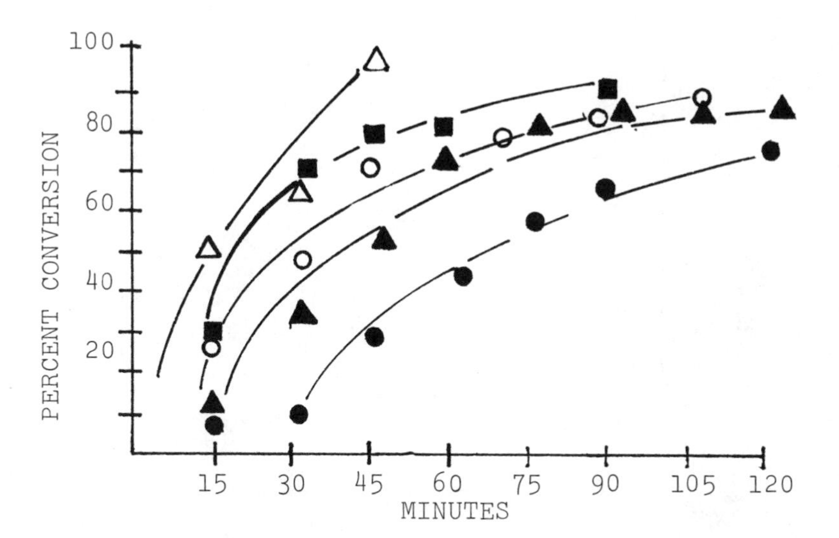

Figure 2. The rate of emulsion polymerization in the presence of Nujol at various concentrations. ○, *standard styrene;* ●, *1:1;* △, *1:0.50;* ■, *1:0.25;* ▲, *1:0.025.*

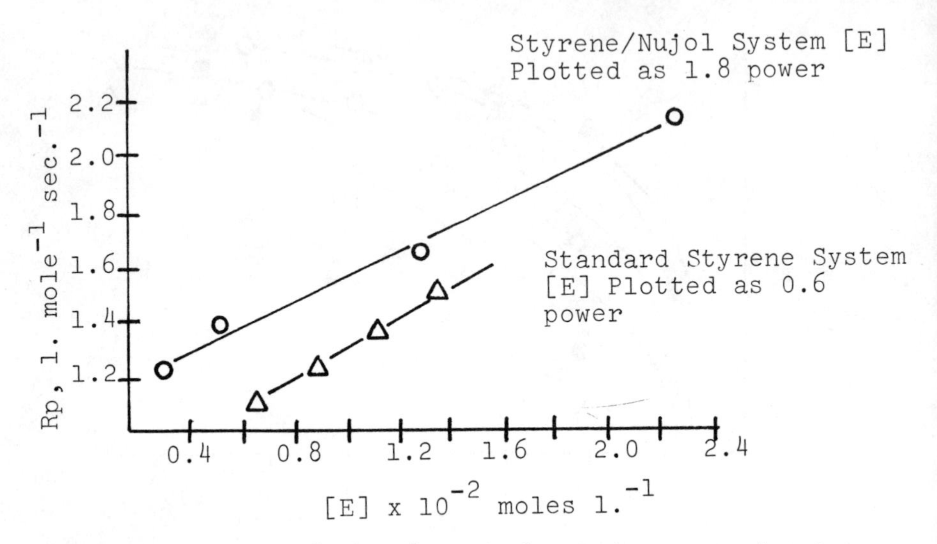

Figure 3. Comparison of the dependency of polymerization rate on surfactant concentration of the standard styrene and styrene/Nujol (1:0.25) emulsion systems

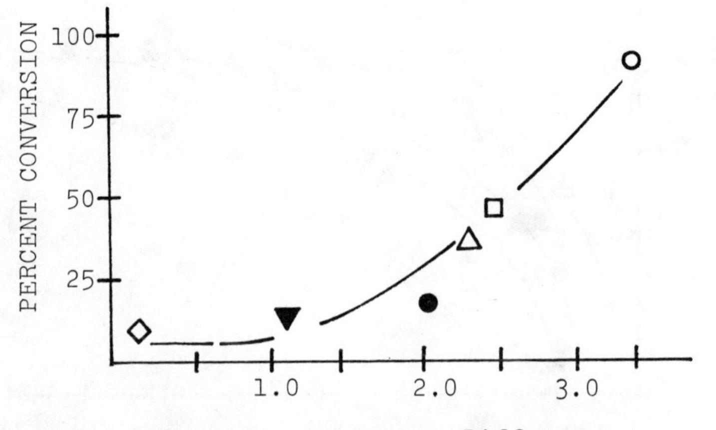

Figure 4. The relationship of solvency to the retardation of emulsion polymerization in the presence of solvents. ◇, benzene; ▼, cyclohexane; ●, octane; △, heptane; ○, Nujol; □, hexane.

polymerization rate (Rp) would be expected to show a 0.6 power dependency on the total surfactant concentration [E]. If the surface of the monomer-polymer particle becomes the loci of polymerization and/or occlusion phenomena become important a large deviation from the control could be expected.

As shown in Figure 3, the standard emulsion formulation Rp exhibited a 0.6 power dependency on [E]. Whereas, at a styrene to Nujol ratio of (1/0.025) a Rp dependency on [E] of approximately 1.8 was observed.

Conclusions

As shown in Figure 4, the rate of polymerization of styrene was retarded by good nonviscous solvents such as benzene, cyclohexane, and octane whose solubility parameters (δ) were within 1.5H of that of polystyrene at styrene to additive ratios of 3 to 1. The absolute rates were slightly increased in poorer nonviscous solvents such as heptane and hexane and were fastest in viscous nonsolvents such as diisoctyl phthalate and Nujol. Rate studies indicated a Rp dependency on [E] substantially greater than unity for the styrene emulsion systems modified with viscous poor solvents.

Literature Cited

(1) Seymour, R. B., Kincaid, P. D., and Owen, D. R., Jour. of Paint Technology, (1973), 45, (580) 33.
(2) "Polymer Handbook", editors V. Brandrup and E. H. Immergut, IV-10, Ref. 51, Interscience, New York (1967).
(3) Seymour, R. B., Kincaid, P. D., and Owen, D. R., Advan. Chem. Ser., (1973), 129, 230.
(4) Trommsdorff, E., Kohle, H., and Lagally, P., Makromol. Chem., (1962), 51, 154.
(5) Seymour, R. B., Stahl, G. A., and Wood, H., Polymer Preprints, 16 (in press).

20

Viscoelastic Properties of Acrylic Latex Interpenetrating Polymer Networks as Broad Temperature Span Vibration Damping Materials

J. E. LORENZ, D. A. THOMAS, and L. H. SPERLING

Materials Research Center, Lehigh University, Bethlehem, Penn. 18015

Polymers form effective damping materials when used for noise or vibration reduction by efficiently degrading mechanical energy into heat. In the vicinity of the glass transition temperature (T_g) polymers have a "loss factor" greater than unity. The loss factor, tan δ, may be defined as the ratio of the out-of-phase portion of the complex Young's modulus E", to the in-phase portion, E'. The glass transition temperature itself is a function of frequency, and individual polymers dampen effectively only over limited ranges of temperature and frequency.

The objective of the present study was to develop damping materials with high lossy mechanical values over wide but controlled temperature ranges in the vicinity of -30°C. to +70°C. The two-stage emulsion polymerized interpenetrating polymer networks (IPN's) (1) were modified by the incorporation of a common comonomer in both stages of the polymerization. The damping properties of the IPN's as well as the mixing characteristics of the two polymers can be studied through dynamic mechanical spectroscopy (DMS). Damping materials may be applied to the vibrating substrate as either one-layer "extensional" coatings or as part of a multiple layered system. The loss factor for damped systems depends on the above system geometry and the mechanical properties of both damping layers and substrate. One of the simpler multi-layered systems consists of a viscoelastic layer and a high modulus "constraining" layer. The constrained layer system has both a higher degree of damping effectiveness and a broader temperature range than the extensional damping coating (2). Models of the loss factor for both constrained and extensional damping are given by Ungar (3). To a first

The authors wish to thank The Human Engineering Laboratory, Aberdeen Proving Ground, Md., and the National Science Foundation for support through Contract No. DAAD05-72-C-0175 and Grant No. GH-40645, respectively.

approximation the effectiveness of extensional damping is proportional to E", whereas the effectiveness of constrained layer damping depends primarily upon tan δ. The damping effectiveness of viscoelastic materials can be readily assessed by studies of E', E", and tan δ as a function of temperature.

Theory of Wide Temperature Span Viscoelastic Damping Materials

The damping behavior of polymers can be altered to optimize either the temperature span covered or the damping effectiveness for particular temperatures. The area under the loss modulus temperature curve tends to be constant for some polymer combination, which has been expressed by the empirical "temperature band width law" of Oberst (2):

$$(E''_{max})\ \Delta T / E'_{\infty} = K \qquad \text{Eq. (1)}$$

where E''_{max} is the maximum value of E"
ΔT is the bandwidth, the temperature range for which $E'' \geq E''_{max}/2$
E'_{∞} is the in-phase complex Young's modulus at infinite frequency or low temperature.
K is a constant that was found by Oberst to be close to 6°K.

As individual homopolymers are effective dampers only over limited temperature ranges, wider temperature range damping materials can be obtained from polymer blends. Blends of incompatible polymers show separate transitions characteristic of the individual polymers, but as the degree of mixing is increased by increasing the compatibility of the polymers, the individual transitions broaden and merge into a single transition. One means of augmenting the compatability of two polymers involves addition of a common comonomer to each of the polymers, thus reducing the positive enthalpic contribution to the free energy of mixing among the polymers.

Copolymerization techniques offer the opportunity to control polymer structure and hence the degree of mixing of the components. The multiphase physical characteristics of polyblends are also observed in graft, block, and heterogeneous copolymers (4, 5). Materials suitable for broad temperature range vibration damping have been prepared from polyblends (4, 6, 7), graft copolymers (2, 8), and IPN's. (1, 10)

IPN's are synthesized by swelling a crosslinked polymer I with monomer II, plus crosslinking and activating agents, and polymerizing monomer II *in situ*. When IPN's are synthesized with emulsion polymerization techniques, each individual latex particle, ideally, consists of a micro-IPN, and contains just two crosslinked molecules(1). In common with most polyblends, grafts, and blocks, IPN's exhibit phase separation. However the insertion of crosslinks into both polymers introduces a new structure-influencing element, because the extent of phase separation and phase domain size are limited. Thus, when the deliberately introduced crosslinks outnumber the accidental grafts, morphologies and properties different from graft copolymers arise. In particular, when polymers of similar solubility parameters are mixed in IPN form, extensive but incomplete mixing may occur, and each volume region within the latex particle may contain a different composition. Thus, each volume element contributes differently to the glass-rubber transition region, which may be made to span the range between the original transitions of polymer I and polymer II.

IPN's and related materials, in fact, have a long history. For example, IPN's were first synthesized to produce smooth sheets of bulk polymerized homopolymers (11). IPN's were next used as solution polymerized ion exchange resins. (12, 13) Further development of IPN's included the syntheses of interpenetrating elastomer networks (IEN's) and simultaneous interpenetrating networks (SIN's) (14). IEN's consist of a mixture of different emulsion polymerized elastomers which are both crosslinked after coagulation. SIN's are formed by the simultaneous polymerization of mixed monomers by two noninterfering reactions (15, 16).

The phase continuity of the individual polymers strongly affects the glass transition of polymer blends. Like most mechanical properties, the DMS behavior of a material is most strongly influenced by the more continuous phase. In a blended polymeric material where one polymer exists as the matrix and another as inclusions, the DMS behavior of the blend resembles that of the matrix polymer, the transition being broadened by the different T_g of the inclusions. High values of tan δ over the widest temperature range occur when the T_g of the inclusions is higher than that of the matrix.

Some two stage emulsion graft copolymer materials synthesized and characterized by DMS: include, the series poly (methyl methacrylate)/poly(n-butyl acrylate) (PMMA/PnBA) synthesized by Dickie (14) and the series poly(ethyl methacrylate)/poly(n-butyl acrylate) (PEMA/PnBA) synthesized by Sperling et al. (1) The present study will continue the development of the PEMA/PnBA damping materials by incorporating a common comonomer, ethyl acrylate (EA), in both stages of the emulsion polymerization.

Creep Behavior of Bulk and Emulsion Polymerized IPN's

The submicroscopic emulsion polymerized form of IPN's would be expected to differ in mechanical properties from the counterpart bulk polymerized form in that: (1) The latex particles are not crosslinked one to another, allowing movement of one latex particle past another. (2) In bulk IPN's (10) it was shown that polymer I forms the continuous or more continuous phase, while in latex IPN's, polymer II tends to form the more continuous phase (1).

As part of this study on IPN compatibility and damping characteristics, the creep behavior of bulk and emulsion polymerized IPN's will be compared to explore the morphological differences expected.

Design of Experiment

The compositions of the IPN latices chosen for study, listed in Table I, represent a 2^3 factorial design matrix suitable for determining first order effects. The three independent compositional variables under investigation were the volume fraction of EMA in monomer charge I (X_1), the volume fraction of nBA in monomer charge II (X_2), and the total volume fraction of seed latex copolymer to total polymer (X_3). The levels of the variables were chosen so that the experimental design could be efficiently extended into an orthogonal second order design by adding six additional data points (17). As in previous papers from this laboratory, the first-mentioned copolymer constitutes the copolymer first synthesized, in this case the seed latex.

Damping Material Syntheses

All latex IPN's were synthesized by two-stage emulsion polymerization techniques, (1, 18) as follows: To 300 ml of deionized, deaerated, stirred water at 60°C were added 50 ml of a 10% (W/V) solution of sodium lauryl sulfate, followed by 5 ml of a 5% (w/V) solution of potassium persulfate. The calculated quantity of comonomer was added at a rate of about 2 ml per minute. When the first monomer was fully added, a minimum of one hour was allowed to elapse. Then a new portion of initiator was added, but no new soap, followed by the second charge of comonomers under similar reaction conditions.

Two important departures from standard procedures should be noted: (1) all monomers contained 0.4% (W/V) of crosslinking agent, tetraethylene glycol dimethacrylate, necessary to form the IPN. (2) No new soap was added for the second stage polymerization, to discourage the formation of new particles. Total polymer concentration of the completed latexes was about 30% (W/V).

The finished latexes were cast as films and vacuum dried

TABLE I

Latex IPN Syntheses

Sample Identification	Composition P(EMA-co-EA)/P)nBA-co-EA)	Reduced Composition Varibles		
		χ_1	χ_2	χ_3
A	P(25-co-25)/P(25-co-25)	.5	.5	.5
B	P(9-co-21)/P(21-co-49)	.296	.296	.296
C	P(9-co-21)/P(49-co-21)	.296	.704	.296
D	P(21-co-9)/P(21-co-49)	.704	.296	.296
E	P(21-co-9)/P(49-co-21)	.704	.704	.296
F	P(21-co-49)/P(9-co-21)	.296	.296	.704
G	P(21-co-49)/P(21-co-9)	.296	.704	.704
H	P(49-co-21)/P(9-co-21)	.704	.296	.704
I	P(49-co-21)/P(21-co-9)	.704	.704	.704
PEA	P(0-co-25)/P(0-co-75)	0	0	.75

EMA = ethyl methacrylate

EA = ethyl acrylate

nBA = n-butyl acrylate

χ_1 = vol. frac. EMA in first stage of synthesis (network I)

χ_2 = vol. frac. nBA in 2nd stage of synthesis (network II)

χ_3 = vol. of network I/ total volume of polymer.

to constant weight to obtain samples suitable for DMS studies.

Creep Specimen Synthesis

IPN's of 54/46 poly(methyl methacrylate)/poly(ethyl acrylate) were prepared by both the latex and bulk (10) routes. Both IPN's contained 0.4% (W/V) tetraethylene glycol dimethacrylate (TEGDM) crosslinking agent in each polymer. Samples of the latex IPN were film formed on glass petri dishes. All samples were vacuum dried at 60°C to constant weight.

Creep Testing

Creep testing was performed on a Clash-Berg Torsion Stiffness Tester. Silicone oil was used as the heat transfer medium. Temperatures were held constant to within ± 1 C°.

DMS Testing

All dynamic mechanical measurements employed a Rheovibron direct reading viscoelastometer, model DDV-II (Vibron, manufactured by the Toyo Measuring Instrument Co., Ltd., Tokyo, Japan.) The temperature range employed was from -100°C to +100°C, with a heating rate of about 1 C° per minute. To better correlate with sound damping experiments conducted in our laboratory, a frequency of 110 Hz was employed. As per requirements of the instrument, the sample dimensions were of the order of 10^{-2} cm x 10^{-1} cm x 2 cm. The quantities E' and E", representing the storage and loss moduli, respectively, and tan δ, which equals E"/E', were obtained.

DMS data was easily obtained but for two limiting conditions: (1) tan δ values above 1.75 could not be determined, and (2) the maximum range of E' for a given sample is about three orders of magnitude. Thus, very soft samples, as well as materials exhibiting very sharp damping peaks, could not be fully evaluated.

DMS Behavior

With the exclusion of formula 'A', Table I, the remaining eight samples are composed of four possible combinations of copolymer in a composition ratio of either 30:70 or 70:30. The values of E', E", and tan δ for 'B', 'D', and 'E' are shown in Figures 1, 2, and 3. These samples represent composition combinations all having an overall composition ratio of seed latex to overcoat polymer of 30:70. The DMS behavior of sample B, which consists of two copolymers each of 70 wt% ethyl acrylate, resembles poly(ethyl acrylate)

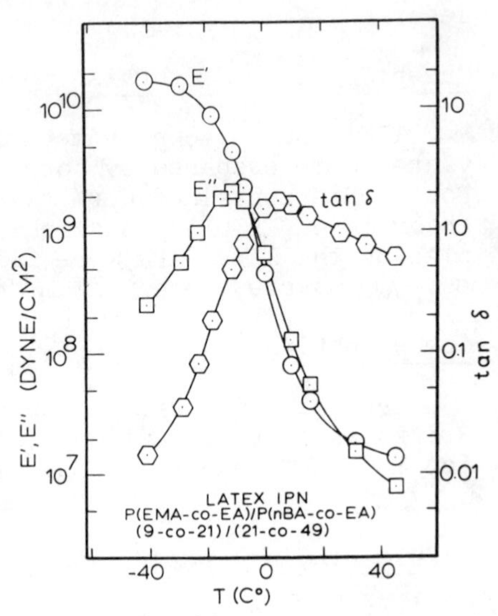

Figure 1. Dynamic mechanical spectroscopy of formula 'B'. This material, containing 70% EA comonomer between polymer networks I and II, displays a mechanical spectrum only slightly broader than would be expected of the corresponding random copolymer.

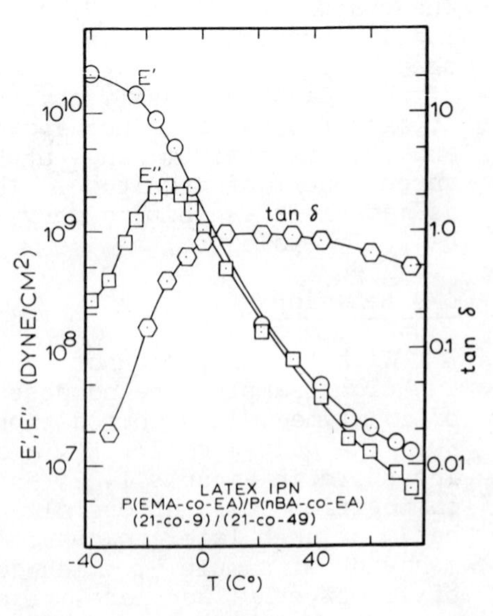

Figure 2. Dynamic mechanical spectroscopy of formula 'D'. With 60% EA the transition is noticeably broader than that shown in Figure 1.

homopolymer behavior. Samples D and E show a broadening of the transition, a flattening of the tan δ versus temperature curve, and a lowering of the maximum value of tan δ as the concentration of ethyl acrylate is reduced.

The DMS behavior of the PEMA/PnBA two stage polymerized latexes (1) are shown in Figures 4 and 5 for comparison. Inspection of the 25/75 PEMA/PnBa composition shows that when the ethyl acrylate is eliminated the tan δ versus temperature curve is bimodal. All three of the PEMA/PnBA materials exhibit low but nearly constant values of tan δ over wide temperature ranges. The strongly bimodal shape of the E" and tan δ curves for 75/25 PEMA/PnBA suggests dual phase continuity. The DMS data of PEMA shown in Figures 4 and 5 was obtained from shear modulus testing at 1 Hz (19). The curves were displaced 15°C upwards in this report to approximate the DMS behavior of 110 Hz. The large discrepancy between the temperature of maximum tan δ for PEMA and the apparent high temperature transition of the PEMA/PnBA materials may be due to either complex morphologies (14) or compatible polymer-polymer behavior, this shifting the T_g to intermediate temperatures.

The reader should note that PEMA homopolymer yields the highest values for E" vs temperature, Figure 4, but the lowest values for tan δ vs temperature, Figure 5. This peculiar feature of damping curves will be discussed further below.

The DMS behavior of all nine materials is shown in Figures 6, 7, and 8. Formulas B and F, both composed of different ratios of the same two copolymers, have similar shaped DMS curves. Another pair of materials composed from the same two copolymers, formulas C and G, each have bimodal tan δ curves. Samples H and I also have similar DMS curves; although composed of two different pairs of copolymers, both samples contain 50 vol% ethyl methacrylate.

The differences in solubility parameters between polymers in a blend strongly influences the degree of compatibility of nearly compatible polymers (20). The solubility parameters (21), calculated on a volume fraction basis, for all six random copolymers are listed in Table II. The differences in solubility parameters between networks I and II for each of the two stage latex IPN's are listed in Table III. The large differences in solubility parameters for the two copolymers of formulas C and G account for their incompatible behavior as evidenced by the strongly bimodal tan δ curves, Figure 8. Comparing formulas C and G with the PEMA/PnBA series and Table II it appears that bimodal tan δ behavior occurs when the glass transitions of the two copolymers are far apart, the copolymers are highly incompatible, or some combination of the two effects.

Due to instrumental limitations values of tan δ at high

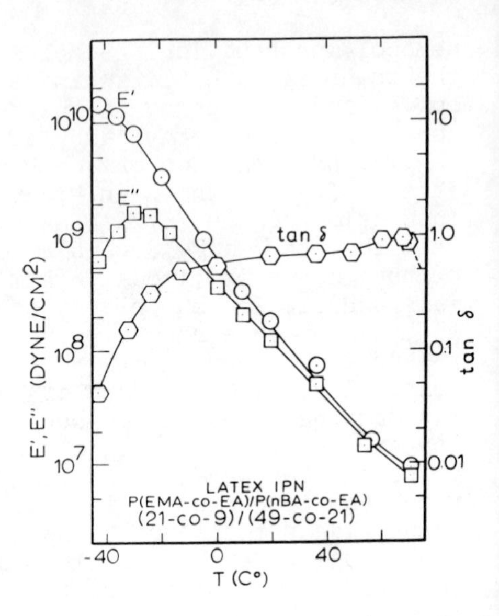

Figure 3. Dynamic mechanical spectroscopy of formula 'E'. Notice the broad temperature span of high tan δ values. With only 30% EA comonomer, the material shows a flat tan δ curve.

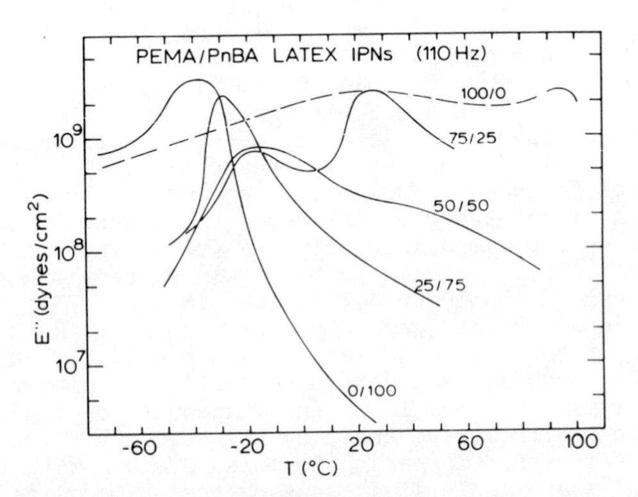

Figure 4. Values of lossy modulus, E''. Data from references 1, 5, and 8. The data for PEMA was obtained from 3G'' at 1 Hz and converted to 110 Hz. The PEMA homopolymer is seen to have a very high E'' value over a broad temperature range brought about by the strong secondary transition. With no common comonomer the loss peaks and also tan δ peaks (Figure 5) of the IPN's tend to be bimodal.

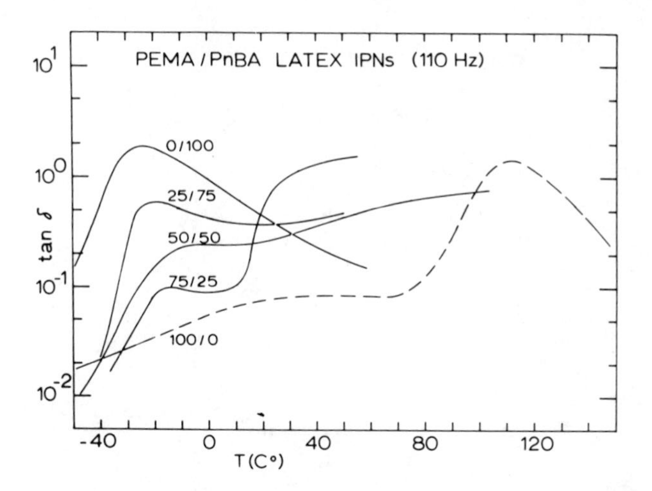

Figure 5. Values of tan δ vs. temperature. Data from references 1, 5, and 8.

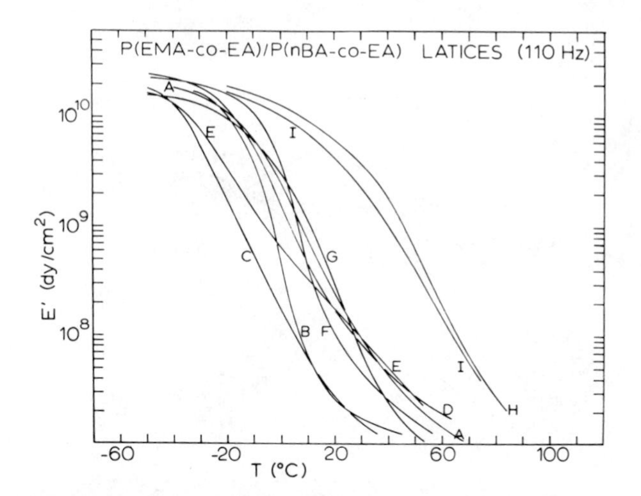

Figure 6. Values of storage modulus (E') vs. temperature. The actual glass transition temperature and the slope of the E' vs. temperature curves depend on overall composition and synthetic detail.

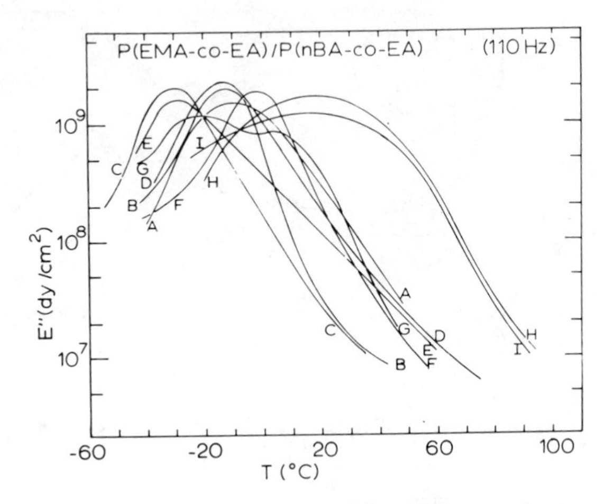

Figure 7. Values of lossy modulus (E″) vs. temperature

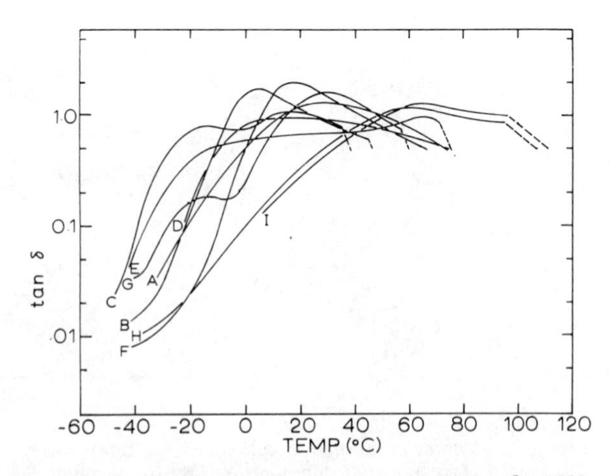

Figure 8. Values of tan δ vs. temperature for the IPN latices P(EMA-co-EA)/P(nBA-co-EA) at 110 Hz

TABLE II

Calculated Solubility Parameters and

Glass Transition Temperatures of Copolymers

Polymer or copolymer	Solubility Parameter[21] δ_p $(cal/cm^2)^{\frac{1}{2}}$	T_g[21] (^oK)
PnBA	8.8	218
PEA	9.35	249
PEMA	8.95	333
P(EMA-co-EA), (30-co-70)	9.23	269
P(EMA-co-EA), (50-co-50)	9.15	295
P(EMA-co-EA), (70-co-30)	9.06	303
P(nBA-co-EA), (30-co-70)	9.19	239
P(nBA-co-EA), (50-co-50)	9.08	233
P(nBA-co-EA), (70-co-30)	8.96	226

TABLE III

Differences in Calculated Solubility Parameters and Glass Transition Temperatures Between Network I and Network II of the Two-stage IPN's

Sample	$\lvert \delta_{p1} - \delta_{p2} \rvert$ $(cal/cm^2)^{1/2}$	$(T_{g_2} - T_{g_1})$ (deg.K)
A	.07	52
B	.04	30
C	.27	43
D	.13	64
E	.10	77
F	.04	30
G	.27	43
H	.13	64
I	.10	77
PEMA/PnBA	.15	115

temperatures could not be determined. For purposes of comparison, the tan δ curves in Figure 8 were extrapolated by assuming that at higher temperatures the value of d(log(tan δ))/dT was constant and equal to the negative of the maximum value of d(log(tanδ))/dT occuring at temperatures lower than the peak. The temperature spans for which tan δ is greater than or equal to 0.5 for all the materials tested are listed in Table IV. The selection of tan δ >0.5 as the criterion for effective constrained layer damping is somewhat arbitrary. If more accurate tan δ versus temperature data were available it would be worthwhile to determine the temperature spans for tan δ values of 0.1 through 0.5.

"Millar" IPN's

Network strain has been suggested as a contributing factor to the widening of the transition between incompatible or semi-incompatible polymers. (8) The materials in this study, synthesized as latex interpenetrating polymer networks, are probably subject to swelling strains. "Millar" IPN's, defined as IPN's of identical copolymers where monomer I and monomer II are the same, are named after the investigator who first prepared such materials. (12) Millar's original IPN's showed an increase in density of 0.5% and a decrease in swelling ability as the number of networks is increased from I to III. Values of E', E", and tan δ were determined on bulk and latex polymerized PEA/PEA "Millar" IPN's. Within experimental error, see Figure 9, no shifting or broadening in the glass transition was noted. (Other results not shown are similar). This negative result indicates that shifts and broadenings of the glass transition in IPN's prepared from two different polymers arises primarily from phase continuity effects and/or thermal motions of one species in the presence of another, but internal molecular strains apparently play only a minor role, if any.

For the present study, "Millar" IPN's of PEA/PEA also served as the limit of total compatibility in the main series of IPN's investigated. One may consider the IPN series herein as "Millar" IPN's to which different comonomers were added to polymer I and polymer II, to drive the glass temperatures apart.

We are now in a position to model the temperature dependence of E". For a completely compatible system, exemplified by the "Millar" IPN, the sharp homopolymer transition behavior is observed, Figure 10a. With the addition of significant non-identical comonomers to polymers I and II, the transition is broadened, but a single peak is maintained, Figure 10b. The semi-compatible PEMA/PnBA system, containing no ethyl acrylate, displays two peaks (see Figures 4 and 5), as modeled in Figure 10c. A completely incompatible

TABLE IV

Loss Tangent and Loos Modulus Temperature Bandwidths

Sample	Loss Tan. Temp. Span $T_{(\tan \delta \geq .5)}$ (deg. K)	Loss Modulus Temp. Bandwidth $T_{(E'' \geq E''_{max}/2)}$ (deg. K)	Loss Modulus Temp. Bandwidth constant, K* (deg. K)
A	72	34	1.9
B	57	20	1.4
C	65	22	1.5
D	82	24	1.8
E	88	26	1.4
F	65	20	1.3
G	54	50	2.1
H	71	54	3
I	76	60	2.8
PEMA/PnBA (27/75)	14	19	1.5
(50/50)	43	48	1.3
(72/25)	40	33	2.7
PEA	58	20	1.2
PnBA	55	34	3.7
PEMA	42	130	11

* Calculated from Eq. 1, assuming that E'_{∞} is 3×10^{10} dy/cm^2

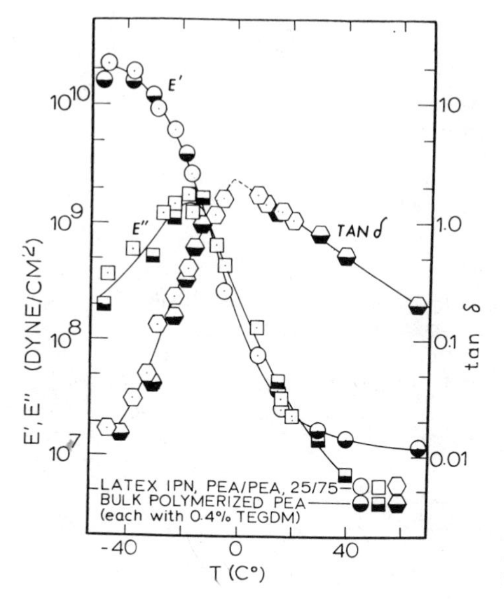

Figure 9. Dynamic mechanical spectroscopy of PEA/PEA IPN's at 110 Hz

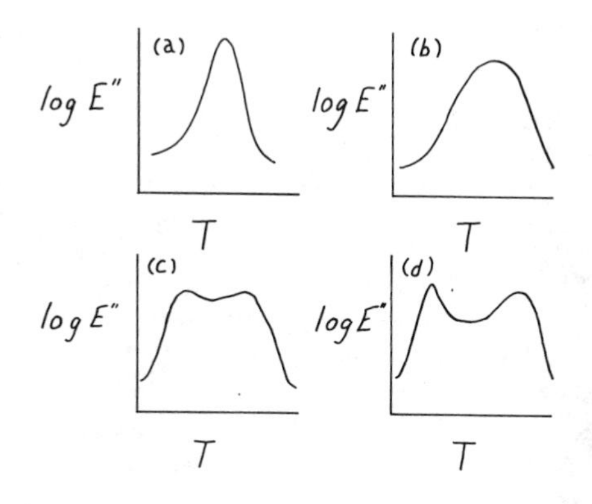

Figure 10. Idealized lossy modulus behavior as a function of polymer I/polymer II incompatibility. With increasing incompatibility, the loss peak first broadens, then forms two distinct loss peaks as the components separate into two distinct phases.

system, such as poly(vinyl chloride)/poly(butadiene), (5) may show two well defined transitions far separated, as shown in Figure 10d. Thus, the total span of possible compatibility ranges may be achieved by controlling the thermodynamic heat of mixing of the two polymers. For similar levels of molecular weight (or crosslinking), the entropy of mixing as such probably does not change significantly. The dependence of tan δ vs temperature or composition will follow a pattern similar to Figure 10 also.

Bandwidth Constant

The loss moduli (E") of the latex IPN's shown in Figures 4, 7 and 10 are lower than that of PEMA with its broad secondary loss maximum. A summary of the loss modulus behavior of all the materials is given in Table IV in the form of the temperature bandwidths and the temperature bandwidth constants, as determined by equation 1. The value of E' was assumed to be the same for all the materials because of the unavailability of accurate E' data and because the calculation of the temperature bandwidth constant is very sensitive to the selection of E'.

The importance of the concentration of polymer having a broad secondary loss maximum upon the loss modulus bandwidth of polymer blend materials is shown in Fig. 11 for the bulk polymerized IPN's of Huelck. (10) The loss modulus temperature bandwidth constant is strongly dependent upon the concentration of methyl methacrylate, irrespective of whether methyl methacrylate is present in the matrix or the inclusions. Overall the Oberst type of analysis indicates that for a given type of polymer blend the area under the E" curve tends to be constant. In other words, one may have height or width but not both.

Modeling of DMS Behavior

The DMS curves of tan δ as a function of temperature for each sample were characterized by four parameters: (1) the temperature bandwidth through which tan $\delta \geq 0.5$, $\Delta T(\tan\delta = .5)$: (2) the maximum value of tan δ, $(\tan\delta)_{max}$: (3) the mean temperature through which tan δ was greater than 0.5, $T_{mean}(\tan\delta > .5)$: and (4) the temperature at which $(\tan\delta)_{max}$ occurs, $T_{(\tan\delta)max}$. Similarily the DMS temperature curves of E" for each sample were characterized by five parameters: (1) the lossy modulus temperature bandwidth, $\Delta T_{(E''>E''_{max}/2)}$: (2) the temperature bandwidth through which $E'' \geq 5 \times 10^8$ dynes/cm^2, $\Delta T(E'' > 5{\times}10^8)$: (3) the maximum value of E", E''_{max}: (4) the mean temperature through which $E'' \geq E''_{max}/2$, $T_{mean}(E'' \geq E''_{max}/2)$: and (5) the mean temperature through which $E'' \geq 5{\times}10^8$ dy/cm^2, $T_{mean}(E'' \geq 5{\times}10^8)$.

The effect of composition on each of the nine parameters describing the shapes of the DMS temperature curves was investigated empirically by fitting these parameters into second order polynomial equations of the type:

$$T = A_1 + A_2X_1 + A_3X_2 + A_4X_3 + A_5X_1^2 + A_6X_2^2 + A_7X_3^2 + A_8X_1X_2 + A_9X_2X_3 + A_{10}X_1X_3$$

The equations, fit to the data by stepwise linear regression, (22) are shown in Table V. In addition to the ten samples of the current study listed in Table I, homopolypolymer data and data for the three latex PEMA/PnBA IPN's, shown in Figures 4 and 5, were also included in the generation of the equations.

The nonlinearity of the empirical equations in Table V indicate that the use of the first order experimental design supplemented by six additional data points was not sufficient to accurately determine the effect of composition upon the DMS behavior of the series studied. The equations, while not quantitatively accurate, do give a qualitatitive description. Equation 2 predicts a maximum in $\Delta T(\tan\delta \geq .5)$ of 76 ^oC. for the IPN PEA/P(nBA-co EA) and a minimum of 37 ^oC. for PEMA. Equation 3 shows that $(\tan\delta)_{max}$ is largest for PEA and for PEA/P(nBA-co-EA) IPN's. The product of $\Delta T_{(\tan\delta \geq .5)}$ and $(\tan\delta)$ max is approximately proportional to the area under the DMS curve of tan δ as a function of temperature and represents a useful measure of damping effectiveness. This product possesses a maximum of 209 degrees for PEA and a minimum of 24 ^oC for PnBA. The product remains high for the IPN series PEA/P(nBA-co-EA) and low for the series P(EMA-co-EA)/PnBA. The composition which maximizes this product is largely due to the arbitrary selection of tan δ $\geq$ .5 as the criteria for the temperature span. Materials which have exceptionally wide temperature spans for lower values of tan δ would not receive adequate attention.

The equations which describe the lossy modulus behavior product and maximum temperature bandwidth for random copolymers of P(EMA-co-EA) and a maximum E''_{max} of PnBA. The product of $T_{(E'' > E''\,max/2)}$ and $E''max(dy/cm^2 \times 10^{-9})$, similar to the temperature bandwidth constant, possesses a maximum for the random copolymer P(EMA-co-EA), (90-co-10). This model of lossy modulus behavior is thus strongly influenced by the exceptionally large secondary transition of PEMA.

Latex VS Bulk IPN Creep Behavior

The time-dependent creep moduli of latex and bulk poly-

TABLE V

Regression Analyses of DMS Behavior

as a Function of Composition

tan δ Versus Temperature Behavior

$$T_{(\tan\delta \geq .5)} = 56+89.5(1.-.1\,1X_2)-11.X_1X_3 \pm 12. \qquad \text{Eq. 2}$$

$$(\tan\delta)_{max} = 2.55-3.9X_1(1.-.89X_1)-.96X_2^2$$

$$+.86X_3(X_2-.8X_1) \pm 0.3 \qquad \text{Eq. 3}$$

$$T_{mean(\tan\delta \geq .5)} = 13.-30.X_2(X_2-.9X_3)+102.X_1X_3 \pm 19. \qquad \text{Eq. 4}$$

$$T_{(\tan\delta)_{max}} = 3.-23.X_2(X_2-X_3)+109.X_1X_3 \pm 17. \qquad \text{Eq. 5}$$

E" Versus Temperature Behavior

$$T_{(E'' \geq E''max/2)} = 22.+153.X_1(1.-X_1)\_11.X_2^2-209.X_3(1.-X_3)$$

$$+99.X_1X_3 \pm 11. \qquad \text{Eq.-6}$$

$$T_{(E'' \geq 5\times10^8 dy/cm^2)} = 24.-36.X_2(1.+.18X_1-1.6X_2)$$

$$+164.X_1X_3 \pm 14. \qquad \text{Eq.-7}$$

$$E''_{max}(dy/cm^2 \times 10^{-9}) = 2.+1.1X_2(1.-1.3X_3)-3.4X_3(1.-.97X_3)$$

$$+.69X_1X_3 \pm 0.4 \qquad \text{Eq.-8}$$

$$T_{mean(E'' \geq E''max/2)} = -15.-27.X_2+26.X_2X_3+55.X_1X_3 \pm 6. \qquad \text{Eq.-9}$$

$$T_{mean(E'' \geq 5\times10^8 dy/cm^2)} = -20.-44.X_2^2 +16.X_1X_2+40.X_2X_3$$

$$+35.X_1X_3 \pm 6. \qquad \text{Eq.-10}$$

merized IPN's of identical composition are shown in Figures 12 and 13, respectively. While the slopes of the lines are very similar, the bulk polymerized IPN is twice as stiff at each temperature tested. The higher modulus of the bulk polymerized material is caused by two structural differences: plastic PMMA is the more continuous phase in the bulk polymerized material (10) whereas elastomeric PEA is the more continuous phase in the latex film; (1) the bulk IPN sample consists ideally of two interwoven molecules whereas the latex film consists of many separate intracrosslinked particles.

The slopes of the two sets of creep curves are nearly identical and therefore the energy of activation is the same for both bulk and emulsion synthesized films.

Conclusions

Acrylic/methacrylic latex IPN's exhibiting two glass transitions (1) can be increasingly compatibilized by the addition of comonomers to match the solubility parameters of both stages. Thus the two transitions can be changed into a single transition having high values of tan δ over a broad temperature range by the addition of a common comonomer of intermediate T_g to both stages of the polymerization. Materials synthesized by this technique are candidates for constrained layer damping whenever damping is required over a wide but controlled temperature range.

The full temperature range of effective damping remains unknown for many of the materials investigated in this study due to instrumental limitations on the Vibron. A more thorough evaluation of materials for constrained damping could be obtained by direct vibration decay testing (2, 6, 23) as a function of temperature. Both sample B and the PEMA/PnBA 25/75 composition have been vibration decay tested in constrained damping applications (24); the effectiveness of vibration decay damping was found to be in good agreement with tan δ as expected.

The empirical temperature bandwidth constant (K), a measure of extensional damping effectiveness, is strongly affected by polymer secondary loss mechanisms. Polymers such as PMMA and PEMA form very effective damping materials because they possess broad temperature span lossy modulus curves.

It was of general interest to investigate the effectiveness of the damping materials in terms of any synergisms that may arise because of special mixing modes. The bandwidth "constant" K indeed is not a constant and depends on the chemistry and other factors of the system, see above. However K is only useful for loss moduli-temperature behavior. No similar theory for tan δ temperature exists, and

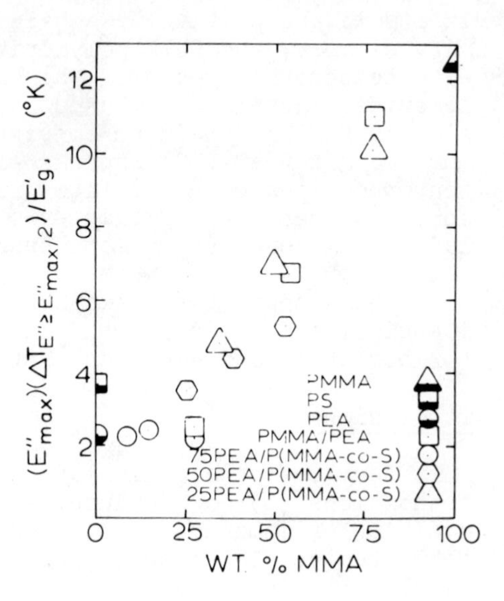

Figure 11. "Temperature bandwidth constant" of Oberst (2) calculated for the bulk polymerized IPN's of Huelck (10)

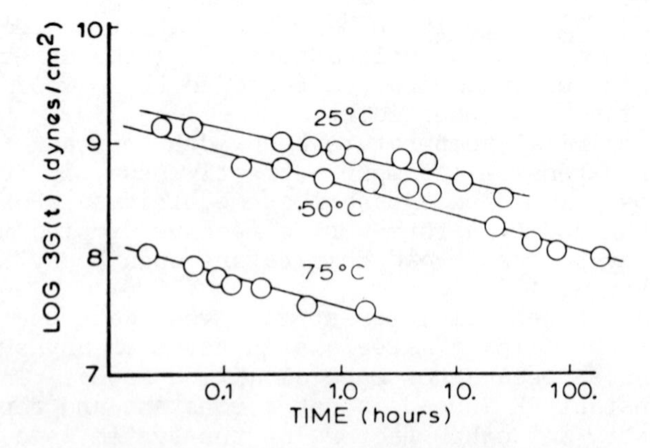

Figure 12. Torsional creep of emulsion polymerized PMMA/PEA, 54/46

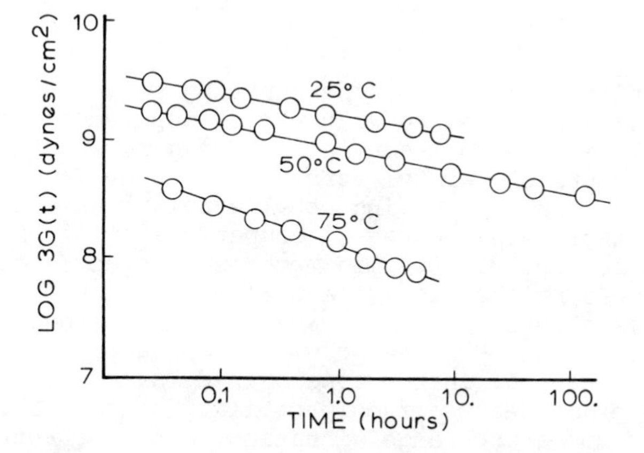

Figure 13. Torsional creep of bulk polymerized PMMA/PEA, 54/46

the very different behavior of PEMA shown in Figures 4 and 5 remains unexplained. Considering similar chemistry, it does appear however, that a particular system may have intensive damping over a narrow temperature range, or more moderate damping over a broader temperature range, the area under the log E"-T curve remaining relatively constant.

Abstract

A series of latex interpenetrating polymer networks (IPN's) were synthesized with random copolymers of ethyl methacrylate and ethyl acrylate forming network I, and random copolymers of n-butyl acrylate and ethyl acrylate forming network II. Having ethyl acrylate as a common comonomer enabled the degree of compatibility of the two networks to be varied. Dynamic mechanical spectroscopy was used to measure the storage and loss moduli and the loss tangent as a function of temperature. The broad but controlled temperature range of vibration damping capability of these materials in the constrained layer mode depends upon the exact degree of compatibility of the IPN. The limited temperature range vibration damping effectiveness of these materials in the extensional mode is described in terms of the Oberst "temperature bandwidth constant." The creep behavior of a latex IPN and its bulk synthesized counterpart was examined. The latex component exhibited greater creep; this result is believed due to the latex material having greater continuity of the elastomer phase, or restriction of the networks to individual latex particles.

References

1. L. H. Sperling, Tai-Woo Chiu, and D. A. Thomas, J. Appl. Polym. Sci., 17, 2443-2455 (1973).
2. H. Oberst, Phil. Trans. Roy. Soc. London, A263, 441-453 (1969).
3. E. E. Ungar, Noise and Vibration Control, L. L. Beranek, ed. N.Y.; McGraw Hill, 1971.
4. A. V. Tobolsky, Properties and Structure of Polymers. N.Y.: John Wiley and Sons, Inc. 1960.
5. M. Matsuo, Japan Plastics, 2, 6 (July, 1968).
6. H. Mizumachi, J. Adhesion, 2, 292 (1970).
7. M. Takayanagi, H. Harima, and Y. Iwata. Mem. Fac. Eng., Kyushu Univ., 23, 1 (1963).
8. R. A. Dickie and Mo-Fung Cheung. J. Appl. Poly. Sci., 17, 79 (1973).
9. R. T. Angelo, R. M. Ikeda and M. L. Wallach. Polymer, 6, 141 (1965).
10. V. Huelck, D. A. Thomas and L. H. Sperling. Macromolecules, 5 348 (1972).
11. J. J. P. Staudinger and H. M. Hutchinson. U.S. 2,539, 376 and U.S. 2,539,377 (1951).
12. J. R. Millar, J. Chem. Soc., 1311 (1960).
13. G. S. Solt. Brit. Pat. No. 728, 508 (1955).
14. K. C. Frisch, H. L. Frisch and K. Klempner. German 2,153,987 (1972).
15. L. H. Sperling and R. R. Arnts, J. Appl. Polym. Sci., 15, 2371 (1971).
16. R. E. Touhsaent, D. A. Thomas, and L. H. Sperling, J. Polym. Sci., 46C, 175 (1974).
17. O. L. Davies. The Design and Analysis of Industrial Experiments, p. 533, Hafner Publishing Co., 1956.
18. F. A. Bovey, I. M. Kolthoff, A. I. Medalis and E. J. Meehan, Emulsion Polymerization (High Polymer Series IX) Interscience, 1955. Pp 280-281.
19. J. Heijboer. Physics of Noncrystalline Solids, (Proc. Inst. Cong. Delft, July 1964, Int. Union Pure Appl. Chem.) Prins, J. A., ed., 1965.
20. S. Krause, J. Macromol. Sci. - Revs. Macromol. Chem. C7, 251 (1972).
21. J. Brandrup and E. H. Immergut. Polymer Handbook. N.Y.: Interscience, 1966.
22. W. J. Dixon, ed. Biomedical Computer Programs, 2nd ed. Berkley: Univ. Calif. Press, 1967. Pp 233-257.
23. G. L. Ball and I. O. Salyer. J. Acoust. Soc. Amer., 39, 663 (1966).
24. L. H. Sperling, J. A. Grates, J. E. Lorenz, and D. A. Thomas. Polymer Preprints, 16(1), 274 (1975).

21

New Developments of Crosslinkable Emulsion Polymers

HENRY WARSON

Solihull Chemical Services, 284 Warwick Rd., Solihull, England, B92 7AF

The application of emulsion thermoplastic polymers into many industries, especially those of coating, textile and paper, saw a major development from about 1958 with the introduction of polymers which could be crosslinked after application. This is achieved by the inclusion of monomers with at least one other reactive group, which is substantially unchanged in the addition polymerisation, but which may later react with itself or another included reactive group either inter- or more probably intra-molecularly. Alternatively the reaction may take place with an added crosslinking agent. Monomers of the first type include N-(methylol)-acrylamide* or its ethers. A typical added crosslinking agent is hexamethoxymethyl melamine. These processes are now well established, and the author's current volume (1) will be taken as a datum line for a survey of current improvements, this being based on the nature of the reactive groups.

Derivatives of Acrylamide

NMAM, $CH_2:CHCONH_2CH_2OH$, available as a 60% aqueous solution, is a highly reactive monomer in copolymerisation, although less so than acrylamide, and copolymerises with most standard monomers, although almost invariably at a high rate. (2)(2A). The ethers, e.g, $CH_2:CHCONHCH_2OC_2H_5$, formed readily in monomeric condition, or by a secondary reaction with units in a polymer, are equally reactive, as the ether group splits readily, and the longer chain ether groups may assist in improving compatibility with other monomers. In this connection N-(iso-butoxymethyl)acrylamide, $CH_2:CHCONHCH_2O\text{-iso}C_4H_9$ has been introduced recently. It is a colorless to light yellow clear oil, purity c . 85%, density 0.98, and BP 99 - 100° at 0.03mm. It is water insoluble, but soluble in almost all organic solvents except for aliphatic hydrocarbons. (2B) Difficulties in obtaining even copolymerisation of NMAM in emulsion are partly due to unbalanced reactivity ratios, and partly due to its high water solubility. These may result in the formation of

*N-(methylol)acrylamide is abbreviated NMAM;

irregular or brittle copolymer films, due to incompatibilities, with accompanying "fisheyes" and other defects. Much current work has been devoted to lowering curing cycles and temperatures.

Difficulties of copolymerisation are most acute with vinyl acetate. Whilst early recommendations include the addition of methanol to improve compatibilities, current disclosures include methods by which NMAM is added to a polymerising emulsion of vinyl acetate. NMAM, although generally about 2.5% of total monomers, is raised to 15% in a process in which the monomers, including 0.9% of acrylic acid, are added to a water phase containing tert-butyl hydroperoxide and a buffered bisulfite redox initiator with a ferrous ion activator, the process taking 2 hours. The final latex is suitable as a base for a plywood adhesive.(3) Terpolymers of this type may be mixed with tin tetrachloride hydrate or other catalyst to produce an adhesive for wood blocks much better than polyvinyl acetate alone.(4)

Alternative processes include the vinyl esters of "Versatic acid" (Shell Chemicals), with vinyl acetate and acrylamide (49:49:2), which are added gradually after conversion to a pre-emulsion, to an anionic surfactant solution containing a persulfate. Transparent waterproof films are formed.

Copolymers of NMAM and vinyl acetate are useful as sizes for fiber glass in conjunction with a silane and methacrylatochromic chloride.

Terpolymers of ethylene, vinyl acetate and NMAM are prepared by adding the latter two under constant ethylene pressure into an initiated emulsifier solution containing polyvinyl alcohol, controlling the amount of unpolymerised vinyl acetate at any one time to be under 3.5%. This method enables ethylene to be copolymerised at relatively low pressures. The NMAM content is 0.5 - 2% of the vinyl acetate. The final latex may be employed as an adhesive with excellent resistance to cold flow, this being probably imparted by slow ambient crosslinking. The latter is assisted by the inclusion of aluminium chloride. (7)(7A)

Acrylic acid, acrylonitrile and optionally triallyl cyanurate are also included in a copolymer which imparts wrinkle resistance to textiles, and also resistant to dry cleaning solvents.(8)

Vinyl acetate - ethylene - NMAM - diallyl maleate copolymers, optionally with itaconic acid, give films which cure rapidly at 132°. (9). Other copolymers with this acid are carpet backsizes.(9A)

N-(isobutoxymethyl)acrylamide (NIBA). NIBA can be copolymerised readily with many standard formulations, since it is soluble in most monomers. Reactivity ratios as determined in a benzene solution copolymerisation are:

methyl acrylate (r_1) : NIBA (r_2)	1.11 : 1.27
ethyl acrylate (r_1) : NIBA (r_2)	0.51 : 4.98
vinyl acetate (r_1) : NIBA (r_2)	0.015 : 34.00

In comparison with NMAM copolymers there is less likely to be premature crosslinking. Crosslinking of NIBA copolymers may occur thermally, at temperatures of 150 - 170°, or under acid conditions. Proposed mechanisms are as follows: (2B)

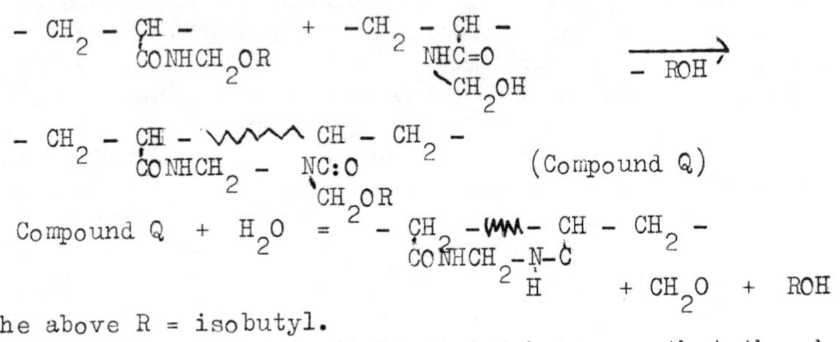

In the above R = isobutyl.

Many workers in this field state, however, that they have no evidence of the liberation of formaldehyde.

A different mechanism is postulated for acid crosslinking, involving the liberation of both isobutanol and its formal, ultimately producing crosslinked units as:

$$\begin{array}{l} -\ \mathrm{CH}-\mathrm{CH_2}-\sim\sim\sim-\mathrm{CH}-\mathrm{CH_2}- \\ \mathrm{O{:}C}-\mathrm{NH}-\mathrm{CH_2}-\mathrm{C{:}O} \end{array}$$

If a reactive substrate is present it is possible that reactive groups, e.g, the hydroxyl in cellulose derivatives, may take part in the condensation.

Latex copolymers including NIBA have found applications in protective coatings, binders for nonwoven fabrics, water and oil repellancy, crease resistance, print pastes, leather finishing, adhesives, paper impregnation, curable thickeners and elastomers. (2B). Comonomers include acrylic and methacrylic esters, acrylonitrile, styrene, and occasionally vinyl acetate or vinyl chloride.

Acrylic Copolymers. Most copolymers in this class are based on acrylic (including methacrylic) esters. These produce most of the required application properties, albeit at a premium cost. A soft, easily integrated film may be first formed, but it is readily cured. Copolymers of acrylic esters with NMAM or methylene bisacrylamide $H_2C(NHCOOCH{:}CH_2)_2$ improve laminates of wool with nylon or acetate fiber. (10)

Acrylic or methacrylic acid, at about 2 - 5% of monomers, is often a component, helping latex stability, curing and adhesive properties. The major monomers are often present in quite complex mixtures to suit specific applications. Thus acrylonitrile assists solvent resistance. (11) Most of the modifications are for various aspects of textile finishing. There is an increasing interest in flocking adhesives. (12 - 14)

Other acrylic latices have a specific application in the coatings industry. One product based on butyl acrylate and acrylonitrile also contains 2.7% of isobornyl methacrylate with 2.84% of NMAM. Self - emulsification takes place with 2-sulfoethyl methacrylate and surprisingly, sodium vinyl sulfonate. Coating to metals is applied from an acid bath containing fluoroboric and hydrofluoric acid with chromic anhydride and potassium ferricyanide. (15) An allied product is based on ethyl acrylate, NMAM and acrylic acid, emulsification being nonionic and polymerisation by redox. A paint at 56% pigment - volume - concentration exhibited high scrub resistance (ASTM-D-2486-66T), and good low temperature coalescence. (16)

Colloidal silica is included in a styrene - methyl methacrylate copolymer with 2.5% each of itaconic acid and NMAM. It is claimed that heating at 180° for 1 minute gives a water resistant coating on aluminium, although some surprise must be expressed at this composition forming other than a powder on drying at ambient or even elevated temperatures. (17)

The ethers, e.g., N-(methoxymethyl)methacrylamide, in conjunction with acrylic or methacrylic acids are alternatives to NMAM in acrylic polymers for nonwoven textiles, giving good wet and dry tensile strength after curing, for which 150° is the optimum temperature, but 110° is effective. (18). Of unusual interest is the disclosure of a conversion of an acrylamide - styrene (11:89) copolymer latex into a Mannich base, including units such as:

$-(CH_2CH)-$, thus affording an ambient crosslink.
$\quad CONHCH_2N(CH_3)_2$

The latex may be added to paper pulp, and after conversion, the resultant paper has improved bursting strength and internal cohesion. (19)

An unusual process features a spun polythene band which has been impregnated with blended acrylic latices, e.g, containing 1,4-butanediol diacrylate, ethyl acrylate and acrylic acid in 1 component, and 1,4-butanediol diacrylate, butyl acrylate, acrylic acid and NMAM in a second. 1% of the photosensitiser, benzoin isopropyl ether is also present. The impregnated band is drawn across an irradiation zone whilst resting on a thin layer of ice, thence forming a continuous leather - like band. (20)

N-formyl-N'-methacryloylmethylenediamine, $HCONHCH_2NHCH_2NHOOC\text{-}C(CH_3):CH_2$ (3.5%), and NMAM (2%) are copolymerised with acrylonitrile and butyl acrylate (approx 1 : 3) to produce a latex which forms tough crosslinked films when dried at 90°. The latex is useful for coatings, binders, and in textile printing. (21)

Fluorinated acrylic monomers, with acrylamide or NMAM are included in compositions intended for imparting oil and water repellancy to textiles. (22)(22A)

External Additives. A methoxymelamine is added to a rather complex acrylic latex including methacrylic acid, acrylamide, NMAM and diallyl phthalate, the product being suitable as a flocking adhesive. (23). An allied product, based on butyl acrylate, acrylonitrile, acrylamide and NMAM is mixed with a polyester resin, and applied to a rayon base, followed by nylon flock. (24)

An interesting variation is afforded by the prior reaction of a fully methylolated urea with N-(2,3-dihydroxypropyl)methacrylamide before copolymerisation takes place, the resultant latex being applied as an impregnant to a nylon fleece. (25) This type of condensate may be worthy of further study. On the other hand, NMAM, with NH_4Br catalyst, may itself be added to a high ethylene copolymer including vinyl acetate and methacrylic acid. The cured latex film enables polypropylene to be heat-sealed to aluminium. (26).

In a variation a polyethylene latex is used with monomeric NMAM and a fiber reactive agent such as dimethylgloxalmonoureine and a latex catalyst, the process being optionally 2-stage. The modified latex features wrinkle and 'wash and wear' properties. There may be some direct grafting on to the polythene. (27)(28).

HO OH HOH$_2$CN NCH$_2$OH O

A reactive acrylic latex, including either acrylamide or glycidyl methacrylate, is applied as an impregnant to nylon and polyester textiles together with glycerol diglycidyl ether or bisphenol A. 3 minutes heating at 150° is required for cross-linking. (29)

Vinyl Halides. In copolymerising vinyl chloride with NMAM, the same precautions to ensure even copolymerisation should be taken as with vinyl acetate.

An NMAM - vinyl chloride copolymer, with a nitrile rubber latex, optionally carboxylated, and trimethylolmelamine, bonds rayon flock to PVC sheet, and PVC to galvanised steel. (31)

A paper coating is based on an ethylene - vinyl chloride - NMAM copolymer, pigmented with TiO_2 and clay. (32) Similar ter-polymers, prepared with delayed addition of vinyl chloride and NMAM, are applied to nonwoven fabrics. Drying at 119° produces a textile of good strength, handle and drape. (33)

The useful properties of vinylidene copolymers in paper coating are further enhanced by including NMAM as a comonomer and subsequent curing. (34). A 15% latex, including vinylidene chloride, acrylonitrile, an acrylic ester, acrylamide and NMAM is useful as a paper impregnant, providing good wet and dry strength after cure at 135°. (35)

Latices. Latices are sometimes defined as emulsion polymers or copolymers containing a diene. If they also contain a carbox-ylic acid, e.g, itaconic and/or methacrylic acids, they are usually of the 'self cure' type. The 'cure' or crosslinking is catalysed by transition metal salts and even by pigments derived therefrom. The action is enhanced by including NMAM as a monomer. (36) Heat curable foam - forming latices are obtained from 1,3-butadiene, styrene and NMAM. (37). Latices including NMAM may be mixed with polyisocyanates for the production of foams and films. (38). However, overall, disclosures including NMAM and allied reactive monomers in diene latices seem to be rather scarce.

Diacetoneacrylamide.

Diacetoneacrylamide, N-(1,1-dimethyl-3-oxobutyl)acrylamide has been the subject of many investigations as a comonomer. It can be modified to a methylol form, but unlike acrylamide, methylolation takes place on the carbon atom adjacent to the ketonic group and the monomer may be written $CH_2:CHC(O)NHC(CH_3)_2$-$CRR'C(O)CH_3$ where R and R' = H or CH_2OH. 3% of this monomer proves optimum in formulations, and addition of HMMM is advised, a pH of 3 being optimum for a cure, which occurs at ambient temperature in 4 - 16 hours. Vinyl acetate and acrylic copolymers are quoted in the literature, the latter even without an auxiliary cross-linking agent. The special interest is in thermosetting coatings, but formulations are suggested for interior gloss emulsion paints. (39)(40).

The Aminimides.

The aminimides (Ashland Chemical Co.), a series of monomers with both positive and negative nitrogen atoms, and which may contain unsaturated groups may be typified by the formula: $CH_2:C(CH_3)CON^- N^+(CH_3)_2CH_2CHOHCH_3$. For preparation see refs. (41 - 44). A feature of these compounds is that they are iso-cyanate precursors, one molecule of triethylamine being lost on heating. They will thus react with compounds containing active hydrogen as an isocyanate, and in the case of a polymeric product will give rise to crosslinking. (45)

Polymerisation and copolymerisation of these monomers is possible in emulsion with redox systems, although styrene will only copolymerise with the hydrochlorides of the aminimides. (47) The major interest seems to be in metal coatings; self - reacting polymers are produced by the copolymerisation of acrylic compounds with active hydrogen, e.g, hydroxyethyl acrylate. At normal curing temperatures the rearrangement of the aminimides is very rapid:

$$RC(O)N^- N^+R''_3 \longrightarrow RNC + R''_3N$$

The development of polymer latices and their pigmentation and application is still at an early stage. However, an emulsion formulation suitable as a base for paints has been given. (46) Here a difunctional aminimide (below) is included at 5% in the

$$(CH_3)_3N^+ N^-OC(CH_2)_8CON^- N^+(CH_3)_3$$

monomers with 10% of 2-hydroxyethylacrylate, the balance being acrylic esters. In some cases the simpler difunctional aminimides can be used to crosslink other acrylic copolymers which include active hydrogen. The water solubility of the aminimides should be noted. The epoxide condensation with difunctional aminimides produces oxazolidone rings which are useful in promoting adhesion and have a suggested application in tire cord adhesives. (45)

Trimethylamine methacrylamide copolymerises well with most standard monomers. The extensive technical literature of the Ashland Co., including patent specifications, gives further details of applications of these compounds, which should have an interesting future. Unfortunately it is understood (1975) that there has been a setback in commercialisation due to restrictions in the supply of unsymmetrical dimethylhydrazine, a key raw material. Some recent developments have been published. (47A)(47B)

Miscellaneous Nitrogen Compounds.

The Aziridine Ring. The aziridine ring .CH(NR)CH is analagous in many ways to the epoxide ring and may be used for crosslinking in rather similar circumstances. Care should be taken both with regard to potential toxicity and hydrolytic stability of any experimental compounds. 2-(1-aziridinyl)ethyl acrylate can be copolymerised directly with acrylic acid using a nonionic surfactant, if the pH of the emulsion is lowered with a pKa below 3.7, e.g., phosphoric acid or an organic acid such as oxalic or malonic acid. (48) The aziridine ring can open quite readily affording many opportunities for crosslinking with other reactive groups.

Ethylene - vinyl acetate - acrylic acid terpolymers have been crosslinked with 2,4,6-tri(2-methyl-1-aziridinyl)s-triazine or tris(2-methyl-1-aziridinyl) phosphite oxide. The improvement in film tensile strength and the reduced solubility in trichlorethylene are very marked. (49)

Compounds such as tris[β-(1-aziridinyl)ethyl] trimellitate react with carboxyl - ended butadiene polymer, and might well be an additive to latices containing carboxyl groups. The original specification quotes 24 hours stoving at 85°. (50).See also(54)

Azide Derivatives. Azidoformates are prepared from a hydroxyl - containing monomer via the chloroformate and sodium azide. Generally hydrolytically stable, they can be used either in direct copolymerisation or as post - additives, functioning by a secondary reaction. In the latter case the reaction is not restricted to azoformates of polymerisable acids. The crosslink at 70° produces a urethance type linkage. (51)

A number of variations exist in which azides, e.g., epoxyazide compounds of general formula R-(epoxide)-QXN$_3$ where R is typically CH_3, Q = CH_2 and X = OOC, are typical crosslinkers for carboxyl - containing polymers, and also assist in the formation of tire cord adhesives, presumably by a crosslinking action. Azides of polycarboxylic acids may also function as photosensitive crosslinking additives, the .NCO group being formed by actinic light. (53) Many coating uses are suggested with these azide derivatives.

Unclassified. Films from latices of butadiene, styrene and fumaric acid may be cured with triethylenemelamine or tris(1-aziridinyl) phosphine oxide, the crosslinked products giving very strong adhesion to glass plates, and having useful textile applications. (54)

N-vinyl lactams, the most wellknown being N-vinylpyrrolidone $CH_2CH_2C(O)NH$, crosslink readily accompanied by ring opening, whilst the polymer has the added bonus of being an emulsion stabiliser. Since ambient crosslinking takes place at pH 2 - 4,

, latices should have the pH raised to maintain stability. Whilst polyvinylpyrrolidone is water soluble, many emulsion copolymers including vinyl acetate and acrylic esters are available. There are useful applications in the paper and textile industries.

A keto acid ester such as allyl levulinate may be crosslinked by adipic acid hydrazide, and some unsaturated diketones or keto-acids would probably react similarly. Polymers have been described in bead form, but conditions could probably be found to ensure emulsion stability. (55)

Mannich bases (see ref. 23), are reported to crosslink in film form at ambient temperature, and are worthy of further application investigation. Polyisocyanates have been used to react with vinyl polymers containing active hydrogen (56), but the subject is too vast for detailed consideration here.

Cationic Compounds - Sulfonium Derivatives

The cationic sulfonium compounds have proved of considerable interest in cationic deposition. Fewer sulfonium groups are needed than anionic groups in the corresponding anionic system; no metal ions are formed at the cathode to cause discoloration, nor is the corrosion rate of the substrate increased. (57)

Associated with this type the very soluble aryl cyclic sulfonium zwitterion should be mentioned. A simple type is illustrated. These compounds form polymers by ring opening, and are capable of curing a range of latices containing carboxyl groups with which they are compatible. Polyfunctional monomers (ACSZ) enable crosslinking to take place, and with an ACSZ monomer containing 3 reactive groups, only 5.9% of the ACSZ compound was required for obtaining water solubility of an acrylic latex including 8% of polymerised acrylic acid. (58)

O^- CH_3 $\cdot 2H_2O$ (+)

Acrolein.

Acrolein CH_2:CHCHO, although often polymerising via the aldehyde group, can be polymerised in emulsion. (59). 1.5% is included in the formulation of a carboxylated latex for carpet backing, also providing a major improvement in wet abrasion resistance when used as a paper coating. (59)(59A) A copolymer of acrylonitrile, styrene, acrylic acid and starch, blended with hydroxyethyl starch, gives mutual insolubilisation on heating, and a suitable pigmented composition gives improvements in the wet rub and IGT pick resistance of coated paper (60)

Acrolein is included at 3 - 20% in a formulation including acrylic esters, styrene, and 2-hydroxyethyl acrylate. The latex is used with tri(methoxymethyl)melamine to provide a solvent and alkali resistant coating. (62)

SILOXANES.

The reactive groups in siloxanes may be utilised to providing crosslinking sites. (63) A disclosure describes a siloxane polymer containing 2% of $CH_2:CHSiO_{3/2}$, which is copolymerised with acrylic esters and acrylic acid in a redox process in which tert-butylaminoethyl methacrylate is added in a final stage. This latex can be cured on a metal plate in 1 hour at 25°, or converted to an emulsion paint with excellent wet adhesion. (64)(64A)

A siloxane emulsion including 1.5% of methylvinyl siloxane units forms, after curing, a soil release treatment for textiles. (66) An unusual method of introducing silicon is to utilise bis-[β-(trimethoxysilyl)propyl] disulfide as a chain transfer agent in the emulsion polymerisation of styrene, partially crosslinking during the reaction, and fully crosslinking by treatment with tin octoate. (67)

Space forbids mention of further recent developments, including many involving the hydroxyl, carboxyl and epoxide groups.

Literature Cited

1. Warson, H. "Applications of Synthetic Resin Emulsions" 584 - 610, Ernest Benn, London, 1972
2. Yocum, H.H., Nyquist, E.B. Eds. "Functional Monomers" Vol. 1 Ch. 1 by Macwilliams, D.C. 197pp. Marcel Dekker, New York 1973
2A. Warson, H. "Reactive Derivatives of Acrylamide & Allied Products" Solihull Chemical Services, 1975
2B. American Cyanamid Co. Tech. Literature
3. Air Products, GP 2,221,690, 1973
4. Croda, GP 2,206,571, 1972
5. Denki Kagaku, JP 73 - 38,231
6. Johns-Manville, GP 2,328,129, 1974
7. Borden, USP 3,714,105
7A. Borden, FP 2,154,837, 1973
8. Japan Synthetic, JP 71 - 07,636; 71 - 10,519
9. Union Oil California, USP 3,714,099-100, 1973
9A. DuPont, BP 1,298,155, 1972
10. Goodrich, B.F., BP 1,309,849, 1973
11. Bar, V., Ver, Z. Hung.P. 7,041, 1972
12. Dearing Milliken, USP 3,616,136, 1971
13. Rohm & Haas, USP 3,705,053, 1972
14. Nissin; Shinetsu, JP 71 - 33,573
15. Celanese, A. Afr. P. 70 - 07,900
16. BASF, GP 2,202,568, 1973
17. Toyo Ink, JK 74 - 04,739
18. Bayer, GP 2,012,287, 1971
19. Am. Cyanamid, GP 2,263,089, 1973
20. BASF, GP 2,202,568, 1973
21. Casella, GP 2,251,922, 1974

22. Dupont, BP 1,303,806, 1973
22A Hoechst, GP 2,115,139 - 40
23. Dainippon Ink, JP 72 - 03,718
24. Rohm & Haas, USP 3,681,108, 1972
25. Nobel Hoechst, GP 2,258,041, 1973
26. Dupont, GP 2,313,021, 1973
27. Daido Maruta, JP 71 - 37,440
28. Kyoto Sarasen, JP 71 - 11,078, 1972
29. Unitika, JK 73 - 09,098
30. No ref.
31. Nissin Chemical, JP 71 - 004,832-3
32. Scott Paper, GP 2,302,044, 1972
33. Monsanto, BP 1,311,256, 1973
34. Solvay, Belg. P. 755,898, 1971
35. Goodrich, B.F., GP 2,145,494, 1972
36. Teikoku Chemical, JP 72 - 50,872
37. USP 3,740,357, 1973
38. GP 2,014,385, 1971
39. Lubrizol Co. USA, Technical Literature
40. Franco, S., Leoni, A., Polymer(1973)14(1) 2-4
41. McKillip, W.J., Clemens, L.M., Haugland, R. Canad. J. Chem. (1967) 45 2613 - 7
42. McKillip, W.J., Slagel, R.C. ibid. 2619 - 23
43. Culbertson, B.M. et al. J. Polymer Sci. A-1,(1968) 6 2197 - 2207
44. Slagel, R.C. J. Org. Chem.(1968) 33, 1374 - 8
45. Technical Literature, Ashland Chemical Co., USA
46. Ashland Chemical, USP 3,527,802, 1970
47. Culbertson, B.M., Randen, N.A. J. Appl. Polymer Sci. (1971) 15, 2609 - 21
47A Langer, H.J., Randen, N.A. Polym. prepr. Am.Chem.Soc.(1975) 16(1) 490 - 7
47B Culbertson, H.M., Langer H.J. ibid. 498-505, 506-13
48. Alcolac, USP 3,719,646, 1973
49. Dainippon Ink, JP 71 - 28,654
50. Dow, USP 3,806,498, 1974
51. Hercules, BP 1,116,728, 1968
52. Hercules, GP 2,035,957, 1971
53. Gavaert, USP 3,278,305, 1966
54. CPC Internat. GP 2,206,966, 1973
55. Hoechst, BP 1,037,131, 1966
56. "Applications" (see ref. 1), pp271 - 2
57. Wessling, R.A. et al. Advan. Chem. Ser.(1973)(119) 110 - 27
58. Schmidt, D.L., Smith, H.B., Broxterman, W.E. J. Paint Technol. (1974) 46(588) 41 - 6
59. Shell Chemical Co. Technical Literature
59A Polymer Corp. GP 2,161,408, 1972
60. Nalco USP 3,733,386, 1973
61. Rohm & Haas, GP 2,205,388, 1972
62. Mitsubishi Rayon, JP 72 - 19,160
63. "Applications" (see ref. 1), pp 261 - 7

64. Dow-Corning, BP 1,297,730, 1972
65. Dow-Corning, GP 2,144,172, 1972
66. Dow-Corning, GP 2,144,377, 1972
67. Dow-Corning, GP 2,142,594, 1972

22

Continuous Emulsion Polymerization—Steady State and Transient Experiments with Vinyl Acetate and Methyl Methacrylate

R. K. GREENE, R. A. GONZALEZ, and G. W. POEHLEIN

Department of Chemical Engineering and Emulsion Polymers Institute, Lehigh University, Bethlehem, Penn. 18015

Continuous reactor processes are used to produce a number of synthetic polymer latexes and ever increasing production demands provide incentives for a broader use of continuous sytems in the future. Research and development studies with continuous reactors are necessary for intelligent commercial application and also for a better fundamental understanding of emulsion polymerization kinetics. A continuous stirred-tank reactor (CSTR), operated at steady state, can provide a quite different perspective for viewing reaction mechanisms. When such a reactor is used without a particle seed stream, one must have a good quantitative understanding of particle formation and competitive growth in order to predict product characteristics.

The major purpose of this paper is to present experimental results for the emulsion polymerization of vinyl acetate (VA) and methyl methacrylate (MMA) in a single CSTR. Both steady state and transient results will be presented and discussed. Possible causes for prolonged unsteady behavior will be outlined and several techniques for achieving steady operation with a CSTR will be described.

Background

Most papers involving the theory of emulsion polymerization begin with a brief discussion of the classical work of Smith-Ewart (1). Their Case II theory, tested extensively with styrene batch reactions, yields the following prediction for polymerization rate (R_p) and particle number (N).

1) $$R_p \approx N = k\ R_i^{0.4}\ (S)^{0.6}$$

where R_i is the rate of initiation, (S) is emulsifier concentration, and k is a constant. The Case II theory is based on the assumption that polymer particles contain one active free radical half of the time and none the other half. Thus $\bar{n}$, the time-average number of free radicals per particle, is 0.5.

Gershberg and Longfield (2) used the Smith-Ewart Case II theory to develop a mathematical model for a CSTR and they showed that experimental rate data for styrene emulsion polymerizations were in good agreement with the model. DeGraff and Poehlein (3) expanded the Gershberg and Longfield theory to include the Stockmayer (4) theory for slow termination in large particles. They also obtained extensive data on particle size distribution, particle number, rate of polymerization, and molecular weight averages. The model equations presented in References (2) and (3) are summarized below.

2) Particle Size Distribution Based on Radius.

$$u(r) = K_2 r^2 \exp\{-K_2 (r^3 - r_o^3)/3\}$$

where u(r) is the particle radius density function, r_o is the particle size at nucleation, and K_2 is a constant which can be determined from literature values of various reaction constants.

3) Particle Number and Polymerization Rate.

$$\frac{N}{R_i \Theta N_A} = \frac{2 R_p}{R_i \Theta k_p [M]} = \left\{ 1.0 + \frac{\alpha_o R_i \Theta}{a_s (S)} \left[\frac{k_p [M] \Theta}{1 - \alpha_1 [M]} \right]^{2/3} \right\}^{-1}$$

where Θ is reactor mean residence time, [M] is monomer concentration in the particles, N_A is Avogadro's number, k_p is the propagation rate constant, a_s is the adsorption area per emulsifier molecule, $\alpha_o = 3.85 \times (M_o V_p)^{2/3}$, $\alpha_1 = M_o V_m \times 10^{-3}$, M_o is the monomer molecular weight, and V_p and V_m are the specific volumes of polymer and monomer respectively.

For most experimental runs the second term in the brackets would be considerably greater than 1.0 and Equation 3) could be simplified to:

4) $$\frac{N}{N_A} = \frac{2 R_p}{k_p [M]} = \frac{a_s (S)}{\alpha_o} \left[\frac{1 - \alpha_1 [M]}{k_p [M] \Theta} \right]^{2/3} \quad \text{or}$$

$$R_p \approx N \approx R_i^{0.0} (S)^{1.0} \Theta^{-0.67}$$

A comparison of this equation with Equation 1) demonstrates that functional relationships between dependent and independent variables can be quite different for CSTR's and batch reactors, even with the same reaction system and the same kinetic mechanisms.

5) Molecular Weight Averages.

$$\overline{X}_n = \frac{N\ G}{R_i\ N} \quad \text{and} \quad \overline{X}_w = \frac{4.84\ N\ G}{R_i\ N}$$

where G is the mean rate of radical propagation. The factor 4.84 includes consideration for the latex size distribution. It comes from a model which assumes that free radicals diffuse into particles at a rate proportional to particle area. No transfer reactions are considered.

Stevens and Funderburk (5) presented theoretical models for particle size distributions based on Smith-Ewart Case II and several other particle growth theories. They concluded that the Smith-Ewart Case II theory containing the Stockmayer modification fit CSTR data for styrene better than other models.

Although theoretical models seem to be quite adequate for styrene emulsion polymerization in either batch reactors or CSTR's, such is not the case with other monomers like vinyl acetate, methyl acrylate, methyl methacrylate, vinyl chloride, etc. One of the early papers to discuss some of the important mechanisms involved with these other monomers was written by Priest (6). He studied the emulsion polymerization of vinyl acetate and identified most of the key mechanisms involved. Priest's paper has been largely overlooked, however, perhaps because of the success of the Smith-Ewart approach to styrene.

A small amount of work has been done with other monomers in continuous emulsion reactor systems. Gerrens and Kuchner (7) studied styrene and methyl acrylate, Berrens (8) used vinyl chloride, Gonzalez (9) studied methyl methacrylate, Senrui et al. (10) examined ethylene, and Greene (11) studied MMA and vinyl acetate. These workers have all presented experimental data but none have offered a complete CSTR model based on fundamental kinetic equations.

Experimental System

The experimental reactor system used for the data presented in this paper is described by DeGraff and Poehlein (3). The temperature control system was changed, however, to include a coil in the reactor. Temperature-controlled water was pumped through this coil to heat and/or cool the reaction media. Some of the experiments were carried out with a tubular pre-reactor located upstream of the CSTR. This reactor consisted of 130 ft of 1/8-inch I.D. teflon tubing. The tubular reactor was divided into three sections of 54 ft, 54 ft and 22 ft to allow sampling at different space times or conversions.

The tube reactor was operated in a plug-flow mode by injecting nitrogen slugs into a three-way valve at the reactor entrance. Thus the polymerizing emulsion moved through the reactor in plugs about 3/4-inch long. Reactor space time could be changed by varying the rate of emulsion flow and/or the rate of nitrogen gas flow.

Vinyl acetate monomer, supplied by Celanese Chemical Company with 3-5 ppm hydroquinone, was distilled under nitrogen through a 5-ft packed column with reflux at 1 atm. The distilled monomer contained less than 0.5 ppm HDQ. Ammonium persulfate (certified grade) was used as the initiator and sodium lauryl sulfate (laboratory grade) was the surfactant. The emulsions were buffered with sodium acetate (certified grade) and acetic acid (reagent grade).

Methyl methacrylate (uninhibited) was used as received from Rohm and Haas. Sodium bicarbonate (certified grade) and sodium chloride were used to adjust ionic strength and buffer the emulsions during the MMA runs. Distilled-deionized water was used for all runs. High purity nitrogen used for degassing solids and for blanketing the reactor and storage vessels, was passed through a bed of heated copper shavings prior to use.

Latex samples used for particle size analysis on the electron microscope were diluted in emulsifier solutions containing HDQ. The MMA latexes were placed on a grid, dried, and platinum shadowed prior to photographing. The PVA latexes were diluted, freeze dried, and shadowed on a Formar support film.

Experimental Results

Vinyl Acetate CSTR. Figure 1 shows the effect of initiator concentration on the conversion in a CSTR at three different values of mean residence time. Conversion is directly related to rate of polymerization (R_p) by the following equation.

6) $$R_p = \frac{X\ [M]_0}{\Theta}$$

where X is fractional conversion of monomer to polymer and $[M]_0$ is the monomer concentration in the mixed feed stream.

Figure 2 shows the influence of emulsifier concentration on reactor conversion at two levels of initiator concentration and two values of mean residence time. An analysis of the data shown yields the following approximate relationship between polymerization rate and the three independent variables.

$$R_p \approx [I]^{0.8} [S]^{0.1} \Theta^{0.0} \quad (7)$$

This dependency is quite different from the predictions of theoretical models based on Smith-Ewart Case II kinetics and also different from styrene data (Equation 1).

Figures 3 through 6 show conversion-time data for a number of vinyl acetate runs. The start-up procedure for these experiments consisted of filling the reaction vessel with degassed water prior to introducing any feed streams. Periodic samples were taken and the monomer conversion measured gravimetrically. As can be seen, some of the conversion transients did not reach a steady state. Tendency toward unsteady behavior and the magnitude of the oscillations seemed to increase with increasing initiator concentration and mean residence time. The influence of changing the emulsifier concentration is not clear.

Figures 7 and 8 show typical particle size distributions for vinyl acetate emulsions produced in a single CSTR. A large number of particles are quite small with 80 to 90% being less than 500 Å in diameter. The large particles, though fewer in number, account for most of the polymer mass as shown by the cumulative volume distributions. Data are also presented on Figures 7 and 8 for the number of particles based on diameter measurements (N_d), the average number of free radicals per particle, and the steady state conversion.

Molecular weight averages for the PVA latexes did not vary significantly with recipe changes or with different mean residence times. The weight-average molecular weights ranged from 1,509,000 to 1,768,000 and the number-averages varied from 524,000 to 729,000. This type of result is expected for low-conversion VA polymerizations because of the high transfer reaction rate with vinyl acetate monomer.

<u>Methyl Methacrylate CSTR and CSTR-Tube Reactors.</u> MMA experiments were conducted with a single CSTR and

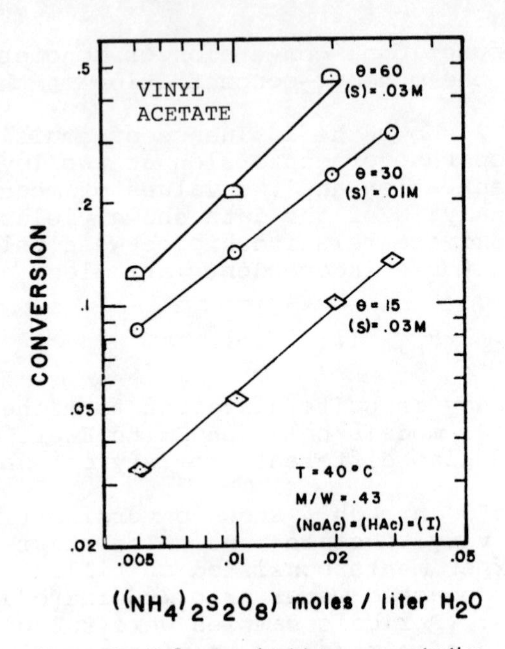

Figure 1. Influence of initiator concentration on conversion

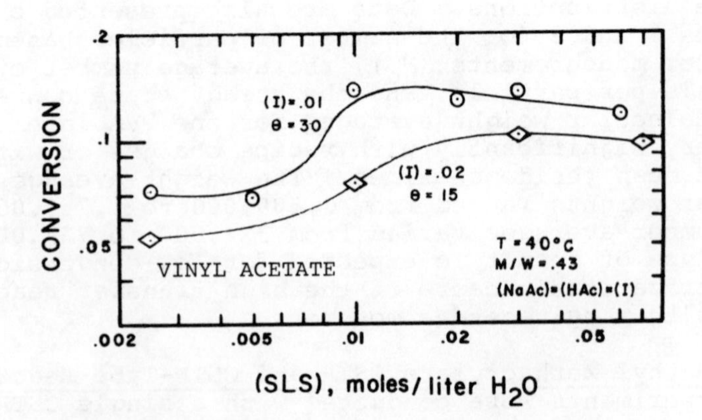

Figure 2. Influence of emulsifier concentration on conversion

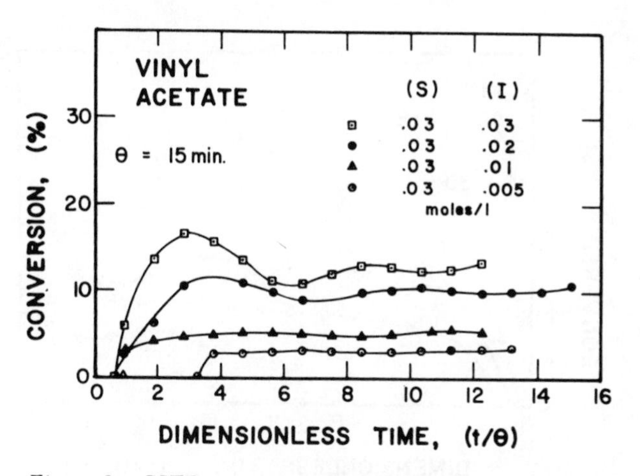

Figure 3. CSTR conversion transients—effect of initiator concentration

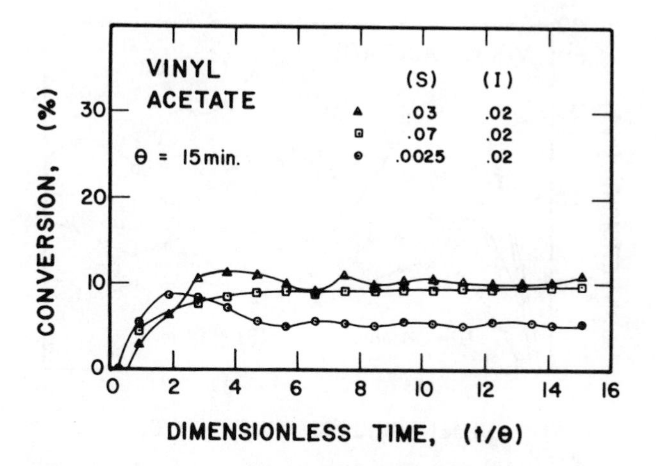

Figure 4. CSTR conversion transients—effect of emulsifier concentration

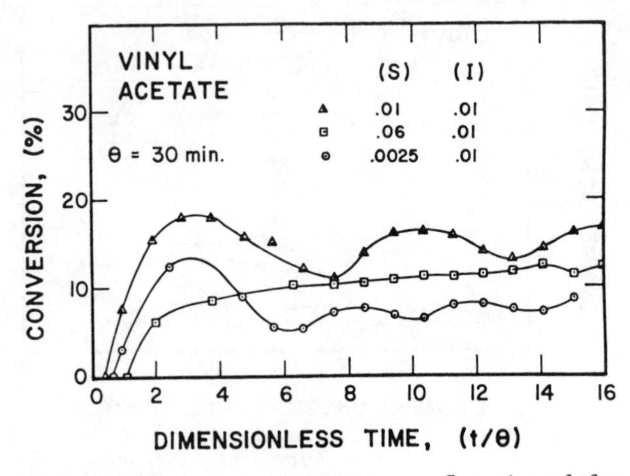

Figure 5. CSTR conversion transients—effect of emulsifier concentration

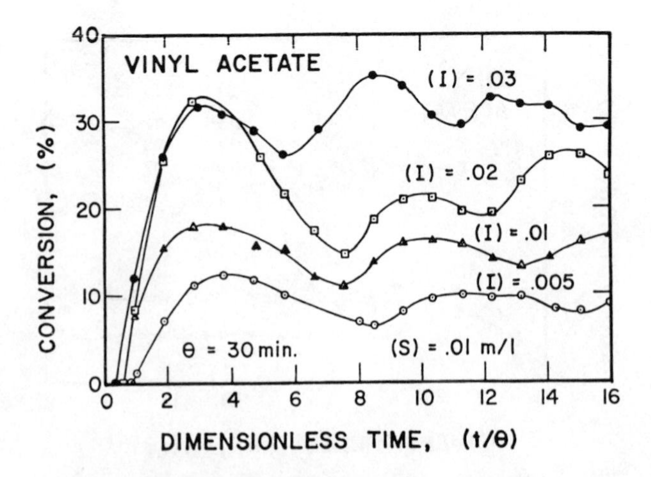

Figure 6. CSTR conversion transients—effect of initiator concentration

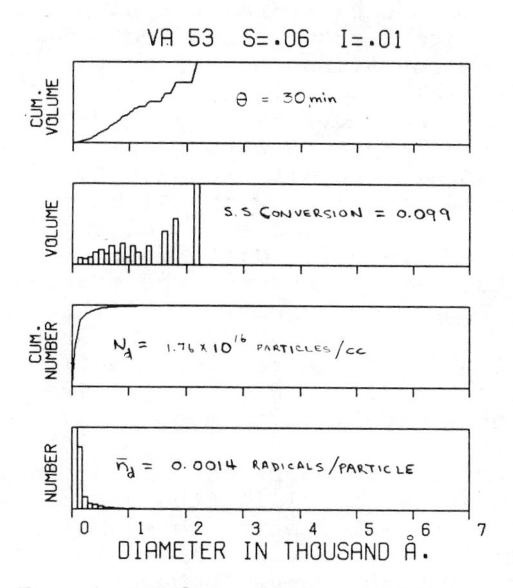

Figure 7. Particle size distribution of PVA latex from CSTR

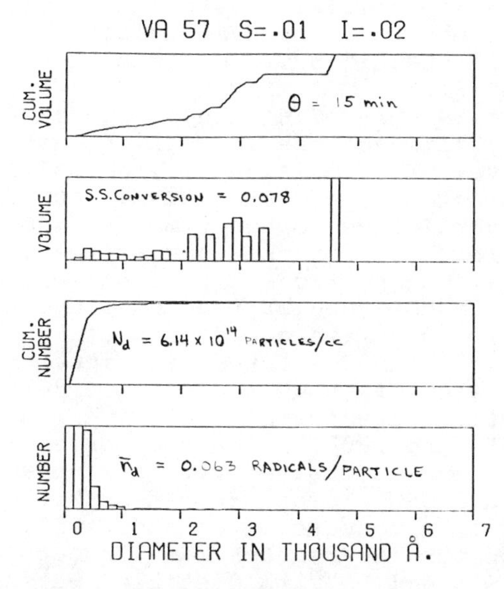

Figure 8. Particle size distribution of PVA latex from CSTR

with a continuous tubular reactor placed upstream of the CSTR. Figure 9 shows the effect of initiator and emulsifier concentrations on the conversion from the single CSTR. The slopes of the initiator and emulsifier lines are about 0.8 and 0.9 respectively. Thus MMA behaves similar to VA with respect to initiator and similar to styrene with respect to emulsifier.

Figures 10 and 11 show transient conversion data for MMA experiments. As one can easily see, the shapes of these curves are quite different. The major difference in the experiments was the start-up procedure. The data in Figure 10 resulted when the reactor was started empty. In this run the feed stream pumps were started at $t = 0$ and the reactor overflow first occurred at $t = \Theta$. As one can readily see, the conversion oscillates widely in an apparently undamped fashion.

The start-up procedure for the experiments shown in Figure 11 involved filling the reactor with distilled, deionized water prior to starting the feed stream pumps. In these runs the conversion seems much more stable. The top curve, however, demonstrates that these apparently stable steady states may be subject to rapid change; perhaps the seeking of a new, more stable, steady state or the beginning of oscillations.

Figure 12 shows a particle size distribution for MMA latex produced in a CSTR. The presence of a large number of small particles is similar to the particle distributions observed for PVA emulsions, but different from the single-peaked broad distributions obtained with styrene (3).

Gonzalez (9) placed a tubular prereactor upstream of a CSTR to eliminate the conversion instabilities reported above. MMA emulsion polymerizations were run with the tube at 60°C and the CSTR at 40°C. A typical curve of conversion vs. space time (residence time in the tubular reactor) for the tubular reactor is shown in Figure 13. The particle size distribution of the PMMA latex from the tubular reactor seemed to depend on the conversion. At a conversion of 3.5% we observed a narrow, single-peaked distribution with a mean particle diameter of about 140Å. At 4.86% conversion the distribution appeared bimodal with peaks at diameters of 95Å and 280Å. When the conversion was 21.9%, the distribution was broader with a single peak centered at $d \cong 390$Å. The maximum number of particles was observed at a conversion of about 11%.

When the tubular reactor was connected to the CSTR (which was filled with distilled water), a

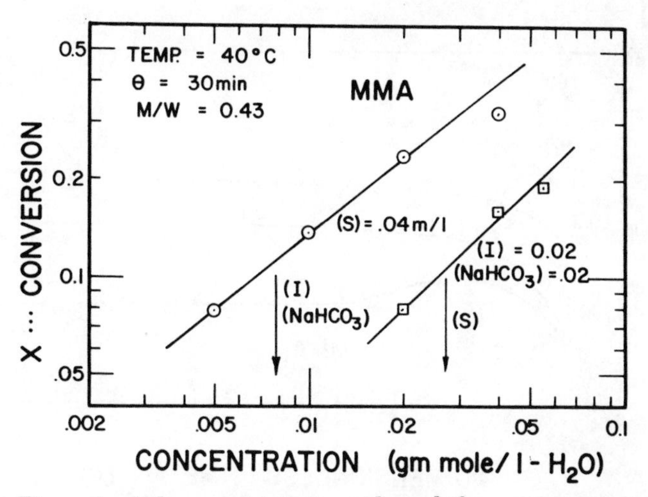

Figure 9. Influence of initiator and emulsifier concentrations on MMA conversion

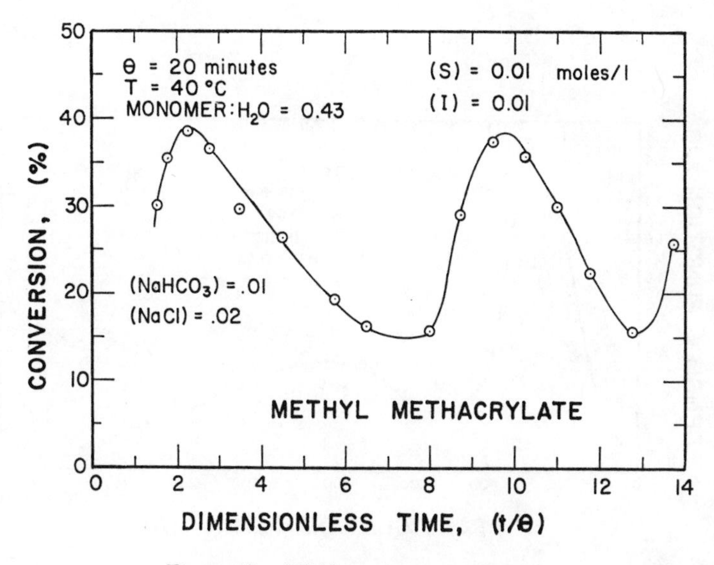

Figure 10. MMA conversion transient

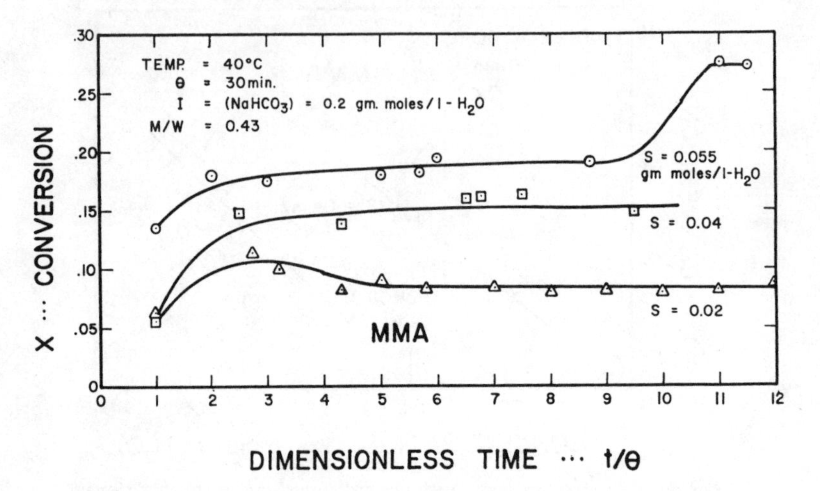

Figure 11. Conversion transients for MMA with slow start-up

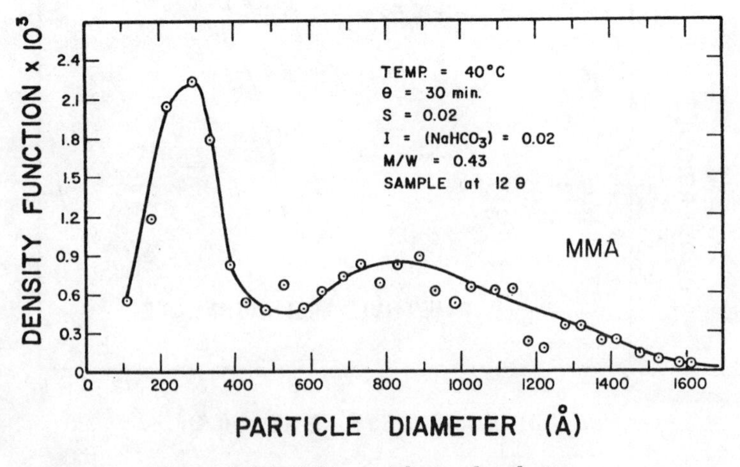

Figure 12. PMMA particle size distribution

smoothly increasing conversion transient was obtained. The steady state seemed to be stable as long as the conversion was modest. Figure 14 is a typical example. If the conversion in the tube was small enough to lead to low conversions out of the CSTR, oscillations were again obtained as shown in Figure 15.

Discussion of Results

Steady-state conversions for VA and MMA polymerizations in a CSTR do not agree with reactor models based on Smith-Ewart Case II kinetics. This is not surprising since such a model does not consider many important phenomena. The particle-formation component of the Smith-Ewart Case II model is based on a simple mathematical relation which assumes that the rate of formation of new particles is proportional to the ratio of free (dissolved or in micelles) surfactant to total surfactant. This equation is based on the earlier concept of particle formation via free radical entry into micelles.

We know that the particle formation mechanisms can be much more complex, especially for monomers such as MMA and VA. The sulfate ion radical formed from persulfate initiator is not likely to be strongly attracted to either micelles or polymer particles due to its charge and hydrophilic nature. Instead, this radical will begin to polymerize monomer which is dissolved in the water phase. As monomer units are added, the aqueous phase ion radical will become less hydrophilic and more surface active. During the period of surface activity, it may well adsorb on a micelle, a polymer particle, or a monomer drop.

As the water-phase polymerization continues, however, the ion radical will become more hydrophobic and less surface active. At this point a large number of paths are feasible; it may move into micelles, particles, or monomer drops; it may self-coil to form a small (primary) particle; it may combine with other aqueous phase oligomers to form a particle; or it may continue to grow. The small primary particles formed by the above mechanisms could become stabilized or they could flocculate among themselves or with other organic particles in the emulsion. In addition, the larger particles in the latex might also be involved in flocculation interactions with other organic components.

It should be obvious that the simple concepts of Smith-Ewart Case II kinetics could not be expected to explain the complex phenomena outlined above. Another

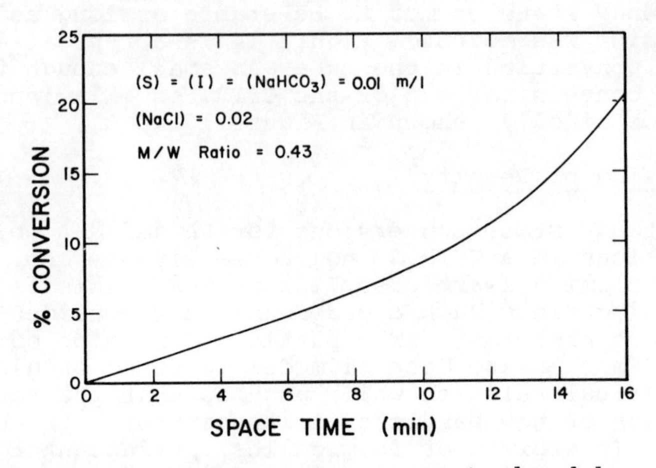

Figure 13. Conversion vs. space time for the tubular reactor

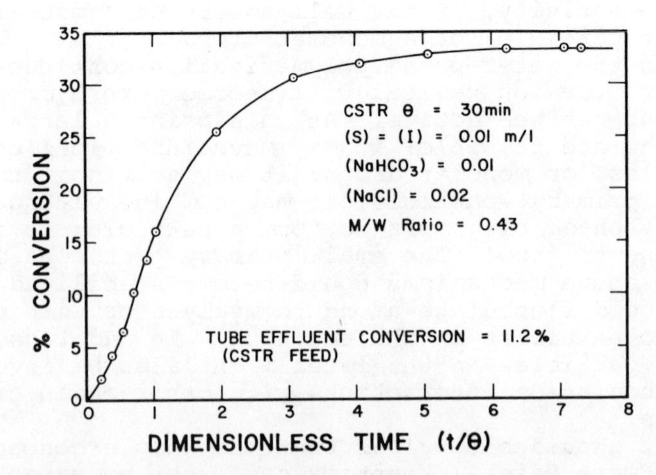

Figure 14. Conversion transient for MMA polymerization in the tube-CSTR reactor system

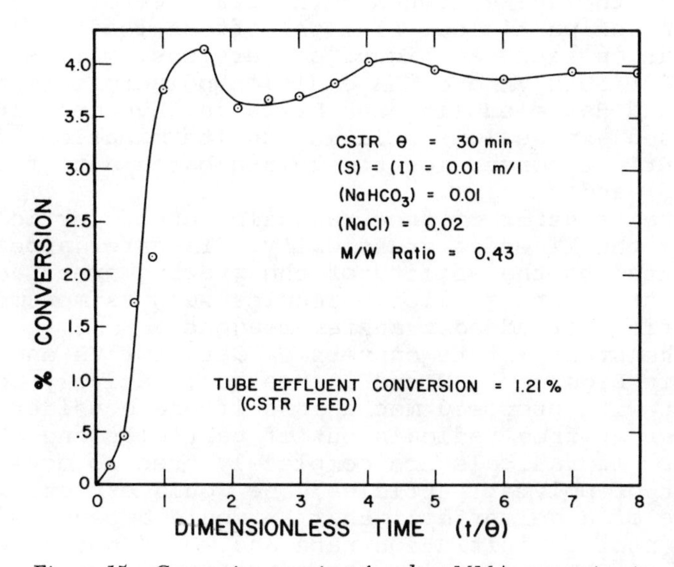

Figure 15. Conversion transient for slow MMA conversion in the tube-CSTR reactor system

weakness of the Smith-Ewart Case II theory is the assumption that all particle growth occurs by polymerization within the particle. Thus growth by capturing smaller particles is not considered. Recent work by Yeliseyeva (12,13) and Furuta (14) suggests that growth by capturing primary particles may be quite significant.

Two other major factors are not included in the CSTR model based on Smith-Ewart Case II kinetics. They are the gel effect and the transfer of free radicals out of particles. The gel effect leads to slower termination rates at higher conversions. It is significant for both VA and MMA emulsion polyermizations. Friis and Hamielec (15) and Friis and Hyhagen (16) have used expressions relating k_t (termination rate constant) to conversion to explain batch data for MMA (15) and VA (16).

The transfer of free radicals out of particles changes the kinetics drastically. Transfer appears to be related to the ability of the growing free radicals to transfer with smaller molecules such as monomer, emulsifier, or added transfer reagents.

The steady-state conversion data for VA and MMA, shown in Figures 1, 2 and 9, are qualitatively consistent with proposed mechanisms if one considers the transfer of free radicals out of particles and the gel effect. If radicals are completely free to move into and out of polymer particles, one would expect, in the absence of a gel effect, that R_p would depend on the square root of initiation rate and would not depend at all on the emulsifier concentration. Ley et al (17) demonstrated that free radicals do transfer out of particles in PVA and PMMA emulsions, and that the transfer rate is considerably higher for vinyl acetate than for MMA.

The fact that the influence of initiator is stronger (0.8) than the square root could be caused by a gel effect, although Stannett, Litt and Patsiga (18) suggest that water-phase termination may be responsible for the higher exponent with vinyl acetate. The lack of a surfactant effect on R_p for VA would seem to confirm the importance of radical transfer. With MMA, however, the surfactant influence is significant. The gel effect is also considerably stronger for MMA and thus we would conclude that this, rather than radical transfer, is the most important factor leading to deviations from Smith-Ewart Case II theory for MMA.

The unsteady behavior of single CSTR's (Figures 3, 4, 5, 6, 10 and 11) is probably caused by a combination of particle formation phenomena and the gel effect.

Matsuura and Kato (19) predict the possibility of multiple steady states under isothermal conditions because of the gel effect, and Gerrens et al (20) experimentally verified these predictions with styrene emulsion polymerization.

Particle formation phenomena could also contribute significantly to oscillations. If a large number of particles are formed during short periods, they could adsorb essentially all of the emulsifier and lead to a prolonged period in which no particles were formed. Since surfactant is continuously added to a CSTR and particles continuously removed, one would eventually return to an emulsifier-excess situation and generate a new population of small particles. If this process is repeated, one could reach a limit cycle situation. Such limit cycles have been reported for other types of systems such as continuous crystallizers (21) and continuous precipitation polymerization (22). Gerrens (23) reported transients in particle size distributions for styrene which illustrates the development of multi-peaked distributions. Berrens' (8) reported similar phenomena for PVC emulsions produced in a CSTR. Ley and Gerrens (24) reported sizable and persistant oscillations in surface tension at nearly constant, high (95+%) con versions of styrene in a single CSTR.

The data on particle size distributions for both PVA and PMMA emulsions suggest that small particles could be quite important in the kinetic scheme, and that the larger particles probably grow by internal polymerization and by flocculation with smaller particles. The experiments with the tubular reactor installed upstream of the CSTR demonstrate a practical way to eliminate uncontrolled transients with continuous systems. We believe that the particles generated in the tube prevent CSTR oscillations by avoiding the unstable particle formation reactions in the CSTR. Berrens (8) accomplished the same results by using a particle seed in the feed stream to a CSTR with PVC emulsion polymerizations.

Acknowledgements

This work was supported by the National Science Foundation on Grant Number GK-36489 and by the Emulsion Polymers Industrial Liaison Program at Lehigh University.

Literature Cited

(1) Smith, W. V. and R. H. Ewart, J. Amer. Chem. Soc. (1947), 69, 1428.

(2) Gershberg, D. B. and J. E. Longfield, paper presented at Symposium on Polymerization Kinetics and Catalyst Systems: Part I, 54th A.I.Ch.E. Meeting, New York (1961), Preprint No. 10.

(3) DeGraff, A. W. and G. W. Poehlein, J. Poly. Sci., A-2 (1971), 9, 1955.

(4) Stockmayer, W. H., J. Poly. Sci. (1957), 24, 314.

(5) Stevens, J. D. and J. O. Funderburk, Ind. Eng. Chem. Process Res. Develop. (1972), 11, 360.

(6) Priest, W. J., J. Phys. Chem. (1952), 56, 1077.

(7) Gerrens, H. and K. Kuchner, Br. Poly. J. (1970), 2, 18.

(8) Berrens, A. R., J. Appl. Poly. Sci., Chem. Ed. (1974), 12, 251.

(9) Gonzalez P., R. A., M.S. Thesis, Dept. of Chem. Eng., Lehigh University (1974).

(10) Senrui, S., A. Kodama and M. Takehisa, J. Poly. Sci. (1974), 12, 2403.

(11) Greene, R. K., Ph.D. Dissertation in Preparation, Dept. of Chem. Eng., Lehigh University (1975).

(12) Yeliseyeva, V. I., Acta. Chim. Acad. Sci. Hung. (1972), 71:4, 465.

(13) Yeliseyeva, V. I., S. A. Petrova and A. V. Ziukov, J. Poly. Sci., Sym. No. 42 (1973), 63.

(14) Furuta, M., J. Poly. Sci., Poly. Letters Ed. (1973), 11, 113.

(15) Friis, N. and A. E. Hamielec, J. Poly. Sci., Poly. Chem. Ed. (1974), 12, 251.

(16) Friis, N. and L. Hyhogen, J. Appl. Poly. Sci., (1973), 17, 2311.

(17) Ley, G. J. M., C. Schneider and D. O. Hummel, J. Poly. Sci., Part C (1969), 27, 119.

(18) Stannett, V., M. Litt and R. Patsiga, J. Poly. Sci., A-1 (1970), 8, 3607.

(19) Matsuura, T., and M. Kato, Chem. Eng. Sci. (1967), 22, 171.

(20) Gerrens, H., K. Kuchner and G. Ley, Chem. Ing. Tech. (1971), 43, 693.

(21) Sherwin, M. B., S. Katz and R. Shinnar, A.I.Ch.E. J. (1967), 13, 1141.

(22) Thomas, M. W. and W. C. Mallison, Petrol. Refin. (1961), 5, 211.

(23) Gerrens, H., Kolloid-Zeitschrift and Zeitschrift for Poly. (1968), 227:1-2, 93.

(24) Ley, G. and H. Gerrens, Unpublished Preprint (1974).

Structural Framework for Modelling Emulsion Polymerization Reactors

K. W. MIN and W. H. RAY

Department of Chemical Engineering, State University of New York at Buffalo, Buffalo, N. Y. 14214

Although emulsion polymerization has been carried out for at least 50 years and has enormous economic importance, the detailed quantitative behavior of these reactors is still not well understood. For example, there are many more mechanisms and phenomena reported experimentally than have been incorporated in the existing theories. Considerations such as non-micellar particle formation, non-uniform particle morphologies, polymer chain end stabilization of latex particles, particle coalescence, etc. have been discussed qualitatively, but not quantitatively included in existing reactor models.

Our purpose in this paper is to present a general modelling framework capable of including these and other possible mechanisms and testing their effect on the model predictions. We shall illustrate the application of this framework through the modelling of a polymethylmethacrylate emulsion polymerization reactor.

The Mathematical Model

As a means of beginning our discussion, let us choose to illustrate the model with a rather standard free radical kinetic mechanism:

Initiation:

$$I \xrightarrow[k_d]{} 2R$$

$$R+M \xrightarrow[k_i]{} P_1$$

Propagation:

$$P_n + M \xrightarrow[k_p]{} P_{n+1} \qquad n \geq 1$$

Chain Transfer:

$$P_n + M \xrightarrow[k_{fm}]{} P_1 + M_n \qquad n \geq 1$$

$$P_n + Tr \xrightarrow[k_{ft}]{} P_1 + M_n \qquad n \geq 1$$

Termination:

$$P_n + P_m \xrightarrow{k_{tc}} M_{n+m} \qquad n,m \geq 1$$

$$P_n + P_m \xrightarrow[k_{td}]{} M_n + M_m \qquad n,m \geq 1$$

We should emphasize that we choose this mechanism only as an example; other mechanisms can be treated in a similar way.

The structure of the mathematical model is shown in Figure 1 where we have divided the model up into general balances, aqueous phase balances, individual particle balances, and particle size distribution balances - all of which exchange information with each other. To give an example of the form of the particle size distribution balances let us consider the total particle size distribution, $F(V,t)$.

For a given set of mechanisms the particle size distribution takes the form:

$$\frac{\partial F(V,t)}{\partial t} + \frac{\partial}{\partial V}(\bar{i}F(V)\mu)$$

[change of total number of particles with time][rate of change by volume increase of growing particles]

$$= \int_0^V k_c(V-v,v,\sigma_p)F(V-v)F(v)\,dv - F(V)\int_0^\infty k_c(V,v,\sigma_p)F(v)\,dv$$

[rate of formation of particles of volume V to V + dV by coalescence] [rate of disappearance of particles of volume V to V + dV by coalescence]

$$+ k_c(V-V_m,V_m,\sigma_p)mF(V-V_m) - k_c(V,V_m,\sigma_p)mF(V)$$

[rate of change of population by coalescence between micelles and polymer particles]

$$+ k_{01} \sum_{n=1}^{\infty} n([M_n]_w)\delta(V-V_0) + \frac{1}{\theta_r}\{F(V,t)|_{feed} - F(V,t)\}$$

[rate of new particle formation of volume V_0 by oligomer precipitation in aqueous phase] [rate of population change by inflow and outflow]

$$+ k_{mm}A_m([R]_w + [P]_w)\delta(V-V_m) \qquad (1)$$

[rate of new particle formation of volume V_m by radical entry into micelles]

Similar partial-differential-integral equations can be written down for the

$f_i(V,t)$ - radical number distribution

$\hat{f}_n(i,V,t)$ - growing polymer MWD

$G_n(i,V,t)$ - dead polymer MWD

etc.

These represent the most general (and most complex) form of the modelling framework. A very detailed description of the modelling equations and mechanisms can be found in (1). An illustration of some of the detailed information made possible by this model is given in Figure 2.

Very often one does not require as much detail as presented in Figure 2 and the model can be simplified considerably. For example, one may only be interested in the first few moments of the latex particle size distribution, F(V,t) so as to get a mean and variance of the distribution. This can be readily calculated from the definition of the jth moment:

$$\overline{F}_V^{[j]}(t) = \int_0^{\infty} (V)^j\, F(V,t)\, dV \qquad (2)$$

which when combined with Eq'n (1), leads to

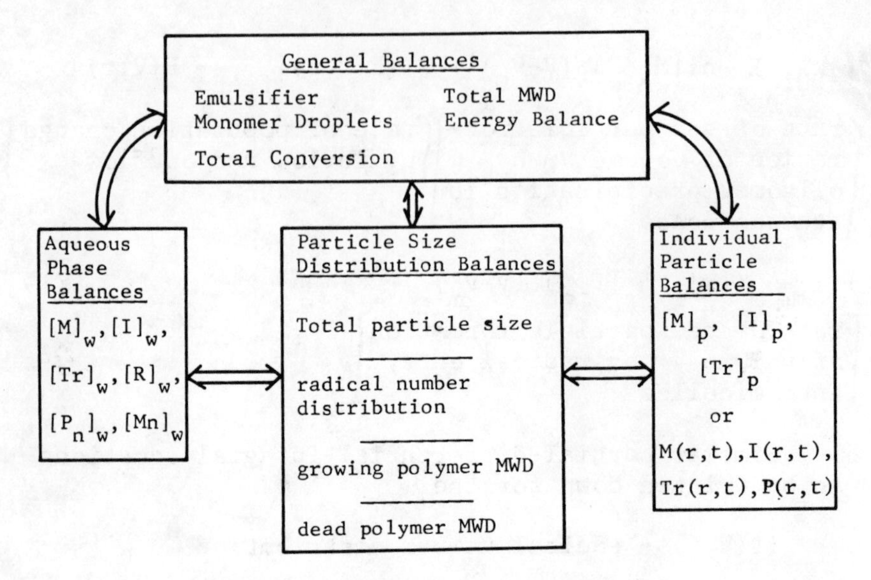

Figure 1. Structure of the mathematical model

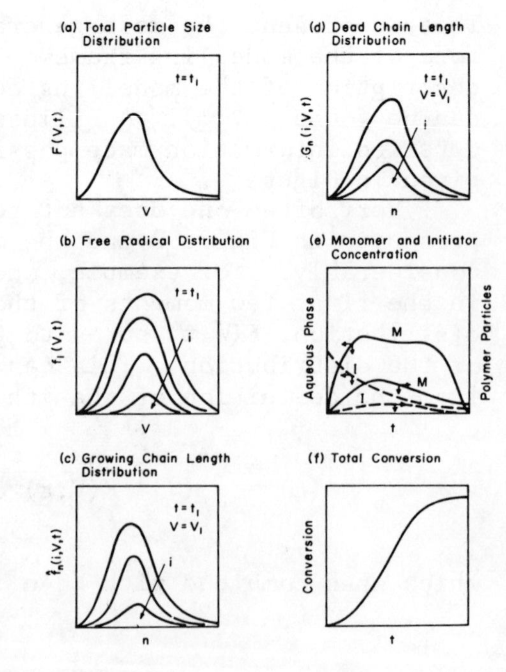

Figure 2. Some possible model predictions

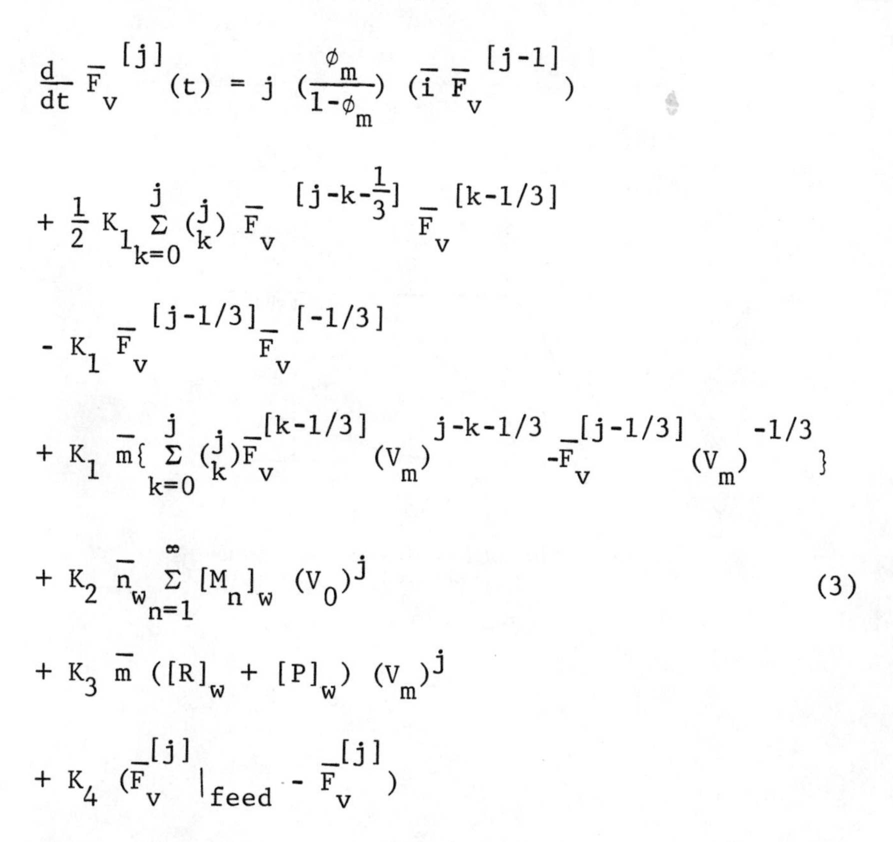

$$
\begin{aligned}
\frac{d}{dt}\bar{F}_v^{[j]}(t) = {} & j\left(\frac{\phi_m}{1-\phi_m}\right)\left(\bar{i}\,\bar{F}_v^{[j-1]}\right) \\
& + \frac{1}{2}K_1\sum_{k=0}^{j}\binom{j}{k}\bar{F}_v^{[j-k-\frac{1}{3}]}\,\bar{F}_v^{[k-1/3]} \\
& - K_1\,\bar{F}_v^{[j-1/3]}\,\bar{F}_v^{[-1/3]} \\
& + K_1\,\bar{m}\left\{\sum_{k=0}^{j}\binom{j}{k}\bar{F}_v^{[k-1/3]}(V_m)^{j-k-1/3} - \bar{F}_v^{[j-1/3]}(V_m)^{-1/3}\right\} \\
& + K_2\,\bar{n}_w\sum_{n=1}^{\infty}[M_n]_w\,(V_0)^j \qquad (3) \\
& + K_3\,\bar{m}\left([R]_w + [P]_w\right)(V_m)^j \\
& + K_4\left(\bar{F}_v^{[j]}\Big|_{feed} - \bar{F}_v^{[j]}\right)
\end{aligned}
$$

where the fractional moments arise because of the form of the coalescence mechanism assumed. Similar moment equations can be written down for the other distributions.

Some Example Calculations

Although a general purpose computer program for numerically solving the modelling equation is still under development, we can provide some example calculations of the total particle-size distribution to illustrate the results. We shall use the kinetic parameters characteristic of the batch polymerization of methyl-methacrylate at 45°C. In Figure 3 are plotted the first few moments of $F(V,t)$ as calculated from Eq'n (3). By expressing $F(V,t)$ in terms of the moments through a series of Laguerre polynominals (2), we can obtain the explicit particle size distribution

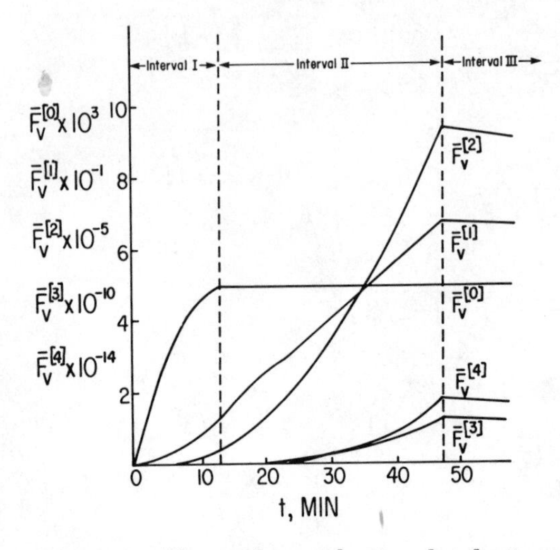

Figure 3. The total particle size distribution moments

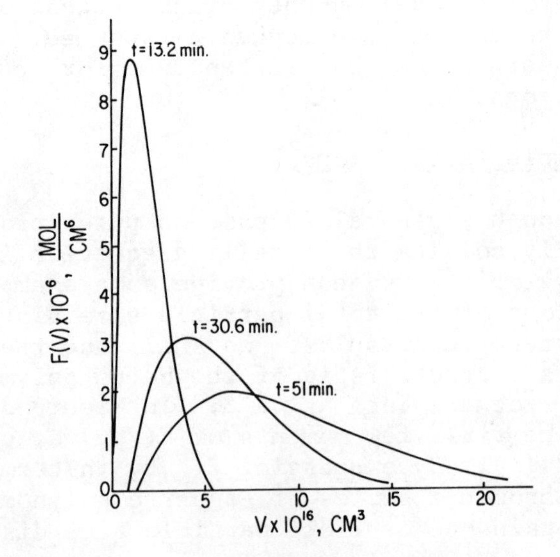

Figure 4. The total particle size distribution

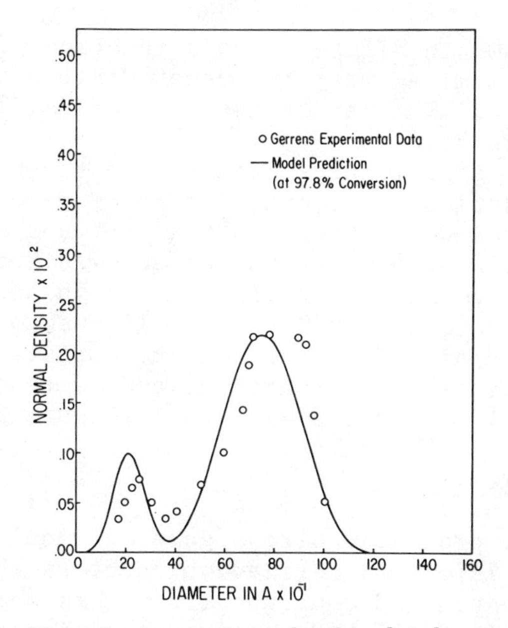

Figure 5. An experimental and predicted particle size distribution

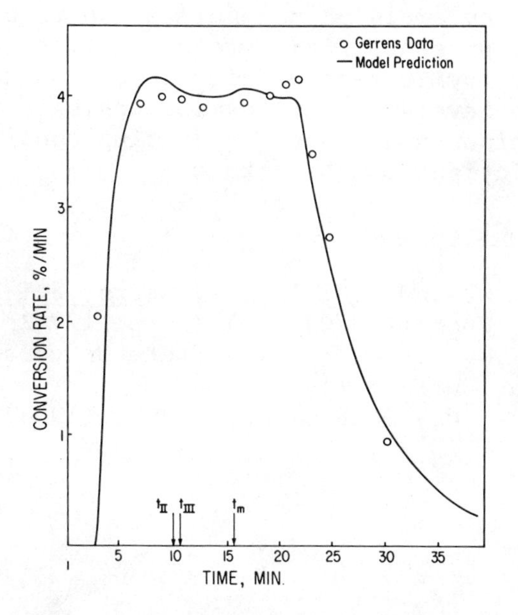

Figure 6. Experimental and predicted rates of conversion

(such as shown in Figure 4) as a function of reaction time. In a similar way, the other distributions can be calculated. The reader is referred to (2) for more complete details.

Gerrens' experimental data (3) on the particle size distribution and conversion rate are compared with our model predictions in Figure 5 and Figure 6. In these calculations the gel effect was included and produces the constant conversion rate in Interval III. Micelle regeneration in Interval III due to the overcrowding of chain end groups on the surface of particles was taken into account and is responsible for the small particle size peak in Figure 5.

Concluding Remarks

We have presented here a general modelling framework for emulsion polymerization reactors which can treat both batch and continuous reactors and includes all previously published models as special cases. Viewing emulsion polymerization through this framework allows one to see the problem in its entirety before proceeding to make simplifying assumptions. The development of a very general computer program for simulating any emulsion polymerization reactor is underway at present. This program will have the ability to provide reactor simulations at several information levels so that by specifying a few modelling assumptions and reactor operating conditions, model predictions can be obtained.

Literature Cited

(1) Min, K.W. and W.H. Ray, J. Macro, Sci-Revs. Macro. Chem. (1974), C11 (2) p. 177-255.

(2) Min, K.W., Ph.D. Thesis, State University of New York at Buffalo (1975).

(3) Gerrens, H., Dechema-Monograph, (1964), 49 (53).

24

Continuous Emulsion Polymerization of Styrene in a Tubular Reactor

MAINAK GHOSH and T. H. FORSYTH†

The University of Akron, Akron, Ohio 44325

Polystyrene can be easily prepared by emulsion or suspension techniques. Harkins (1), Smith and Ewart(2) and Gardon (3) have described the mechanisms of emulsion polymerization in batch reactors, and the results have been extended to a series of continuous stirred tank reactors (CSTR)(4). Much information on continuous emulsion reactors is documented in the patent literature, with such innovations as: use of a seed latex (5), use of pulsatile flow to reduce plugging of the tube (6), and turbulent flow to reduce plugging (7). Feldon (8) discusses the tubular polymerization of SBR rubber with laminar flow (at Reynolds numbers of 660). There have been recent studies on continuous stirred tank reactors utilizing Smith-Ewart kinetics in a single CSTR (9) as well as predictions of particle size distribution (10). Continuous tubular reactors have been examined for non-polymeric reactions (11) and polymeric reactions (12,13).

The objective of this study was to develop a model for the continuous emulsion polymerization of styrene in a tubular reactor, and to verify the model with experimental data.

Experimental

Figure 1 shows the equipment used. The tubular reactor was 240 ft (73m) long, 0.5 inch (1.27cm) OD, Type 316 stainless steel. The reactor was placed in an agitated, constant temperature water bath. Two gear pumps were used to give metered flow of the two feed streams- an emulsion of styrene in an equal volume of water, and a solution of potassium persulfate in water. Table 1 shows the recipe used for polymerization.

† Author to whom correspondence may be directed.

Table I

Polymerization Recipe

Component	Standard	Range Studied
Uninhibited Styrene %	33	33
De-Ionized Water %	67	67
Sodium Lauryl Sulfate (Technical) %*	0.8	0.4-1.2
Potassium Persulfate (Reagent) %*	1.0	0.21-1.5
Temperature °C	60.	40 - 90
Total Flow Rate, cc/min	3.0	0.5-5.1
Residence Time, min	60	10 - 102
Viscosity Average Molecular Weight	$2.0 \cdot 10^6$	$0.7\text{-}8.1 \cdot 10^6$

* wt % based on water

Unsteady start-up behavior has been reported in some continuous reactors (14). In this work, the reactor was initially filled with water or partially polymerized latex, and the % conversion asymptotically approached a constant steady-state value, which required less than twice the residence time in the reactor.

In early tests, the temperature of the emulsion was measured by thermocouples. The thermocouples were removed for later tests, because no significant exotherm was observed and because tube plugging sometimes occurred near the thermocouple probe.

Plugging of the reactor occurred when soap concentration was less than 0.4% and at high temperatures. Plugging was often preceded by a pulsating flow of two-phase liquid. Intensely agitating the emulsion feed in the storage tank helped in preventing plugging also.

The per cent conversion was measured after short-stopping with 200 ppm hydroquinone, by coagulating in isopropanol, and then drying to constant weight. The molecular weight was obtained from the viscosity of a 10% solution of coagulated polymer in toluene, as measured in an Ostwald Capillary Viscometer (size 50). The per cent conversion was reproducible to within 15%, while the molecular weight was reproducible to within 50%.

Mathematical Model

The flowing emulsion was assumed homogeneous, so that the continuity equations could be used. Additional assumptions were: the fluid is an incompressible Newtonian with constant properties; the flow is laminar at the maximum experimental Reynolds number of 210 and less; there is negligible viscous heating; flow is at steady state with no entrance effects or radial velocity components; body forces are neglected; axial heat conduction is small compared to radial conduction; Region I of Smith-Ewart kinetics (i.e., when micelles are first forming) is neglected and the initiator concentration is constant. The model may be summarized as:

$$V_z = \bar{V}_z[1-(r/R^2] = \frac{Q}{\pi R^2}[1-(\frac{r}{R})^2] = \frac{\Delta P}{L}\frac{R^2}{8\mu}[1-(\frac{r}{R})^2] \qquad (1)$$

$$\rho C_p \bar{V}_z[1-(r/R)^2]\frac{\partial T}{\partial z} = \frac{k}{r}\frac{\partial}{\partial r}[r\frac{\partial T}{\partial r}] + \Delta H R_p M \qquad (2)$$

$$2\bar{V}_z[1-(r/R^2]\frac{\partial C}{\partial z} = \frac{D}{r}\frac{\partial}{\partial r}[r\frac{\partial C}{\partial r}] - R_p\frac{M}{\rho} \qquad (3)$$

$$N = C(C_S)^{0.6}(C_I)^{0.4} \qquad (4)$$

$$R_p = \frac{[M]N}{2N_a} A \exp(-E/R_g T) \qquad (5)$$

where [M] is the concentration of monomer in the swollen droplets and is constant up to 50% conversion. Between 50 and 100% conversion [M] is the bulk monomer concentration.

Further combinations of terms simplify the above equations and make it possible to analyze in terms of dimensionless numbers. The energy equation is:

$$(1-y^2)\frac{\partial T}{\partial z} = \frac{G_1}{R}[\frac{\partial^2 T}{\partial y^2} + \frac{1}{y}\frac{\partial T}{\partial y}] + \frac{G_2 T_w}{R}\exp[-\frac{G_3 T_w}{T}] \qquad (6)$$

where: $y = r/R$

$$G_1 = \frac{k}{2\bar{V}_z \rho C_p R} = \frac{1}{(\text{Reynolds})(\text{Prandtl})}$$

$$G_2 = \frac{(\Delta H)(N)(A)(R)}{2\bar{V}_z(C_p)(N_a)(T_w)}[C]$$

$$G_3 = \frac{-E}{R_g T_w}$$

The equation for monomer concentration becomes:

$$(1-y^2)\frac{\partial C}{\partial z} = \frac{G_4}{R}[\frac{\partial^2 C}{\partial y^2} + \frac{1}{y}\frac{\partial C}{\partial y}] - \frac{G_5}{R}\exp - \frac{G_3 T_w}{T} \qquad (7)$$

where: $G_4 = \frac{D}{2\bar{V}_z R} = \frac{1}{(\text{Reynolds})(\text{Schmidt})}$

$$G_5 = \frac{(N)(A)(R)[C]}{2\ \bar{V}_z\ N_a}$$

Solution of the differential equations was by Gauss-Seidel iteration (with the fluid property values given in Table II) on an IBM 370 digital computer using implicit difference equations of the Crank-Nicholson type. The program was convergent and stable for all conditions tested.

Table II

Constants Used in Computer Program for Styrene

A	cm^3/mole-sec	$2.24 \cdot 10^{17}$
[C]	wt fraction	0.572
C_p	cal/gm-°C	1.0
D^p	cm^2/sec	10^{-6}
E	cal/gmole	17,570
ΔH	cal/gm	160.6
k	cal/sec-cm-°C	0.0015
M	gm/gmole	104
N_a	molecules/mole	$6.02 \cdot 10^{23}$
R	cm	0.546
Rg	cal/gmole-°K	1.987
μ	gm/cm-sec	0.01
ρ	gm/cm	1.0

Using Gardon's (3) expression for N, the number of particles per unit volume, gave conversion rates two orders of magnitude larger than were experimentally observed. To obtain accurate estimates of the conversion, N was empirically evaluated (at about 10^{14} particles/cm^3) from one data set at each temperature.

Results and Discussion

Experimental. Figure 2 compares molecular weight data reported by Gardon (10) for batch reactors and by Poehlein for CSTR reactors (5), with the data obtained in this study for a tubular reactor. The solid lines are predicted by Gardon's theory (10). The molecular weights obtained in this experimental study were predicted within a factor of 3 by Gardon's theory. No direct comparison can be made with the data of other workers, yet this molecular weight data is consistent (at least within experimental error) with data obtained in other types of reactors.

Figure 3 compares the experimental rate of monomer loss in Gardon's batch reactor, the tubular reactor, and

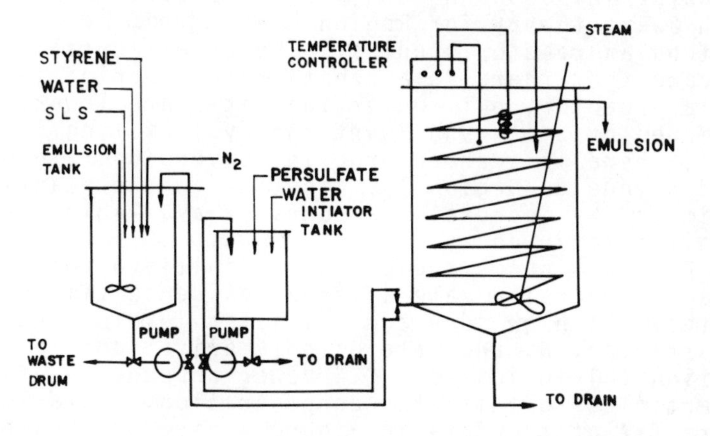

Figure 1. Sketch of equipment used for the continuous emulsion polymerization of styrene in a tubular reactor

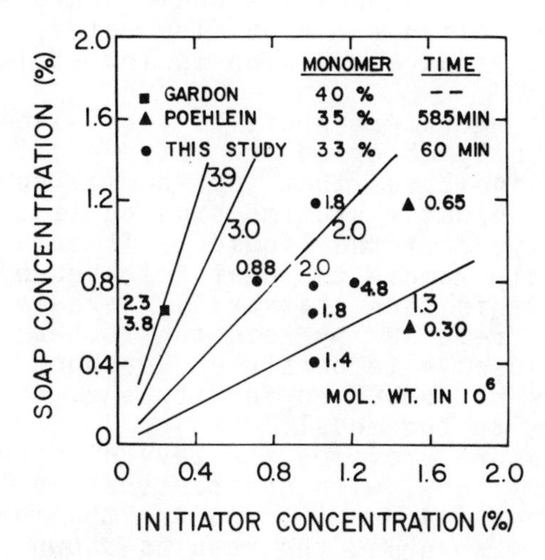

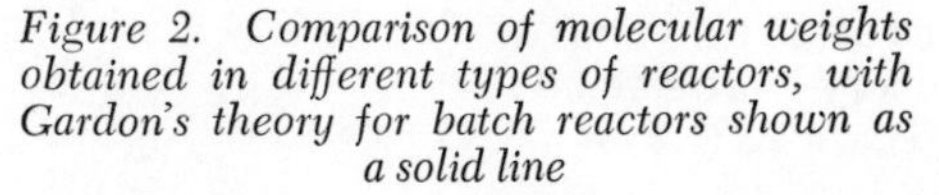

Figure 2. Comparison of molecular weights obtained in different types of reactors, with Gardon's theory for batch reactors shown as a solid line

two CSTR. Also shown is a predicted curve based on Smith Ewart theory for Region II for standard conditions of the tubular reactor studies. The experimental results indicate that there is a constant rate period in the tubular reactor, but conversion rates are lower than predicted by Smith and Ewart theory. At long residence times, there is a sharp drop in conversion rate which could be due to Region III kinetics (where reaction rate is limited by decreasing amounts of available monomer) or to initiator decay.

Figure 4 shows the effect of initiator on the average conversion rate after a residence time of 60 minutes. At high soap and initiator levels, the number of particles, N, and rate of polymerization are high. Equation (4) indicates a 0.4 power dependency of number of particles to initiator concentration, and a least square fit of the data in Figure 4 gave this same dependency for rate of polymerization.

Model Figure 5 shows the effect of the dimensionless variable, G_1, on conversion at different positions down the tube. The curve marked X is for the standard conditions of Table 1. It can be seen that there is little change in conversion as a function of G_1 if heat conduction in the fluid is high. Figure 5 shows that the temperature for curves B and X reach an asymptote slightly higher than the wall temperature, such that heat generation is numerically equal to heat transfer. When the rate of reaction decreases at higher conversion, the average fluid temperature would decrease to 60°C. With high convection (or high flow rates) the fluid is not heated, so that conversion is low at low values of G_1.

Radial temperature profiles were almost constant as experimentally observed and later predicted by the model. This indicates that the thermal conductivity for the styrene/water systems studied is sufficiently high to quickly heat the liquid as it enters the tube, and efficiently remove the heat released by the reaction. In Figure 6 it is seen that as G_2 increases, a high ratio of heat released by reaction to heat removed by convection causes a temperature overshoot. Such an overshoot was not observed for styrene, either experimentally or with the model. Figure 7 presents predicted axial temperature profiles for several monomers at standard conditions, with constants as in Table III. While equations (4) and (5) require the monomer to be only partially soluble, the results shown in Figure 7 indicate a strong exotherm for vinyl chloride, acrylonitrile and vinyl acetate.

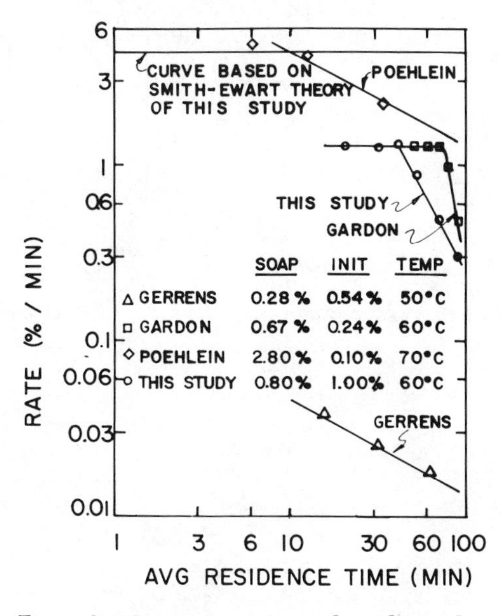

Figure 3. Comparison of rate data obtained in different types of reactors

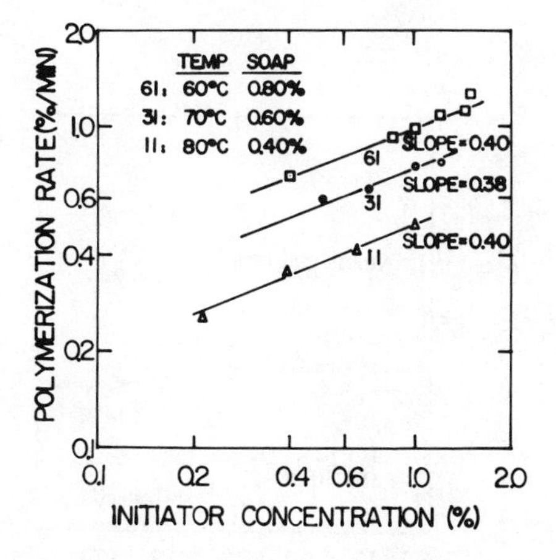

Figure 4. The dependency of reaction rate on initiator concentration

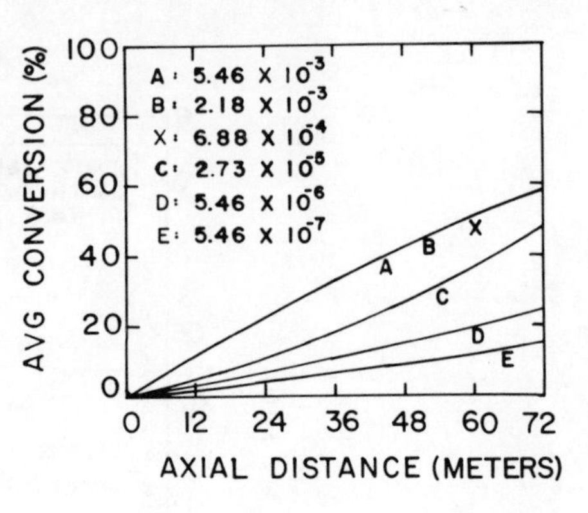

Figure 5. Computed axial conversion profile with changes of variable G_1 (ratio of conductive to convective heat transfer)

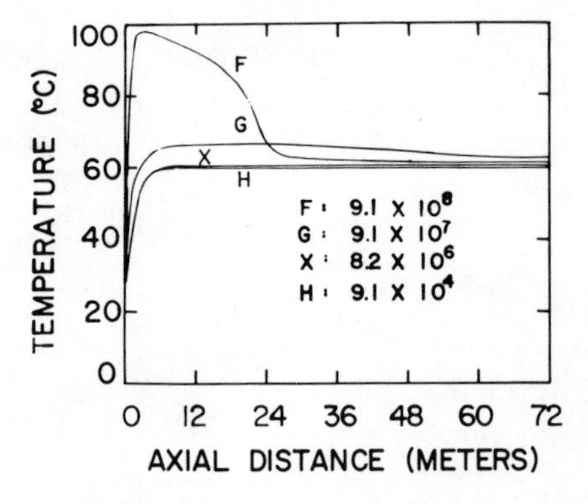

Figure 6. Computed axial temperature profile with changes of variable G_2 (ratio of heat released by polymerization to convective heat transfer)

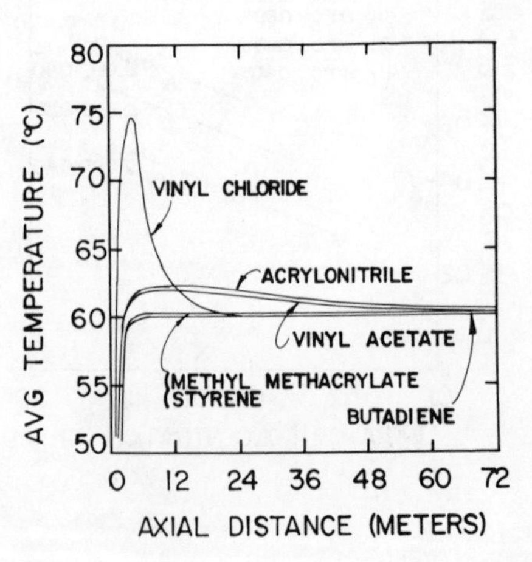

Figure 7. Computed axial temperature profiles for different monomers that can be emulsion polymerized

Table III

Polymerization Constants for Several Monomers Commonly Prepared by Emulsion Polymerization (15)

Monomer	Heat of Polymerization ΔH	Frequency Factor A	Activation Energy E
Methyl Methacrylate	132.0	$8.7 \cdot 10^{8}$	4700
Styrene	160.6	$2.2 \cdot 10^{17}$	17570
Vinyl Acetate	246.5	$3.2 \cdot 10^{10}$	6300
Butadiene	326.0	$1.2 \cdot 10^{11}$	9300
Acrylonitrile	340.7	$7.6 \cdot 10^{8}$	3900
Vinyl Chloride	366.4	$3.3 \cdot 10^{9}$	3700

Figures 8-10 show the effect of changes in G_3, G_4 and G_5 on average conversion. Small changes in G_3, the activation energy group, lead to large changes in conversion. Larges changes in G_4, the ratio of mass transfer by Diffusion to mass transfer by convection, cause small changes in conversion. Although monomer diffusion to the polymerizing particle is required, diffusion of the polymer is effectively non-existent for styrene, so that the first term on the right of Equation (3) could have been neglected. Changes in G_5, the ratio of mass reacted to convective mass transfer, does affect conversion strongly.

Comparison. Figure 11 shows the experimentally measured average conversion, and average conversion predicted by Equation 1 to 5. At low conversions, the experimentally measured values are low, probably because of significant Region I effects (i.e. micelle formation occurring, so that the actual particle number is lower than predicted by the model). Figure 12 shows the effect of wall temperature on experimental and predicted conversion rate, and the greatest error is again at low conversion. There is significant error at 80°C, too, probably because of colloidal instability at high temperatures.

Conclusions

This study on the continuous tubular emulsion polymerization of styrene leads to the following conclusions:

1. Although tube plugging occurred at other conditions, the reactor could be operated continuously at 50-80°C, with high initiator and soap concentrations.

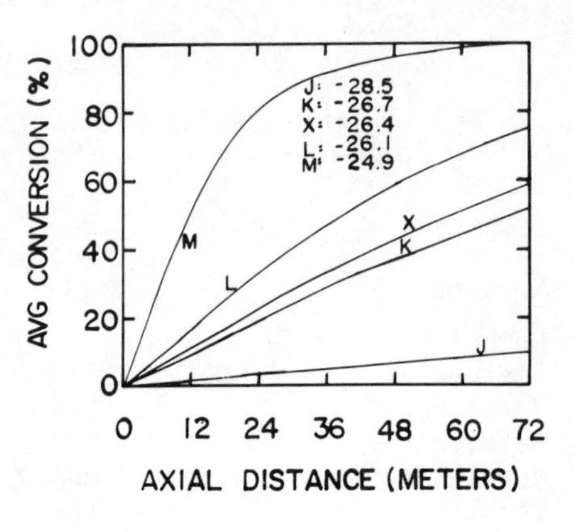

Figure 8. Computed axial conversion profiles for changes of variable G_3 (activation energy group)

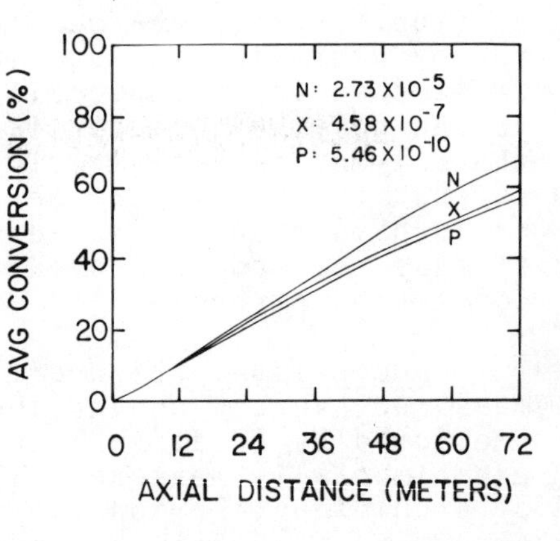

Figure 9. Computed axial conversion profiles for changes of variable G_4 (ratio of mass diffusivity to mass transfer by fluid motion)

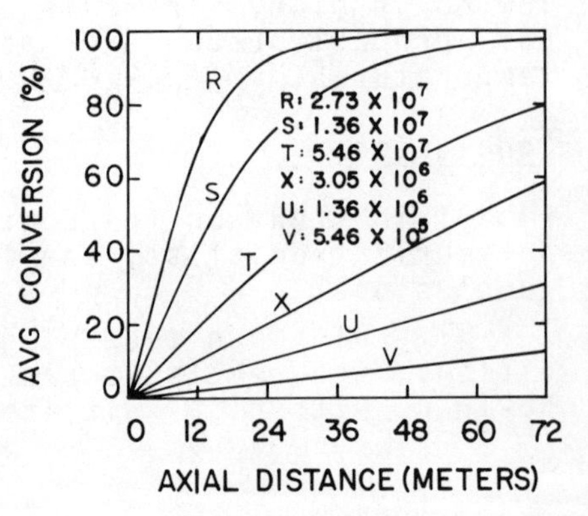

Figure 10. Computed axial conversion profiles for changes of variable G_5 (ratio of reaction ratio to mass transfer by fluid motion)

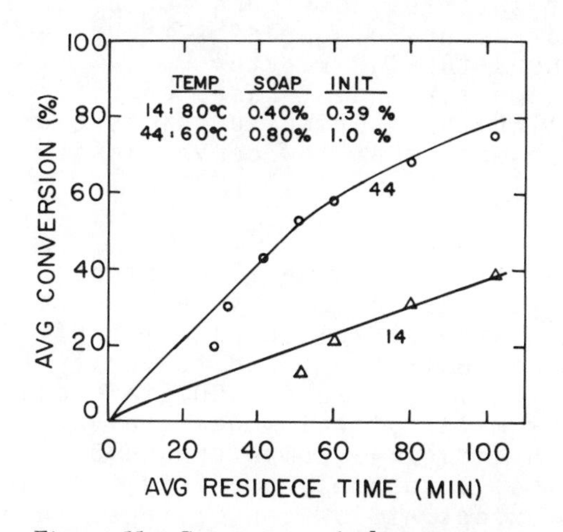

Figure 11. Comparison of the conversion predicted by the mathematical model with the experimental data at different times in the reactor

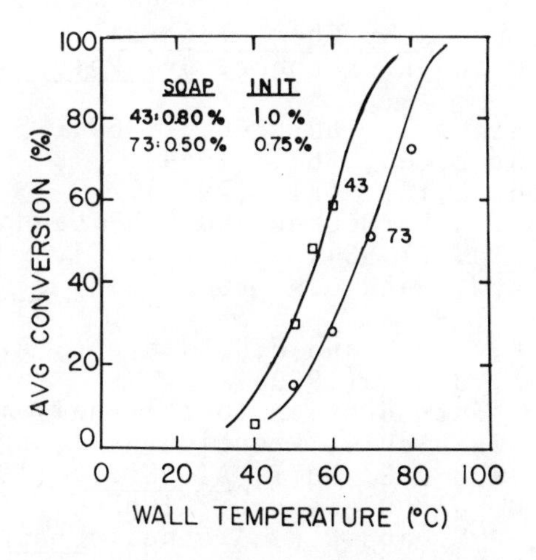

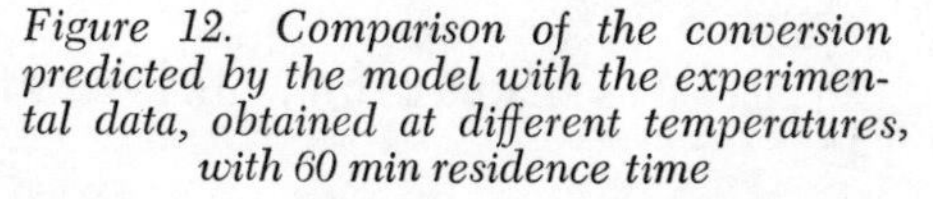

Figure 12. Comparison of the conversion predicted by the model with the experimental data, obtained at different temperatures, with 60 min residence time

2. The theories developed for batch and CSTR reactors do not accurately predict the rate data obtained in a continuous tubular reactor.

3. A computer model, based on Smith-Ewart kinetics and the continuity equations predicts experimental conversion data, except at low conversions.

Abstract

The objective was to develop a model for continuous emulsion polymerization of styrene in tubular reactors which predicts the radial and axial profiles of temperature and concentration, and to verify the model using a 240 ft. long, 1/2 in. OD Stainless Steel Tubular reactor. The mathematical model (solved by numerical techniques on a digital computer and based on Smith-Ewart kinetics) accurately predicts the experimental conversion, except at low conversions. High soap level (1.0%) and low temperature (less than 70°C) permitted the reactor to perform without plugging, giving a uniform latex of 30% solids and up to 90% conversion, with a particle size of about 1000 Å and a molecular weight of about 2 x 10^6.

Literature Cited

1. Harkins, W.D., J. Am. Chem. Soc., (1947), 69, 1428.
2. Smith, W.V., and R.H. Ewart, J. Chem. Phys., (1948) 16, 592.
3. Gardon, J.L., Rub. Chem. Technol., (1970), 43, 74.
4. Gerrens, H and K. Kuchner, Br. Poly. J., (1970), 2, 18.
5. British Patent 1, 168, 760, (1969).
6. U.S. Patent 2,465, 363, (1949).
7. U.S. Patent 2,161, 481, (1939).
8. Feldon, M., R.F. McCann and R.W. Laundrie, India Rubber World, (1953), 128, (1), 51.
9. DeGraff, A.W., and G.W. Poehelein, J. Pol. Sci.-A-2, (1971), 9, 1955.
10. Funderburk, J.O., and J.D. Stevens, AIChE, reprint, Washington, D.C., (1969).
11. Aris, R., "Introduction to the Analysis of Chemical Reactors, Chapter 9, Prentice Hall, New York, 1965.
12. Lynn, S., and J.E. Huff, AIChE J., (1971), 17, 475.
13. Smith, J.M. and D. Sandru, AIChE J., (1973), 19, 558.
14. Poehlein, G.W. and R.K. Greene, J. Appl. Pol. Sci., in press.
15. M. Ghosh, Ph.D. Dissertation, University of Akron, 1974, p. 146.

25

Modification of Polystyrene with Polybutadiene and the Method of Studying Morphology of the Obtained Multiphase High Impact Systems

V. D. YENALYEV, N. A. NOSKOVA, and B. V. KRAVCHENKO

Donetsk State University, Donetsk, 340055, U.S.S.R.

High Impact Polystyrene (HIPS) obtained by means of copolymerization of styrene with rubber represents heterogeneous system consisting of polystyrene matrix and the particles of rubber phase dispersed in it; the particles in their turn keep the graft copolymer and a great number of occluded polystyrene (I). Physical and mechanical properties of HIPS are defined by its morphology (2).

Detailed investigation of mechanism of formation of HIPS microstructure and its properties is done in the works of many researchers: the influence of the agitation speed (3,4) graft copolymer (1,5), the degree of grafting (6), thickness of the intermediate layer (7). It was shown that mechanical properties of HIPS depend in the size of rubber particles and the increase of rubber phase volume raises rubber efficiency (2,8). In the work (9) the attempt was made to find the connection between dynamic-mechanical properties, rubber concentration and impact strength of different commercial samples of HIPS. However size of low-temperature damping peak in general does not appear to be related to the actual degree of toughness, except within a fairly narrow family of similarly prepared polymers.

In literature the following methods of preparation of polymer material specimen and studying their morphology are described: methods of ultrathin section and films with contrasting of osmium tetroxide (10,11), method of replication of the brittle fractured surface (11,12) oxygen and chemical etch of the polished surface or the fractured surface with the following replication for electron microscopy (11, 14).

The most spread methods for studying HIPS are the one of ultrathin section and films as well as the replica technique (without etch). These methods supplement one another to a certain degree, but give limited information regarding HIPS morphology. At the section when contrasting the structure of rubber phase is seen well, but the information of polystyrene matrix structure is inevitably lost. When replic of films and section is used the structure distortion is out, nevertheless identification of rubber and polystyrene phases of HIPS becomes more difficult because of strongly developed polystyrene matrix relief. Besides more often than not changing of rubber particles forms takes place (their "smoothing") while being prepared.

The present paper deals with working out electron microscope method of HIPS studying, less difficult than above mentioned but letting get fuller information about material morphology; in choosing the way of formalized quantitative interpretation of the obtained structure; in establishing dependence between morphology of HIPS and their physical and mechanical properties as well as in making exact the mechanism of microstructure formation.

Investigation of HIPS morphology.

For working out a new electron microscope method of studying HIPS microstructure selective swelling of different parts of HIPS being influenced by the vapour of the solvents was used, selectively influenced either the rubber or polystyrene phases. Compression-molded specimens of HIPS obtained by block copolymerization of styrene with polybutadiene synthetic rubber in the form of bar in the size of 100x6x4 mm, were cooled in liquid nitrogen for 30 minutes, then with the help of a knife made a fractured surface and after heating specimen to the room temperature they were placed above the solvent (at 2-3 cm from the surface distance) for 0.5, 1.2, 3.5, 10 and 15 minutes. From the worked in this way fractured surface carbon-palladium replic was made in a usual manner (12) and was studied with electron microscope UEMV - 100. For solvents ethyl-acetate, dimethylformamide, doixan, benzene were used. The vapour action of solvents was in relief of polystyrene matrix and polystyrene occlusions becoming smoother (for because of selective swelling) without changing rubber phase structure and as a result of it better revealing rubber particles on the surface of the HIPS fracture. Best results were achieved when the surface was worked out with ethyl acetate for 1 minute, dimethylformamide for 2-3 min., benzene and dioxan for 0.5 min. When time is being prolonged for vapour influence of solvents it gives the contrast diminishing and washing out of rubber particles contours. Not sufficient time of working out with vapour of solvents doesn't smooth out the polystyrene matrix relief and makes difficult microgramms interpreting . Rate of polymer swelling is known to depend on its molecular weight and composition and molecular structure. That's why for achieving best contrast (relief) when working with different copolymers the optimal time of working out with the solvent vapour is to be chosen experimentally for each type of materials.

Microgramms of HIPS morphology revealed without working out with the solvent vapour and by means of above described method are given in figure 1,a,b. When comparing them with the photos obtained with the help of generally known Kato method (10), one can see beside similarity of general picture of rubber distribution also difference: membranes between occluded polystyrene and rubber particles in our picture are much thicker. In our opinion it can be explained by the fact that when being worked out one can see only contracted with osmium-tetroxide rubber and in fig. 1 b membranes between occlusions represent rubber together with the intermediate layer, consisting of graft polymer of poly(styrene-gr-butadiene).

Side by side with good visial qualitative observation of HIPS morphology obtained by means of above described method we applied the method used in metallography (14) and adapted for polymer material for optical microscopy (15) for getting distribution characteristics of rubber phase in the material and quantitative value of microstructure elements in electron microscope pictures.

When applying this method according to micrograms one can calculate: volume fraction of HIPS rubber phase Vf in per centage; Intermediate surface of rubber and polystyrene phases S_V in mm^2/mm^3 ; Mean cord spheres $\bar{c}$ which is proportioned to diameter of rubber particles $\bar{c}$ in μ ; Mean free distance among the particles MFD in μ .

The mentioned structural characteristics of HIPS at electron-microscope micrograms are defined with greater exactness than at optical because only rubber particles located in one surface can be calculated while at phase contrast optical microscope particles located in all volume of the observed film are taken into account.

It was found that for quantitative characteristics of HIPS morphology it is sufficient to make calculation according to 4 micrographs from the square of the surface 20x30 , obtained from different samples of the material.under general increase 5000x. While doing it the relative mistake of defining for all structural characteristics doesn't exceed 9 %.

The above described method was applied for characteristics of HIPS both commercial and obtained in laboratories by means of different methods using different types of polybutadiene rubbers. Microstructure elements of this material are compared to their physical and mechanical properties in table I.

Table I.

Physical and mechanical properties and structural parametres of HIPS being synthesized under different conditions. Rubber contents in all specimens is 5 %.

Polymers	Mechanical properties			Structure properties			
	Notched Izod kg cm/cm^2	Tensible strength kg/cm^2	Elongation at break %	volume fraction of rubber phase V_f, %	Mean cord $\bar{c}$, μ	Mean free distance between particles MFD, μ	Dispersion of mean cord S^2
A	12.o	290	32	21.1	1.4	5.7	2.91
B	11.0	310	30	18.4	1.1	4.6	3.14
C	8.8	300	28	13.1	0.7	7.0	1.14
D	5.9	314	16	10.2	0.6	5.3	1.04
E	5.6	265	22.8	8.7	1.6	19.0	7.2
F	4.5	450	20.0	4.8	1.5	32.0	6.15

The data of table I. witness the qualitative correlation between morphology and properties of HIPS. E.g. regularly toughness increases with the increase of volume fraction of rubber phase. The conclusion made in literature about the dependence of physical and mechanical properties of HIPS on the size of rubber particles (8,16,17) is confirmed.

However the stated optimal size of the particles (1-10 μ) is a necessary but not sufficient condition for providing high properties of the material (polymers E,F, table I). Due to analysis of many data analogous to given in table I. one can come to conclusion that for achieving maximal firmness of HIPS under the minimal rubber contents, i.e. at the least expenditures it is necessary by means of different technological procedures to obtain the material to satisfy the following requirements:

a) optimal size of rubber phase particles (1-3 μ);

b) not a wide distribution of the rubber phase particles on the size- quadratic dispersion of particle diameter must not exceed 4;

c) volume fraction of rubber phase must not be less than 13-15%.

Definition of the residual internal microtension in HIPS.

Besides the known from literature on HIPS morphology elements of microstructure in electron microscope photos obtained by means of the described above method a new phenomenon is obtained which is not seen in the pictures made by Kato method (10) and in the pictures of the fractures which were not worked out by the solvent vapours - on the smoothed surface of polystyrene matrix a net of crazes is clearly seen. These crazes have internal striped structure analogous to microcrack structure which appear in HIPS. When the external tension is applied . The appearance of these crazes on the polymer material surface under the influence of solvents and their vapours is due to existence of residual tension in them (19, 20) which appear during forming specimens or tensions caused by stress.

In our method applying the electron microscope there arose the possibility to observe crazes which appeared under the influence of solvent vapour on the surface of the HIPS fracture that witness the existence and distribution of the residual internal microtension in the material itself, these crazes playing in our mind a great importance in toughning polystyrene by means of rubber and they are not discovered up to the present time.

For clearing out the nature of these microtensions we microscoped HIPS examples according to the above described method: a) before and after extrusion and molding under pressure; b) after annealing for 24 hours at 60°-80° C before and after obtaining fracture; c) after heating with the field of high frequency 10^{10} Hz for 15 minutes. In all the described cases the character of distribution of crazes was not changed. It proves that revealed crazes have no nature of tensions which appear in HIPS while forming specimen disappearing during annealling material as well as they don't appear at the moment of destroying specimen when the surface is being prepared for the replication.

For comparing we microscoped in the described manner the specimen of homopolymer of styrene where there is no rubber. In the not annealled specimens of the material the crazes were formed which were the same as those in HIPS (according to their sight). In those annealled at 80° C for 24 hours crazes don't exist. It shows that the cause for forming residual microtension in HIPS is presence of rubber particles in it.

If the surface of the fracture is worked with the solvent vapours when applying the external stress (strain or compression) in the microgramm a great number of wider crazes appear (fig.2). The external loading was 30 % less than that of destroying one, i.e. at these loadings without the solvent vapours influence don't appear either "whitening" or "silver". However in the HIPS specimens the effect of orientation of crazes perpendicular to the direction of the applied tension doesn't appear. In the homopolystyrene specimen under the influence of solvent vapours a great many of crazes appear orientated perpendicularly towards the direction of the applied tension (fig. 3). In the last case the working out of specimen under stress with the solvent vapour lasted for 1 minute (i.e. less than the minimal time) to define the direction of the applied tension according to the direction of orientation of the microstructure elements on the fractured surface as it is difficult to do in the other way because of many-stage process of making replic from the fracture of polystyrene. All this makes possible to give the statement on the nature of the revealed internal microtensions by means of the above described method. HIPS represents a heterogeneous system in which rubber phase particles and polystyrene matrix are connected together due to "seizure" of grafted chains of copolymer both rubber and homopolystyrene phases. When cooling HIPS different deformation of polystyrene and rubber phases takes place and as a consequence microtensions appear. The obtained results give proof to suppositions made by Schmitt (21) and Sternstein (22) stated on the basis of indirect data. The sizes of tension regions of polystyrene matrix stretch from the interface surface for several micrones into matrix. The external stress applied to

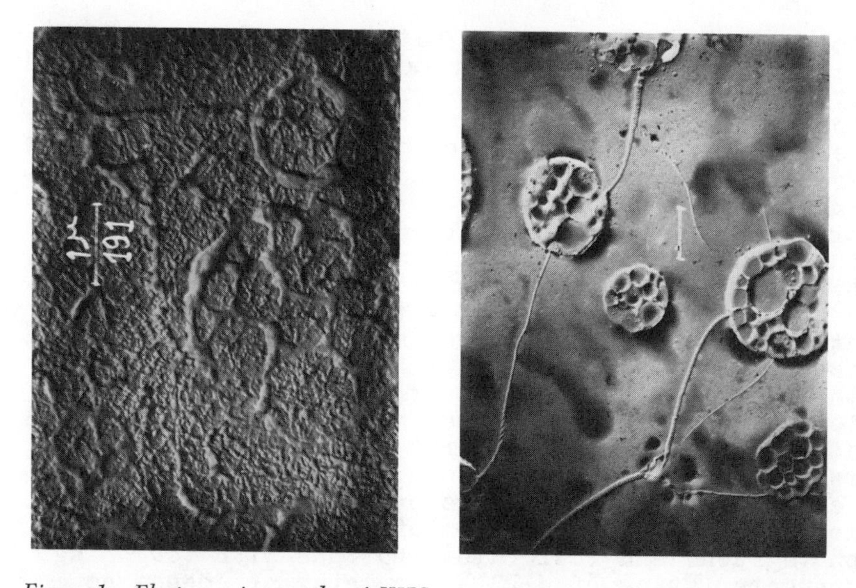

Figure 1. Electron-micrographs of HIPS microstructure. a. (left) fracture without treatment by the vapors of solvent, increase of ×12000. b. (right) fracture, treated by the vapors of ethyl acetate for 2 min, increase of ×7000.

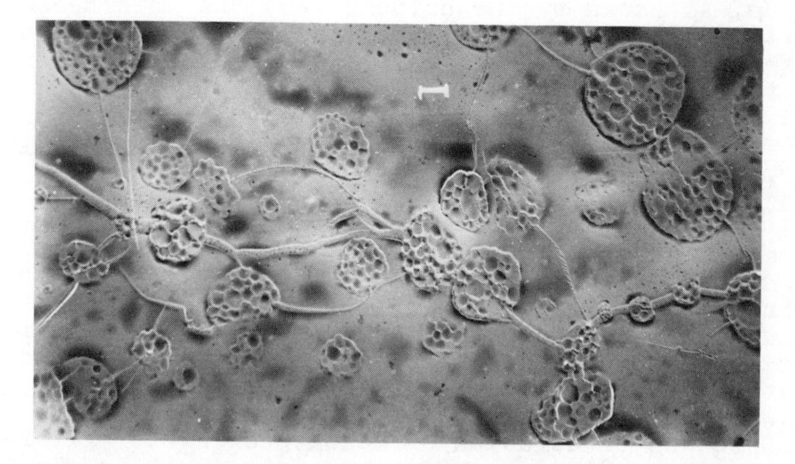

Figure 2. Surface of the HIPS fracture treated by the solvent vapors while applying the external straining stress, increase ×3500

the specimen is summed up with internal microtensions according to the scheme of summing, described by authors (23) and cause formation of a lot of crazes being developed in directions which are not perpendicular to the direction of the applied stress. From this point of view we come to understand cracking in not oriented with respect to the stress applied which was observed by Schmitt and Keskkula by the optical microscope (24) and by the authors (23) by the electron microscope when the material was destroyed being effected by the cycle sign-variable stress. The internal microtensions found by the above mentioned method in HIPS finally determine the distribution of the destroying stress applied to the specimen. Therefore the character of the distribution of these microtensions in the material is of importance in toughning polystyrene by rubber. We find experimental corroboration of the said above in qualitative correlation of sizes of crazes on photographs and their distributions with toughning characteristics of HIPS - the material possesses higher toughness if it has a sufficiently great number of crazes connecting the particles of rubber phase.

Thus it follows that to obtain HIPS with high physico-mechanical properties independent of the manner of its obtaining, it is necessary to create conditions for forming discrete particles of rubber phase, which are at most filled with polystyrene occlusions and which have an optimal external diameter as well as possessing sufficient connection with polystyrene matrix due to the intermediate adhesion layer. As shown by the authors (25) the thickness of this layer essentially depends on the number and size of polystyrene occlusions inside the rubber particles.

The investigation of mechanism of HIPS microstructure formation.

The bases of HIPS microstructure, as it is known, are laid at the stage of prepolymerization, when polymer-polymer emulsion (POO-emulsion) is obtained due to incompatibility of homopolystyrene being obtained and polybutadiene in styrene solution. Due to the fact that the particles of the discrete phase of emulsion and layers of continuous phase between them have sizes not more than tens of μ , the process of defusion of monomer owing to agitating is proceeding at sufficient rate. That is why the composition of polymer emulsion phases approaches the equilibrium. With the purpose of finding the factors influencing the formation of this or that internal structure of rubber phase, the phase equilibria of multicomponent polymer systems of polybutadiene-polystyrene-styrene and polybutadiene-polystyrene-graft copolymer-styrene.

The phase equilibria were studied on model emulsions with various ratios and concentrations of polymers modelling different degrees of copolymerization conversion. The emulsions have been prepared by mixing equal volumes of styrene solutions of ISR polybutadiene rubber "Intene - 55" NFA,(PB) and non-fractionated polystyrene (PS), which has been obtained by free radical polymerization in the presence of benzoyl peroxide(Bz_2O_2) and also by their simultaneous dissolving in styrene. After centrifuging (separating factor G= 5000) polymers from each phase have been separated by precipitation with the help of methylalcohol. Molecular weight of homopolystyrene was determined by viscosity measurement. Studying the process of microstructure forming at prepolymerization stage was carried out with the help of phasecontrast microscopy and by separating polystyrene and by rubber phases by centrifuging POO obtained by copolimerization of styrene with polybutadiene.

While studying polymer distribution between the emulsion phases it was found that in the systems mentioned above obtained both by copolymerization of styrene with polybutadiene rubber and mixing styrene solutions of polymers when the composition is far enough from the critical mixing point, thermodynamic equilibrium is reached. At this thermodynamic equilibrium the ratio of polymer concentration (C_p) in rubber (index') as well as in polystyrene (index '') phases is practically constant (table II).

The state of equilibrium in the investigated system is proved to be true, thermodynamic by the fact that the phase distributing coefficient of polymers does not depend on the following:

1.the way of preparing emulsion: styrene and rubber copolymerization (lines 1,2 of the table II) and the presence of graft copolymer in the system as a result of it; simultaneous dissolving of polymers in styrene (lines 12-14); mixing equal volumes of solutions of PS and PB (the concentration of initial solutions varied widely);

2.Summed concentration of polymers in the emulsion (6 - 32%);

3.Molecular weight of homopolystyrene (lines

Table II.
Equilibrium concentrations of polymers in POO-emulsions polybutadiene-polystyrene-styrene.

№№	Total concentration of polymers in emulsion, % w/w	T°C	C_{PB}/C_{PS} in emulsions	$\bar{M}_v \cdot 10^{-5}$ PS	Type of emulsion*	Concentration of polymers in phases, %, w/w: rubber C'_p	polystyrene C''_p	$K = \frac{C'_p}{C''_p}$
			Polymerization emulsions					
1.	16.0	30	0.4	1.3	PB/PS	13.o	17.6	0.74
2.	8.1	30	1.0	1.1	"	7.4	10.2	0.72
			Model emulsions					
3.	19.2	30	0.1	3.0	PB/PS	15.0	21.4	0.70
4.	12.0	30	0.2	3.0	"	9.3	13.3	0.70
5.	32.0	30	0.3	0.3	"	25.0	34.8	0.72
6.	8.0	30	0.3	3.0	"	6.4	8.9	0.72
7.	12.0	30	0.5	3.0	M	10.3	14.1	0.72
8.	6.0	30	0.5	3.0	"	4.7	6.6	0.71
9.	8.0	30	0.6	3.0	"	6.5	9.5	0.68
10	8.1	30	0.6	2.3	"	6.6	9.5	0.69
11	8.0	30	0.6	0.9	"	6.7	9.2	0.73
12	6.0	20	1.0	2.9	"	5.0	7.3	0.68
13	6.0	40	1.0	2.9	"	5.0	7.3	0.68
14	6.0	60	1.0	2.9	"	5.0	7.3	0.68
15	6.0	30	2.0	3.0	PS/PB	5.4	8.0	0.67
16	7.0	30	2.5	3.0	"	6.4	9.2	0.70

* PB/PS - preinversion emulsion
M - inversion emulsion
PS/PB - post inversion emulsion

9,10,11);

4.Temperature (lines 12-14);

5.Rubber and polystyrene concentration ratio . The latter influences only the type of the emulsion, the presence of the graft copolymer influencing only its stability.

While analysing the composition of phases of the full separation of the emulsion it was found that the experimentally found concentrations of polymers in these solutions differ from those calculated for the case when each polymer is present in one phase only. It can be supposed that due to the partial compatibility in both emulsion phases there are both polymers present, but the "rubber" phase is a polybutadiene solution with the admixture of small quantity of PS, and the "polystyrene" phase represents a polystyrene solution with the admixture of PB. On the basis that in model emulsions of equal compositions the volume of rubber phase increases as the molecular weight of polystyrene decreases, and $\bar{M}_V$ of homopolystyrene in the polystyrene phase increases (table III) we can draw a conclusion that low-molecular fractions of polystyrene migrate into the rubber phase.

Table III.

Dependence of a volume fracture (V) of the rubber phase of POO-emulsion on $\bar{M}_V$ of polystyrene. Composition of emulsion: PS- 5%; PB - 3%;

M_V 10^{-5} of PS used to prepare the emulsion	3.0	2.2	1.0	0.7
$\bar{M}_V$ 10^{-5} of PS in the polystyrene phase	3.05	2.30	1.15	0.78
V of the rubber phase; %	50	53	58	67

Rubber in its turn is distributed so that a part of it defuses into the polystyrene phase. However, the rubber concentration in the polystyrene phase is not high due to non-equal solvent distribution between phases and use for investigation of polybutadiene obtained by ionic polymerization with narrow MWD and high medium value of molecular weight, the concentration of rubber in polystyrene phase is small.The selective disolvent method found it to be equal to 0.05 - 0.1%.

The equilibrium composition of phases can be represented vividly with the help of the diagram for ternary mixture. As seen in fig.4 points 1',2',3' of intersection of connecting line with binodales describing the composition of the rubber phase of POO-

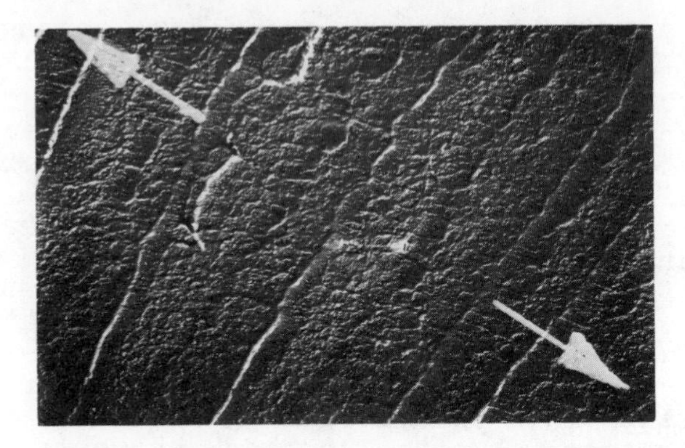

Figure 3. Surface of the homopolystyrene fracture treated by the vapors of ethyl acetate for 1 min while applying the external straining stress. The arrows show the direction of the stress applied.

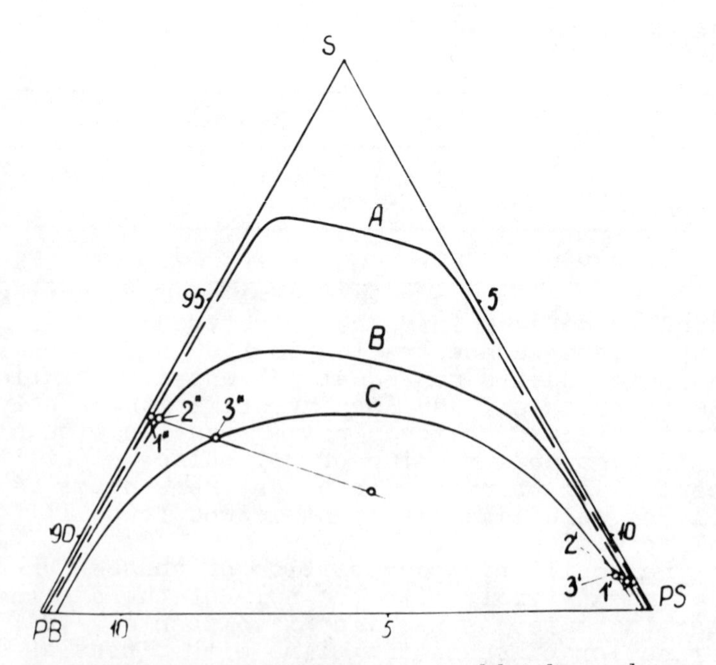

Figure 4. Diagram of ternary system of polybutadiene–polystyrene–styrene at t = 20°C for PS, the values of the molecular weight being: A—3.1 · 10^5; B—0.7 · 10^5; C—0.3 · 10^5

emulsion with polystyrene of different molecular weight lie near each other in the field of low concentrations of rubber, while points 1", 2" and 3" characterizing the composition of the polystyrene phase move considerably into the triangle as the $\bar{M}v$ PS decreases.

The composition of emulsion phases changes also when the temperature increases. For example, concentration of polystyrene in the rubber phase increases. This is proved by its turbidity when being cooled in case of separation of the emulsion at high temperatures.

Proceeding from said above the coefficient of polymer distribution between the phases of POO-emulsion can be written down in the following way:

$$K = \frac{C'_{PB} + C'_{PS}}{C''_{PS} + C''_{PB}} \simeq 0.7 \qquad (1)$$

where C'_{PB} and C'_{PS} - rubber and polystyrene concentrations in the rubber phase;

C''_{PB} and C''_{PS} - rubber and polystyrene concentrations in the polystyrene phase.

After complete POO separation in the centrifuge rubber and polystyrene phase volumes have been measured. The results of these measurements are represented in fig.5, it being known that the region of existence of multiphase emulsions (corresponding to the phase inversion) has been shaded. From the data given it is clear that the determining conditions of phase inversion of polymer emulsion is not the PS content but the complex of factors, providing the definite volume ratio of discrete and continuous phases - the content of both polymers, their molecular weights and MWD. The emulsion multiphase character is kept in the wide interval ratios of volumes, i.e. condition for the inversion of POO-emulsion phases may be represented as follows:

$$V'/V'' = 0.7 \div 1.4$$

Proceeding from the above mentioned, exactly it is possible to define the mechanism of HIPS morphology formation. According to Molau (1,5) the process of rubber particles formation takes place during and after the phase inversions (formations of polystyrene occlusions of I and II-type).

In case of the formation at prepolymerization of low molecular homopolystyrene fractions, compatible with the rubber in the rubber phase, the polymer concentration in the system is rather high by the end of inversion. During the following polymerization the homopolystyrene concentration in the rubber drop will exceed the compatibility limit, but the high PB solution viscosity makes the polystyrene macromolecules diffusion more difficult from the drop of rubber phase, homopolystyrene separates inside the rubber phase as drops of a new polystyrene phase. During the polymerization the volume of these drops grows. The rubber solution volume naturally decreases without the change of the total rubber phase volume. We suggest to call such polystyrene particles inside the rubber ones formed according to such mechanism - the occlusians of the III-type. The II-type occlusions partially degenerate in such a case. The III-type occlusions formation may be illustrated by the following model experiment. 5% homogeneous styrene solution of polybutadiene containing 1.55, of polystyrene with the molecular weight of $4 \cdot 10^4$ is added to 50% polystyrene solution with the molecular weight of $9 \cdot 10^5$ without being mixed. To reach the equilibrium in the system the styrene would migrate into the solution of high-molecular polystyrene from the rubber one , the latter would concentrate. Due to this the lowmolecular polystyrene becomes incompatible with the rubber and the rubber phase becomes turbid, drops of a new phase, representing the lowmolecular polystyrene solution in styrene are formed inside it.

To observe the formation of the III-type occlusions is also possible at high stages of the thermal polymerization without the agitation of the rubber solution, which has been obtained by complete separation in the centrifuge of polymer with lowmolecular homopolystyrene. Heterogeneity in the rubber phase appeares immediately after the beginning of polymerization. The momentary turbidity of the solution witnesses this. Small particles of the polystyrene phase are seen in microscope. The electron-microscope photograph of microstructure of the final product is represented in fig.6.

The influence of the molecular weight of homopolystyrene POO on the formation of polystyrene occlusions is vividly seen in fig.7, where photographs of microstructures of HIPS specimens are represented; The specimens of HIPS are obtained by the polymerization of model emulsions, prepared by mixing solutions of rubber - 8% and polystyrene - 30% in ratio 1:1. MW PS varied at $0.7 \div 3 \cdot 10^5$. As seen in fig.7 the rubber particles, formed from multiphase model emulsion as a result of redistribution of monomer, differ greatly from each other. MW of polystyrene

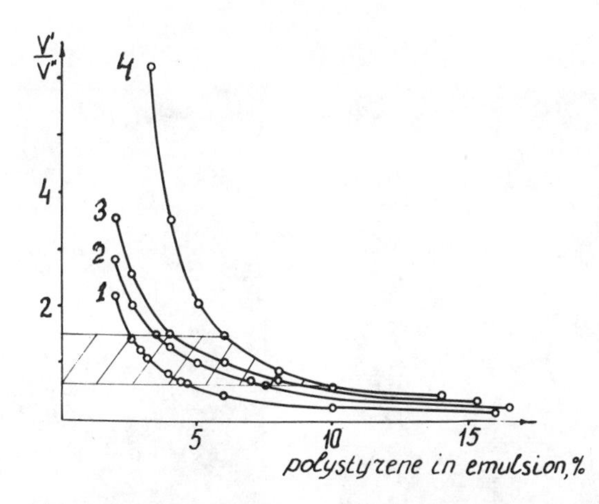

Figure 5. Dependence of the ratio of volumes of the rubber and polystyrene phases in polymer emulsion on PS concentration ($\overline{M}_v = 3 \cdot 10^5$); C_{PB} being: 1—2%; 2—3%; 3—4%; 4—$\overline{M}_v$ PS being $0.5 \cdot 10^5$ its concentration—4%

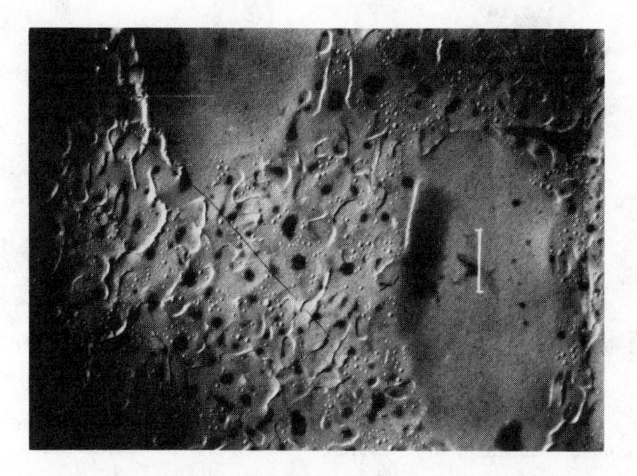

Figure 6. Microstructure of polymer, obtained by the polymerization of the rubber phase, separated from the prepolymer, increase ×8000. Occlusions of the II-type are seen.

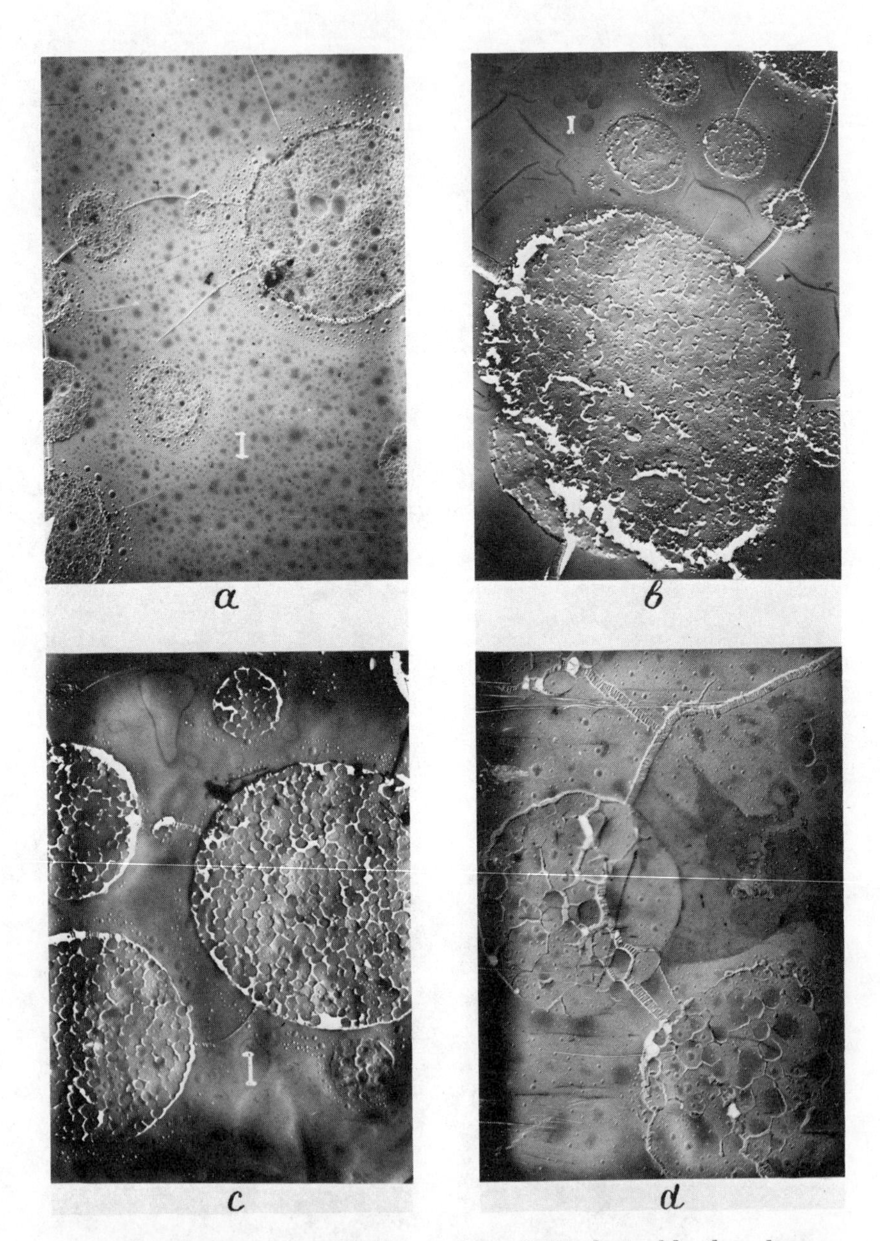

Figure 7. Structure of the rubber particles in HIPS, obtained by the polymerization of model emulsions, MW of homopolystyrene in which being: a) $2.9 \cdot 10^5$ $\times 3500$, b) $1.7 \cdot 10^5$ $\times 2500$, c) $1.17 \cdot 10^5$ $\times 5000$, d) $7.0 \cdot 10^5$ $\times 7000$.

being $1.7 \cdot 10^5$ and $2.9 \cdot 10^5$ (fig. 7, a,b) when polystyrene is practically incompatible with polybutadiene in styrene solution, the rubber particles with high concentration of PB and occlusions of II-type only are formed. With the decrease up to MW PS of $1.17 \cdot 10^5$ the sizes of the occluded particles increase up to 0.3 - 0.5μ , i.e. the internal structure of particles is analogous to HIPS, obtained by copolymerization in block. $\bar{M}_v$ of polystyrene being $0.7 \cdot 10^5$ the contents of PB in the rubber particles is minimal and the sizes of occlusions sometimes exceed 2μ, and this considerably increases the total volume of the rubber phase.

Summing up the above mentioned, the mechanism of formation of the internal microstructure of the "rubber" phase particles after the inversion of phases can be represented in the following way:

a) in case of absence of lowmolecular weight fractures of homopolystyrene the process proceeds according to Molau (1), i.e. with the formation of occlusions of I and II-types;

b) in case of presence of lowmolecular weight homopolystyrene after the inversion of phases a part of it remains within the "rubber" phase, separating during the further polymerization into the new polystyrene phase, forming the occlusions of the III-type. Molecules of PS appearing during the process of homopolymerization inside the rubber solution diffuse into the occluded drops of the polystyrene solution, increasing in such a manner their sizes. Thus,the possibility of forming new occlusions of the II-type decreases or becomes impossible.

LITERATURE CITED

1. Molau G.E., Keskkula H.J. J. Polymer Sci.,(1966), A 1, 4, 1596.
2. Wagner E.R., Robeson M.M., Ruber Chem. and Technolog, (1970), 43 , 1129.
3. Frugward G.W. Polymer, (1972), 13, 366.
4. Frugward G.W. Karmarkar M., J. Appl. Polymer Sci., (1972), 16, 69.
5. Molau I.E. Koll- Z u Z für Polymer, (1970), Bd.238, 493.
6. Bongardt J., Plaste und Kautschuk, (1973), 20 , 265.
7. Pohe G., Plaste and Kauschuk, (1973), 23, 340.
8. Bender B.W., J.Appl. Polym.Sci, (1965), 9, 2887.
9. Keskkula H., Turley S.G., Boyer R.F., J.Appl. Polymer Sci., (1971), 15, 351.
10. Kato R., Elektron Microscopy, (1965), 14 , 220.

11. Williams R.Y., Hudson R.W., Polymer, (1967), 8, 643.
12. Пилянкевич А.Н.Практика электронной микроскопии, Машгиз,Москва,1961.
13. Keskkula H.J. Tranlor P.A., J. Appl. Polym. Sci., (1967), 11, 2361.
14. Underwood E., Amer. Soc.Metals. Eng. Quart., (1962), 11, 62.
15. Cigna G., J. Appl. Polym. Sci., (1970),14, 1781.
16. Енальев В.Д.,Зайцев Ю.С.,Зайцева В.В. и др., Пласт.массы,(1969),10,11.
17. Кулезнев В.Н. в кн.Многокомпонентные полимерные системы, 27,Химия,Москва,1974.
18. Kambour R.P. and Rusel R.R., Polymer, (1971), 12, 237.
19. Ziegler E.E., SPE - journ., (1965),10, 12.
20. G.A. Bernier, R.P. Kambour, Macromolecules, (1968) 1, 393.
21. J.A. Schmitt, J. Appl. Polym. Sci., (1968), 12, 533.
22. S. Sterstein, J. Paterno und Ongchink., Intern. Conf. on Weld, Deformation and Fracture of Polymers, Cambridge, 31 March-3 April (1970).
23, J.A. Manson, R.W. Hertzberg, J.Polymer Sci., (1966), A 1, 4, 1595.
24. J.A. Schmitt, H.J. Keskkula, J. Appl. Polym. Sci., (1960) 3, 132.
25. T.O. Graig, J Polym. Sci.,: Polym. Chem. Ed., (1974), 12, 2105.

INDEX

F

G

H

I

T